INTERPRETATION OF GEOLOGICAL MAPS

LONGMAN EARTH SCIENCE SERIES

Edited by Professor J. Zussman and Professor W. S. MacKenzie, University of Manchester

B. C. M. Butler and J. D. Bell: Interpretation of Geological Maps
P. R. Ineson: Introduction to Practical Ore Microscopy
P. F. Worthington: An Introduction to Petrophysics
B. W. D. Yardley: An Introduction to Metamorphic Petrology

INTERPRETATION OF GEOLOGICAL MAPS

B. C. M. Butler and J. D. Bell

Copublished in the United States with
John Wiley & Sons, Inc., New York

Longman Scientific & Technical,
Longman Group UK Limited,
Longman House, Burnt Mill, Harlow,
Essex CM20 2JE, England
and Associated Companies throughout the world.

Copublished in the United States with
John Wiley & Sons, Inc., 605 Third Avenue, New York, NY 10158

First published 1988

British Library Cataloguing in Publication Data
Butler, B. C. M.
Interpretation of geological maps. —
(Longman earth science series).
1. Geology — Maps
I. Title II. Bell, J. D.
551.8 QE36

ISBN 0-582-30169-6

Library of Congress Cataloging-in-Publication Data
Butler, B. C. M., 1932–
Interpretation of geological maps.

(Longman earth science series)
Bibliography: p.
Includes index.
1. Geology — Maps. I. Bell, J. D. (John David)
II. Title. III. Series.
QE36.B83 1988 912′.1551 87–3151
ISBN 0-470-20844-9 (Wiley, USA only).

Set in 10/12pt Linotron 202 Ehrhardt Roman
Produced by Longman Singapore Publishers (Pte) Ltd.
Printed in Singapore

CONTENTS

PREFACE

Geological maps are the means of recording and storing information about the distribution, composition, and structure of the rocks of the surface of the Earth. They are used at every stage from the initial observations of the field geologist to the published geological survey map which may remain for years the definitive document about an area. Geological maps are thus a primary source of data about rocks and structures, a record of the environments and processes of the past, and an important element in the teaching of modern geology.

Some students and teachers may prefer to start the study of maps with synthetic examples designed to show idealised geometrical relationships; for those who prefer this approach, good simple textbooks are already available. We have found, however, that it is not only possible to teach map interpretation from the beginning using published geological maps, but by so doing, to introduce the student to maps as presentations of factual data about geological situations from the very earliest stage. Moreover this approach avoids the conceptual hurdle that some students find when idealised constructions are replaced by published maps.

In this book we explain both the geometrical and the geological interpretation of maps. We take the view throughout that maps are a cartographical representation of real rocks whose shapes and relationships must be understood as the product of geological processes operating through geological time. We make use of modern geological concepts relating to the environments of formation of sedimentary, igneous, and metamorphic rocks and mineral deposits, and of geological structures. These are summarised in Tables 2.1, 3.2, 3.3, 3.4, 7.1, and 10.1, and we recommend them to the reader for detailed study. We illustrate the determination of the rates of processes through the use of chronometric and radiometric ages and show how actualistic analogies can be made with present-day processes. We also introduce limited but meaningful interpretation of plate-tectonic environments from the evidence available on maps.

The book is designed for use by those beginning to study geology for the first time, whether at school or at university. It provides material covering the equivalent of about the first two years of instruction in most geology courses. Chapters 1 to 7 describe the basic techniques of map interpretation, using a step-by-step explanatory method, with worked examples at each stage. Chapter 8 deals with aspects of ocean-floor geology, and includes examples which are directly illustrative of plate tectonics. Chapters 9 and 10 deal in greater depth with the details of geological interpretation of maps and apply them with a comprehensively worked example.

We have drawn our examples from an international range of recently published maps which can, if the user wishes, be made into the nucleus of a teaching package

(Appendix 6). We decided to restrict interpretation largely to what can be derived directly from the printed maps, even where we recognise that further interpretation is possible in the light of other data. This is because the stated intention of the book is to explain the methods of map interpretation rather than a discussion of the geology of individual areas.

We have chosen to refer to recent textbooks in each field of geology rather than to list large numbers of original papers, in the belief that this provides a more useful introduction to the literature of geology for the readers for whom the book is primarily intended.

Regretfully we have had to omit from consideration the numerous other types of map used by the practising geologist – geophysical, geochemical, hydrological, tectonic, palaeontological, palaeocontinental, etc. We have also omitted the interpretation of cross-sections produced directly from geophysical traverses. The principles of interpretation of these maps and sections are based on the techniques set out in this book, which forms the necessary ground-work for more specialised studies.

Acknowledgements

It is a pleasure to record the help given by our colleagues Drs P. A. Allen, W. S. McKerrow, J. P. Platt, H. G. Reading, and P. N. Taylor in reviewing parts of the manuscript. Professor E. H. Francis read the whole of Chapters 9 and 10, and his constructive comments were of great value. We are grateful to Dr N. J. Snelling for sending us a preprint of the revised geological time-scale. Mr D. S. Paterson and Mr P. J. T. Barbary greatly assisted our search for maps in the library of the British Geological Survey, and similarly Mrs Joan Morrall in our own library. Mrs Vivien Chamberlain and Mrs Andria Fowler typed the manuscript with patience and skill.

The authors wish to make special acknowledgement to the Directors and other officers of the geological surveys whose maps we have used as illustrations in this book. In particular we are grateful to the Director of the British Geological Survey for permission to reproduce Plates 1, 3, and 4 from BGS maps, to the Director General of the Department of Mines and Energy, Geological Survey of South Australia for permission to reproduce Plate 2 from the South Australia Geological Atlas Series, and to the State Geologist, Pennsylvania Geological Survey for permission to reproduce the front cover of the book from the Geologic Map of Pennsylvania.

All the line diagrams of maps in this book were traced from published maps (see Appendix 6), and we acknowledge the following sources:

British Geological Survey; Bundesanstalt für Geowissenschaften und Rohstoffe, West Germany; California Division of Mines and Geology; Commission for the Geological Map of the World and UNESCO; Elsevier Applied Science Publishers, Ltd.; Geological Survey of Canada; Geological Survey of Japan; Geological Survey of South Australia; Landeshydrologie und -Geologie, Geologische Landesaufnahme, Switzerland; Ministry of Petroleum and Minerals, Sultanate of Oman; The Open University, England; Parc Naturel Régional des Volcans d'Auvergne; Pennsylvania Geological Survey; Service Géologique National, Orléans, France; Servizio Geologico d'Italia; United States Geological Survey; Virginia Division of Mineral Resources.

We also acknowledge satellite and aerial photographs from the following sources: Aerofilms, Ltd, Borehamwood; British Geological Survey, Edinburgh; National Aeronautics and Space Administration, Houston, Texas; Royal Aircraft Establishment, Farnborough.

LIST OF COLOUR PLATES

between pages 92–93 and 180–181

Plate 3 Extract from B.G.S. Sheet 39W (Stirling) reproduced by permission of the Director, British Geological Survey: N.E.R.C. copyright reserved. Scale 1 : 50 000. The map shows Carboniferous lavas and sediments and Upper Old Red Sandstone sediments resting unconformably on Lower Old Red Sandstone sediments in the west of the area. The west end of the Ochil Fault forms the boundary between these units in the east of the area. See Table 9.1 for abbreviations of the names of lithological units. Arrows show dip directions and amounts. A — B is the line of section shown in Fig. 9.6.

Plate 4 Extract from B.G.S. Sheet 39E (Alloa) reproduced by permission of the Director, British Geological Survey: N.E.R.C. copyright reserved. Scale 1 : 50 000. The map shows the Ochil Fault separating Lower Old Red Sandstone lavas in the north from folded and faulted Carboniferous sediments with coal seams in the south. See Table 9.1 for abbreviations of names of lithological units. Arrows show dip directions and amounts. C — D is the line of section shown in Fig. 9.7.

Front cover map

Part of the Geologic Map of Pennsylvania (1980), scale 1 : 250,000, showing the area around State College (40°48′N, 77°52′W), reproduced by permission of Donald M. Hoskins, State Geologist, Pennsylvania Geological Survey. The stratigraphical succession is Cambrian and Ordovician marine limestones and dolomites, followed by Late Ordovician to Devonian marine sandstones and shales, and then by Mississippian and Pennsylvanian shallow-water marine and terrestrial sediments including cyclic sequences of limestone, shale, sandstone, and coal.

The Allegheny Front, running northeast-southwest through the northwest quarter of the map area, marks the sharp transition from the strongly folded and faulted Cambrian to Pennsylvanian rocks of the Valley and Ridge Province in the southeast to the mildly folded Upper Silurian to Pennsylvanian of the Allegheny Plateau in the northwest corner.

The Cambrian and Ordovician rocks represent carbonate shelf deposits on a passive continental margin on the southeast side of the early North American continent, followed by clastic sediments derived from erosion of the Late Ordovician Taconian orogenic belt to the southeast. Compression in the foreland during the Alleghanian orogeny in the Late Carboniferous produced the prominent folds and thrusts of the Valley and Ridge Province, and the low amplitude folds and strike-slip faults in the Middle and Upper Paleozoic rocks of the Plateau.

LIST OF TABLES

1 ACTUALISTIC INTERPRETATION OF GEOLOGICAL MAPS

1.A INTRODUCTION

The study of rocks starts where they can be seen – at the surface of the Earth. There are two principal sources of data:

1. The situations in which rocks are forming at the present day and the processes that operate within them.
2. The rock types and structures that were formed in the past and are now exposed by erosion at the surface of the Earth.

Interpretations of the two kinds of data are inter-related. By observing environments and processes at the present day we can develop the techniques for understanding the rocks and structures of the geological past. Conversely, the evidence of ancient rocks and structures known to have been formed below the surface of the Earth leads to the understanding of processes within the Earth which are not accessible for direct observation at the present day.

The techniques of geological mapping are described by Lahee (1961) and Barnes (1981). A geological map is a record of the distribution of rocks of different ages and compositions at the surface of the Earth. It is also, and more importantly, a source of evidence about how the Earth, or some small part of it, has operated as a physical, chemical, and biological system through the millions of years of its history. There are two stages in the interpretation of a geological map. The first is to show how the map reveals the two- and three-dimensional geometry of rock shapes. The second and more important stage is to show how the rock shapes are used to reconstruct the environments and processes of the past – to convert geometrical interpretation into geological understanding. We shall explain principles of both kinds of interpretation in this book. In particular, we shall show how knowledge of the dates of geological events enables us to evaluate the rates of geological processes, and so to begin to make realistic dynamic interpretations of how the geology of an area developed in the course of geological time.

In this chapter we shall briefly review some of the environments and processes that are observable at the present day as a framework for the interpretation of geological maps in subsequent chapters.

Table 1.1 Major Earth environments

	Deep oceans	*Continental shelves*
Geographical location	Oceanic regions generally deeper than 4000 m	Shallow submarine extensions of continents and island arcs
Topography	Active spreading ridges; aseismic ridges; large areas of gentle relief with isolated sea mounts or chains of sea mounts and fracture zones; deep active trenches	Generally even; gentle slope down towards deep oceans
Composition	Thin pelagic sediments covering basic igneous oceanic crust; terrigenous sediments transported by slumps from continental slope and by floating ice; metal-rich muds in some spreading rifts; manganese nodules	Clastic and carbonate sediments, evaporites and/or volcanic rocks
Age of exposed rocks	Mesozoic to Recent	Tertiary to Recent
Rate of deposition	Very low except locally near continental areas with major rivers	Variable: greatest in areas near large river systems
Rate of erosion	Very low	Generally low but locally and temporarily high in regions subject to sudden storms
Mobility	Generally very low except in spreading ridges and in trenches	Generally low; epeirogenic movements typical
Examples	Shallower linear areas: active spreading ridges in Atlantic and eastern Pacific. Aseismic ridges: 90 East Ridge in Indian Ocean. Abyssal plains in all oceans. Sea mounts in Pacific Ocean. Ocean trenches: West Pacific, west coast of South America, Caribbean.	Partially enclosed seas: Persian Gulf; North Sea; Gulf of Mexico. Recently formed oceans: Red Sea, Gulf of California. Open coastlines: Africa, Europe, Australia, Americas. Offshore areas of island arcs: western Pacific, Caribbean.

Lowland areas	*Orogenic belts*	*Shield areas (cratons)*
Inland areas of continents adjacent to orogenic belts; coastal plains bordering stabilised orogenic belts and shields; intracontinental rifts	Destructive plate margins: island arcs; edges of continents; inner parts of continents where continental collision has occurred	Inner parts of continents; edges of continents subjected to geologically recent tectonic spreading
Low and relatively even but with large volcanic piles in some areas	Very variable, from ocean trenches (−11 000 m) to highest mountains (+ 8000 m)	Low but irregular relief
Continental, fluviatile and lacustrine sediments; geologically recent glacial deposits in some areas; volcanics in some areas	Folded volcanic and sedimentary rocks in the younger belts. Metamorphic and plutonic igneous rocks exposed in the older more deeply eroded belts	Basement of high-grade metamorphic rocks formed at depths of 10–20 km, with thin veneer of sedimentary rocks in places
Palaeozoic to Recent	Precambrian to Recent	Precambrian (or later cover)
Variable: greatest in basins of subsidence and in volcanic areas	Very high to low, depending on position in orogenic belt and topographical location	Generally low
Variable: greatest during sudden storms and volcanic explosions	Very high to low, depending on age and topographical location	Generally low
Generally low; epeirogenic movements characteristic but lowland areas are liable to be involved in orogeny later; faulting in rifts	All show intense folding; young belts very mobile with strong seismic and tectonic activity; older belts more stable with seismic activity associated with faults	Resistant to folding; epeirogenic movements with block faulting in some areas
Inland depositional desert basins of Africa, Australia and Asia; Indo-Gangetic plain of India; Great Valley of California; glaciated plains of northwestern Europe; East African and Lake Baikal rifts	*Present-day*: western margins of North and South America; island arcs of the west Pacific, including Japan. *Tertiary*: Alps, Himalayas, Atlas Mountains. *Late Palaeozoic*: Ural Mountains. *Early Palaeozoic*: Appalachians of eastern North America; Caledonides of north-western Europe; eastern Australia	1. Exposed shield areas: Baltic Shield (Scandinavia), Central Australia, Canadian Shield, SW Greenland. 2. With thin cover of younger rocks: Columbia Plateau (western USA), Russian and Siberian platforms; Deccan plateau (India); much of Africa; Brazil

1.B ENVIRONMENTS

The interpretation of geological environments of the past depends on matching the rocks and structures recorded on the map with the products of analogous situations

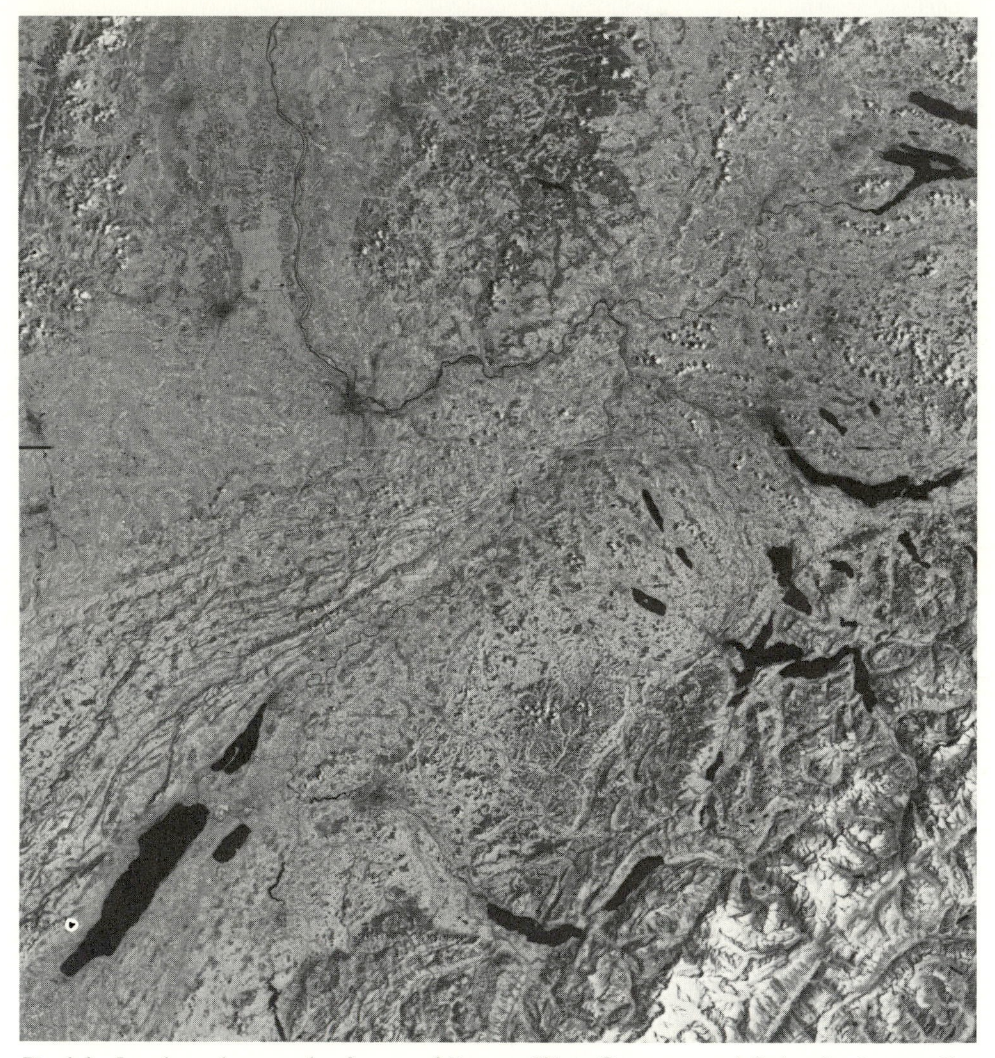

Fig. 1.1 Landsat photograph of parts of France, West Germany, and Switzerland, extending from Lac de Neuchâtel in the southwest to the west end of the Bodensee in the northeast. The area shown is approximately 115 km × 115 km. The following large-scale geographical and geological features are visible:

The Precambrian and Palaeozoic rocks of the Vosges (northwest, dark and partly cloud-covered) and of the Black Forest (north-central, dark) are separated by the Quaternary rocks of the Rhein Graben (pale); to the northeast are the flat-lying Mesozoic and Tertiary sediments of the Schwabische Alb.

The Mesozoic rocks of the Jura Mountains, folded during the Alpine orogeny, extend from the southwest corner to the centre of the photograph (striped light and dark) and are separated by the Tertiary molasse basin of the Franco – Swiss plain (pale, with numerous lakes) from the orogenic belt of the Alps (southeast corner, rugged topography, with snow on the higher ranges).

See Figs 4.17 and 7.4 for geological maps of parts of this area.

of the present day. Five kinds of broadly defined major Earth environments are set out in Table 1.1, each distinguishable by the kinds of rock that underlie them and the processes that operate within them (cf. Fig. 1.1). Finer divisions of environments based on distinctive rock types and structures are summarised in Tables 2.1, 3.2, 3.3, 3.4, 7.1, and 10.1.

1.C PROCESSES

All parts of the surface of the Earth are moving and changing (Table 1.2). We must distinguish movements of the crust as a whole, resulting from plate tectonic movements, uplift, and subsidence, from changes at the surface caused by erosion, deposition of sediments, and volcanic activity. The processes are distinct but inter-related – horizontal plate movement can produce vertical uplift and erosion in one area, while an adjacent area undergoes subsidence and deposition.

Because the rates of present-day geological processes are extremely slow it is easy to get the impression that their effects are trivial, with the exception of sporadic violent volcanic eruptions and earthquakes. However, both continuous and periodical processes are significant over the course of geological time and produce extensive changes. An average rate of deposition of only 0.1 millimetre of sediment per year persisting for 1 million years produces a pile of sediments 100 metres thick; the 3000-kilometre width of the North Atlantic Ocean between the continental shelves of Spain and Newfoundland was produced by a relative movement of the North American and Eurasian plates of 30 millimetres per year for the last 100 million years since the Cretaceous when the rifting of the two continents began.

It is perhaps helpful to think of the view of the Earth that we acquire during our short lifetime as like a single frame from a continuously running sequence. Details may vary – for example, there are no Phanerozoic analogues of Precambrian banded iron formations or of anorthosite intrusions, and the rates of some present-day surface processes are more vigorous than in preceding geological time because of after-effects of the Pleistocene ice age. But the fundamental processes – plate movement, uplift, subsidence, erosion, deposition, volcanism, and those processes that can be inferred to be operating in the interior of the Earth – form a continuum with the past. This is the basis of the **actualistic** view of geology – that the rocks and structures of past geological time were produced by essentially the same processes as those that are observable today (Holmes 1978, p. 30).

The same view of geology needs to be kept in mind when studying geological maps. A sedimentary rock unit shown with a uniform colour on a map seems at first sight to represent a single discrete event; in fact such a unit is the product of thousands or millions of years of sedimentation, and its boundaries are a convenient and artificial subdivision of essentially continuous processes. Similarly, igneous intrusions, faults, folds, and unconformities represent the products of processes that are extended over prolonged, and often determinable, durations of time. It is particularly important to think of the geology of an area as having a history which is continuous from the date of the oldest rocks (and earlier) up to the present day. Even during times when there is no solid record, in the form of rocks or structures, the area was involved in processes of some kind. Throughout the book we shall show how it is possible to evaluate the rates of geological processes by relating observations on the map to the geological time-scale (Appendix 2), and to relate them to the rates of comparable processes at the

(*Text continues on p. 10*)

Table 1.2 The rates of some present-day and recent movements of the surface of the Earth.

Every part of the surface of the Earth has its own unique history and environment, which determine the patterns of local movement of the crust of the earth and of erosion and sedimentation. This table lists some examples of the net rates of movement and of individual events which have been recorded or deduced in particular areas. The rates of movement in analogous situations at the present day may well differ by an order of magnitude or more; comparable variation can be expected in past geological time.

Note that three different units of measurement have been used in this table – thicknesses in millimetres and in millimetres per year, lateral movements in centimetres and in centimetres per year, and volumes in cubic kilometres and in cubic kilometres per year.

References: Many of the data are from Elder (1976), Ollier (1981), Leeder (1982) and Keller (1982). Other sources are: American Association of Petroleum Geologists, 1981, *Plate-Tectonic Map of the Circum-Pacific Region*; Blatt, H., 1982, *Sedimentary Petrology*, Freeman, San Francisco; Booth, B., 1979, *Jl geol. Soc. London*, **136**, 331–40; Bridge, J. S. & Leeder, M. R., 1979, *Sedimentology*, **26**, 617–44; Coleman, J. M., 1976, *Deltas: processes of deposition and models for exploration*, Continuing Education Publication Company; Decker, R. & Decker, B., 1981, *Scientific American*, **244** (3), 52–64; Dewey, J., 1982, *Jl geol. Soc. London*, **139**, 371–412; Donovan, D. T. & Jones, E. J. W., 1979, *Jl geol. Soc. London*, **136**, 187–92; Evans, G., 1979, *Jl geol. Soc. London*, **136**, 125–32; May, S. K. *et al.*, 1983, *Episodes*, **6**, 31; Molnia, B. F., 1980, *Pacific Coast Palaeogeography Symposium 4* (eds Field, M. E. *et al.*), 121–41; Ollier, C. D., 1981, *Tectonics and Landforms*, Longman, New York; Reading, H. G., 1982, *Proc. Geol. Assoc*, **93**, 321–50; Reineck, H-E. & Singh, I. B., 1973, *Depositional Sedimentary Environments* (2nd edn), Springer-Verlag, Heidelberg; Riecker, R. R. (ed.), 1979, *Rio Grande Rift: Tectonics and Magmatism*, Amer. Geophys. Union, Washington, D. C.; Schlager, W., 1981, *Bull. Geol. Soc. Amer.*, **92**, 197–211; Snelling, N. J. (ed.), 1985, *The Chronology of the Geological Record*, Memoir No. 10, Geological Society, Blackwell, Oxford; Tungshen, L. *et al.*, 1985, *Episodes*, **8**, 21–8; Williams, H. & McBirney, A. R. 1979, *Volcanology*, Freeman Cooper & Co., San Francisco.

Process	*Example and typical average rate*		*Magnitude of typical event*	
Horizontal movements of the crust				
EXTENSIONAL				
Ocean-floor spreading	East Pacific Rise	10–15 cm/y	Width of individual sheeted dykes in ophiolite complexes	50–100 cm
	Mid-Atlantic Rift	2 cm/y	Width of open fissures in Icelandic active volcanic zones	400 cm
Stretching of continental crust	Rio Grande Rift, USA	1.2 cm/y		
COMPRESSIONAL				
Relative plate movement at convergent plate margins:				
1. oceanic : oceanic	Island arcs, western Pacific	6–10 cm/y		
2. oceanic : continental	Nazca : South American plates in Chile–Peru trench	9 cm/y		
3. continental collision	Australia–India : Eurasia plate in Himalayas	1 cm/y		

Process	*Example and typical average rate*		*Magnitude of typical event*	
SHEAR				
Relative movement at transform plate margin	San Andreas Fault, California	5–6 cm/y	Movement in San Francisco earthquake, 1906	500 cm
Vertical movements of the crust				
UPLIFT				
Uplift of orogenic belt	Pamir and Tienshan Himalayas	80 mm/y 7 mm/y		
Uplift of active continental margin	Gulf of Alaska	5–10 mm/y	Movement in Alaska earthquake, 1964	5000 mm
Epeirogenic uplift	Southern South America	2 mm/y		
Uplift in active volcanic area			Campi Flegrei, southern Italy	Irregular movements up to 1000 mm/y
Isostatic uplift after melting of ice sheet	Scandinavia (last 10 000 y)	up to 20 mm/y		
SUBSIDENCE				
Decrease in height of newly formed oceanic crust	(i) in first 10^4 y (ii) after 10^6 y	4 mm/y 0.4 mm/y		
Subsidence in area of active deposition	Mississippi Delta (Cenozoic)	0.2 mm/y		
Subsidence due to volcanic activity			Tenerife, Canary Islands (prehistoric)	300 000 mm
Subsidence in area adjacent to former ice-sheet	Southeast England	2 mm/y		
Subsidence due to dewatering of sediments			Venice Mexico City (since 1940)	4 mm/y up to 180 mm/y
Subsidence due to extraction of oil			Long Beach, California (1940–1974)	up to 260 mm/y
Changes at or near the surface of the earth				
EROSION				
Average rate of erosion of continental land-mass		0.01 mm/y		
Average rate of erosion of world's mountain regions		0.09–0.8 mm/y		
Average rate of erosion of world's plains		0.01–0.1 mm/y		

(*Table 1.2 cont.*)

Process	*Example and typical average rate*		*Magnitude of typical event*	
Erosion by post-depositional slumping			Submarine slump from upper continental rise, NW Africa	600 km^3
Erosion by subaerial landslip			Downslope movement in Portuguese Bend landslip, Los Angeles	up to 250 cm/y
Erosion by volcanic activity			Volume of rock displaced in eruption of Mt St Helens volcano, USA, 1980	2.7 km^3
DEPOSITION				
Continental areas				
Deposition in intracratonic basins		0.02 mm/y		
Deposition of loess	China	0.07 mm/y		
Deposition of peat	Okefenokee swamp, SE Georgia, USA	0.8 mm/y		
Deposition of alluvium on flood plain	Mississippi River (last 30 000 y)	1 mm/y	Ohio River, 1937 flood	up to 560 mm
Deposition in glacial lake	Weistriztal, Sudeten Lake Constance (last 8000 y)	60–100 mm/y up to 100 mm/y		
Accumulation of volcanic products			Thickness of ash fall, Mt St Helens, 18 May 1980	up to 2000 mm
			Volume of flood basalt flow, Lakagigar, Iceland, 1783	12 km^3
Coastal environments				
Deposition of deltaic sediments	Mississippi delta (last 26 500 years)	8 mm/y		
Continental shelves and platforms				
Deposition on passive continental margins		up to 0.1 mm/y		
Deposition in subduction-related basins		up to 1 mm/y		
Deposition in strike-slip basins		up to 1 mm/y		
Deposition of mud on clastic shelves		3–5 mm/y		

(*Table 1.2 cont.*)

Process	*Example and typical average rate*		*Magnitude of typical event*	
Formation of limestone in coral reefs		1–10 mm/y		
Precipitation of $CaCO_3$ Bahamas		0.25 mm/y		
Deposition of halite by evaporation		10–50 mm/y		
Deposition in submarine fan	Amazon	1.7 mm/y (max)	Grand Banks, Newfoundland, turbidite flow, 1929	100 km^3
Deposition on continental rise	Greater Antilles outer ridge	0.06–0.3 mm/y		
Oceanic areas				
Sedimentation in landlocked deep sea	Black Sea	0.1 mm/y		
Deep ocean pelagic sediments		0.0001–0.01 mm/y		
Formation of manganese nodules		3×10^{-6} mm/y		
Accumulation of volcanic products	Average rate of eruption, Hawaii	2.6×10^{-2} km^3/y		
LATERAL MIGRATION OF SHORELINES				
In area of active deposition	Mississippi delta	100–2000 cm/y		
In tectonically active area	Liguria, Italy	1800 cm/y		
In tectonically stable area	Atlantic coast of USA average erosion rate	80 cm/y		
Growth in area affected by post-glacial rebound	Maine coast, USA	100 cm/y		
EUSTATIC CHANGES OF SEA-LEVEL				
Rise of sea-level caused by increase in volume of oceanic ridges		0.01 mm/y		
Rise of sea-level caused by melting of Pleistocene ice sheets		10 mm/y	Total rise of sea-level caused by melting of Pleistocene ice sheets	130 000 mm

(*Table 1.2 cont.*)

Process	*Example and typical average rate*	*Magnitude of typical event*	
Fall of sea-level caused by resubmergence of dried-out marine basin		Total fall of sea-level caused by infilling of Mediterranean basin (Miocene)	10 000 mm
MISCELLANEOUS MOVEMENTS			
Rise of exposed salt domes	1–5 mm/y		
Rise of granite batholiths	20 mm/y		
Rise of magma		Kilauea, Hawaii (1959)	2.5×10^7 mm/y
Growth of volcanic structures		Paricutin cinder cone	5×10^4 mm/y
		Mauna Loa, Hawaii (since 1831)	0.17 km^3/y
Man-made erosion		Bingham Canyon Copper Mine, USA	0.08 km^3/y
Retreat of Pleistocene ice sheets of Scandinavia, 10 000–9000 BP	1.2–4 $\times 10^4$ cm/y		
Flow of Alpine glaciers	up to 7500 cm/y		

present day (Table 1.2). Chapter 10 will describe the integration of all the observations about environments and processes to create an account of the geological evolution of an area.

1.D A NOTE ON CONVENTIONS

For the examples in this book we have made use of a small number of modern maps published by various geological surveys (see Appendix 6). Different maps use different symbols, notably the one to indicate the direction and amount of dip. In the interests of consistency in this book we have standardised the symbols on diagrams re-drawn from original maps (see the list on page 230); in particular we have chosen to use the strike-and-dip symbol throughout.

1.E ADVICE TO STUDENTS

Geological maps present a large amount of information, often in a very condensed form. The Sections in the ensuing chapters of this book illustrate how the data needed to answer a particular problem can be selected out from all the other information on the map; in some cases, for instance, it may be the geometrical relationship between the topographical contours and a lithological boundary that is significant, in others the relative ages or time interval between the ages of the units adjacent to a boundary. The reader will find it instructive to compare the line diagrams in this book with the original maps so as to see which items of the printed map information are needed to answer a particular question. It will always be necessary to look closely at the detail printed on the map. However, we would like to add five further pieces of advice:

1. Determine the regional setting by referring to a smaller-scale map that shows the relationships of the area in question to surrounding areas.
2. Read the legend and other marginal information of the map to be studied, noting in particular the conventional symbols that are used.
3. Base your conclusions of detailed study on the evidence of more than one area or part of a map – it is quite common for a particular outcrop or exposure to represent a local anomaly or deviation from the general pattern.
4. Study the map as a whole, for instance by viewing it from a distance of a few metres, in order to recognise the broad pattern of rock relations that are shown in the area.
5. Relate your interpretation of the map to the environments and processes that created the rocks and structures that you are studying. In this way the geology of the area shown by the map will 'come alive', and it will be possible to picture the development of the area in actualistic terms as the continuously evolving product of processes with which you are familiar at the present day.

2 PARALLEL-SIDED LITHOLOGICAL UNITS

An area of uniform colour on a geological map of sedimentary rocks represents a **stratigraphical unit**. This is a sequence of individual beds of the same or related lithological type which are grouped together for convenience of mapping, description, and interpretation. The lateral extent of a stratigraphical unit indicates the geographical area over which the environmental conditions were more or less constant. The thickness of a unit is related to the period of geological time during which these conditions persisted. Individual units may be a few metres to several thousand metres thick (formed during a few thousand to several million years) (Fig. 2.1).

Fig. 2.1 An area in the Sulaiman Range, Pakistan, showing part of a thick succession of Eocene sediments. They can be subdivided into three mappable lithological units; green and brown shales on the left, the white limestone (10 metres thick) which forms the scarp- and dip-slope in the centre, and grey-green shales, in part covering the limestone dip-slope, on the right. In the background are ridges of Cretaceous and Jurassic limestones. Photograph: Professor W. D. Gill.

Deposition of most sediments occurs on the nearly flat surface of lowland areas of continents or on the sea floor. Individual beds form irregular, thin, elongate lenses or wedge shapes, but deposition of successive beds leads to the formation of stratigraphical units which are *essentially parallel-sided layers of constant thickness which were originally horizontal at the time of deposition.* Exceptions are localised deposits formed in restricted environments (Table 3.1). Apart from these, the **constant-thickness principle** is a powerful guide to understanding the shape of the outcrop of a lithostratigraphical unit on the present-day topographical surface as shown on a geological map (Sect. 2.A).

Volcanic rocks (lavas and volcaniclastic deposits), though ultimately of very different origin from sediments, are deposited as layers on the surface of the Earth and must be regarded as part of the stratigraphical succession. The constant-thickness principle applies to some volcanic rocks (e.g. flood basalts) as much as it does to sediments (Fig. 2.2). Certain intrusive igneous rocks form parallel-sided units (Sect. 2.I), and the geometrical methods of this chapter apply also to them, though it must be remembered that intrusive rocks are later than the stratigraphical units into which they are emplaced.

Observations of the distribution and shapes of outcrops of stratigraphical units on a geological map lead to the determination of the **stratigraphical succession**, or sequence of rocks from the oldest to the youngest (Sect. 2.B). The **stratigraphical thickness** of a unit is calculated from its outcrop width and dip (Sect. 2.C). The two kinds of information can be combined to construct a **stratigraphical column**, summarising the succession and thickness of the units in an area (Sect. 2.D).

Determination of the thickness of a single unit at a number of points in an area allows the construction of a **thickness-contour** (isopachyte or isopach) **map** showing the regional variation of the thickness (Sect. 2.E).

Fig. 2.2 An area in the Western Ghats, India, showing nearly horizontal flood basalts of the Deccan Traps (Eocene); the escarpment is approximately 1400 metres high. Photograph: Dr K. G. Cox.

Table 2.1 Lithologies and facies

		Oceanic environments		
		Pelagic oceanic environments	*Clastic oceanic environments*	
			Active continental margins and island arcs	*Passive continental margins*
Clastic sediments	Conglomerate	O	–	–
	Sandstone	O	X	X
	Siltstone	O	X	X
	Mudstone	X	X	X
	Turbidite (alternating sandstone and mudstone)	X	X	X
	Sedimentary mélange	O	X	O
	Boulder clay	O	O	O
Predominantly biogenic sediments	Coal	O	O	O
	Chert	X	O	O
	Marl	x	–	–
	Limestone	X	O	O
Predominantly chemical sediments	Dolomite	X	O	O
	Oolite	O	O	O
	Evaporite	(x)	O	O
	Phosphorite	O	O	O
	Ironstone	x	O	O
	Banded iron formation	O	O	O
	Metalliferous sediment	X	–	O
Igneous	Ultrabasic and basic rocks	x	x	O
One or two lithologies usually predominate				
Characteristically two or more lithologies alternate		X	X	X
Facies can be laterally extensive		X	X	X
Facies usually laterally restricted				
Present-day example		Mid-ocean ridges and deep ocean basins	Pacific coast of USA, island arcs of west Pacific, Japan and China seas	Continental shelf and abyssal plain off eastern USA

Notes: This table, based on Reading (1986), Blatt, Middleton, & Murray (1980), Hallam (1981), and Leeder (1982), provides a simplified guide to the correlation of lithology, as commonly recorded on published geological maps, with facies or environment of formation of sedimentary rocks.

The identification of facies and the determination of the environment of formation of an assemblage of rock-types usually depends on the recognition of characters on a scale smaller than that of a mappable litho-stratigraphical unit (e.g. details of sedimentary structures, chemical composition, fossils). Conversely the names of lithological units and the terms used to describe the lithology on maps are commonly not directly correlatable with facies. Before attempting to identify facies, additional information should always be sought from descriptions or memoirs accompanying the published map, or from general reference books on stratigraphy.

Coastal and shelf environments						Continental environments			
Carbonate shelves	*Clastic shelves*	*Sub-aqueous evaporites*	*Arid shore-lines*	*Clastic shore-lines*	*Deltas*	*Glacial environments*	*Lakes*	*Rivers*	*Deserts*
O	–	–	–	–	–	X	–	X	x
–	X	–	–	X	X	X	X	X	X
–	X	–	–	X	X	X	X	X	–
–	X	–	–	X	X	X	X	X	–
O	O	O	O	O	–	O	x	x	O
O	O	O	O	O	O	O	O	O	O
O	O	O	O	O	O	X	–	–	O
O	O	O	O	–	X	O	x	x	O
–	O	–	O	O	–	O	–	O	O
x	–	–	x	–	–	O	–	–	–
X	–	x	X	–	–	O	–	–	–
X	O	–	X	O	O	O	–	O	–
X	O	–	X	–	–	O	–	O	–
x	O	X	X	O	O	O	(x)	O	(X)
(–)	(–)	O	O	O	O	O	O	O	O
(X)	(X)	O	O	(X)	(x)	O	(x)	O	O
(X)	(X)	O	O	O	O	O	(x?)	O	O
–	–	–	–	–	–	O	–	–	–
O	O	O	O	O	O	O	O	O	O
X	X	X	X	X		X		X	X
					X		X		
X	X	X			X	X			X
			X	X			X	X	
Bahama Platform, Great Barrier Reef, Arabian Gulf	Continental shelf of northwest Europe	No modern examples	Trucial Coast of Arabian Gulf	Atlantic Coast of USA	Deltas of Mississippi, Niger and Ganges–Brahmaputra	Pleistocene and Recent glacial deposits of northern Europe	Lakes of East African rift valley	Alluvial plain of Ganges and Brahmaputra	Desert of North African Sahara

Notice that most lithologies and many associations of two or more lithologies occur in more than one environment. Facies may persist, with a small change in the proportions of rock-types, over more than one mapped lithostratigraphical unit.

X Distinctive lithology of a facies
x Minor lithology, which may be locally abundant
– Lithology which could be present in appropriate palaeogeographical conditions
O Lithology which is unlikely in the facies
( ) Lithology which is not a necessary diagnostic feature, but which is highly distinctive if present.

The lithology of a stratigraphical unit or of an assemblage of units indicates the **facies and environment of formation** of the rocks, though only within broad limits (Table 2.1, Sect. 2.F). Additional information from descriptions of the rocks in the margin of the map or from an accompanying report or memoir allows a more precise determination.

In most environments deposition (including the formation of volcanic rocks) is episodic, with long time intervals between the formation of individual beds. During such an interval erosion of immediately preceding beds may occur. However, uniformity of the type of sediment or facies through a series of layers indicates persistence of the same environmental conditions through an extended period of geological time. This uniformity may include the height of the surface of deposition relative to sea-level, as shown most strikingly in deltaic facies where all the varied lithologies are formed within a few tens of metres above or below sea-level. This leads to the important proposition that, for most sediments, subsidence is necessary to create the conditions for sedimentation to continue over time. Without subsidence, sedimentation leads quickly to a change of facies and ultimately to cessation of sedimentation.

An approximate determination of the **average rate of sedimentation** can be made using the measured stratigraphical thickness and the chronometric ages (Appendix 2), interpolated if necessary, of the base and top of a unit (Sect. 2.G). Generally it is found that terrestrial and near-shore marine sediments formed in high-energy environments were deposited more rapidly (with rates up to about 1 mm/y) than deep marine sediments (commonly less than 0.1 mm/y) – cf. Table 1.2.

The construction of a **structure contour map** is an aid to understanding the three-dimensional shape of a geological surface, such as the boundary between two lithological units; it enables the shape of the surface to be more easily visualised, free from the intricacies of the outcrop pattern resulting from the intersection of the geological surface with an irregular ground surface. The method of construction is described in Section 2.H, and is further extended in Section 9.D.

2.A TO RECOGNISE LOW, MEDIUM, OR HIGH DIPS IN DISSECTED TOPOGRAPHY

The dip is indicated on most maps by conventional symbols. However, it is useful to be able to recognise the effect of the dip from the outcrop pattern.

2.A.1 LOW OR HORIZONTAL DIPS

Geological boundaries are nearly or exactly parallel to topographical contours. They form a contour-like pattern, far apart on gentle slopes, close together on steep slopes. Boundaries form sharp 'V' shapes in steep-sided valleys. If the dip is up the valley the 'V' points up the valley; if the dip is down the valley the 'V' points down or up the valley according as the dip is greater or less respectively than the slope of the valley floor. More resistant rocks (e.g. limestones, sandstones, lavas, etc.) may form broad outcrops on plateaus or dip-slopes.

Fig. 2.3

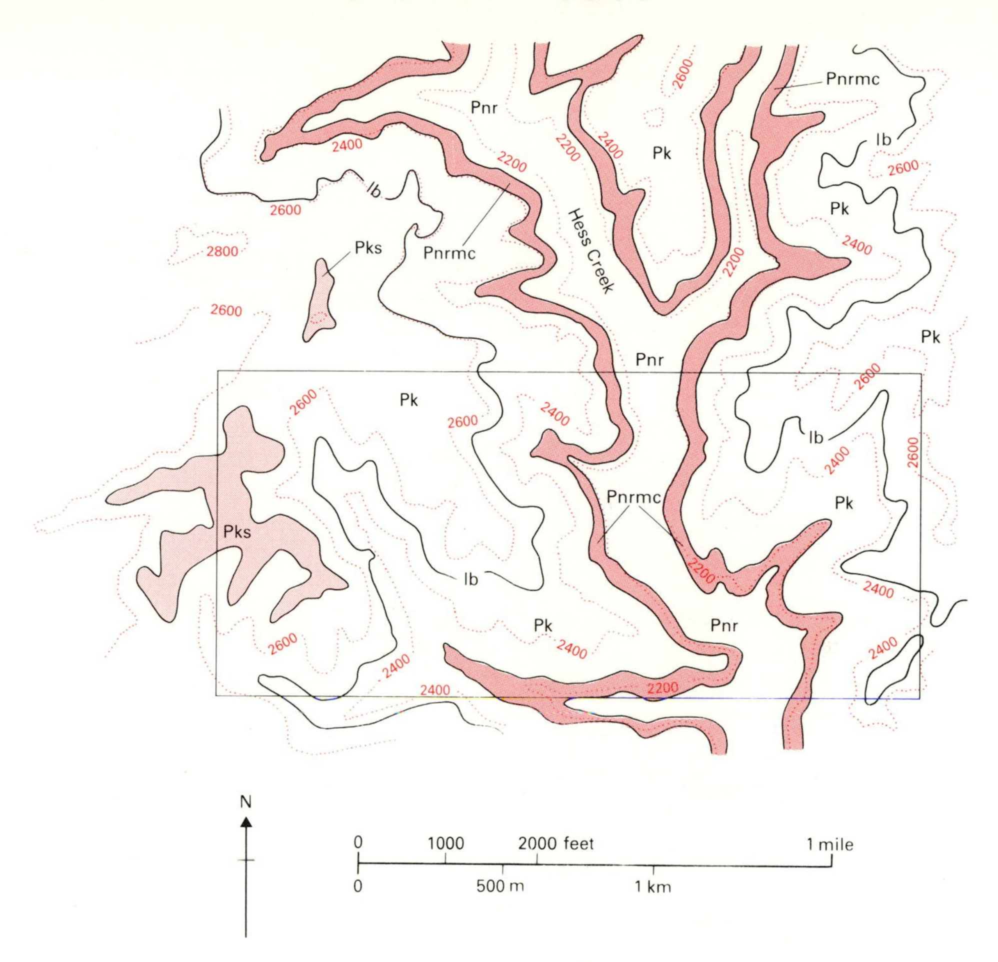

Fig. 2.3 Part of the central area of U.S.G.S. map GQ 1542 (Honaker Quadrangle, Virginia), showing nearly horizontal rocks of the Lower and Middle Pennsylvanian.

Pks	Unnamed sandstone member of the Kanawha Formation	Middle
Pk	Kanawha Formation (including lb: Lower Banner coal bed)	Pennsylvanian
Pnrmc	McClure Sandstone Member of the New River Formation	Lower
Pnr	New River Formation	Pennsylvanian

The rectangle shows the area of the block diagram (Fig. 2.6).

2.A.2 MEDIUM DIPS

Geological boundaries are nearly parallel to each other. In areas of level topography or where boundaries are exactly parallel to topographical contours, the boundaries are **strike lines** (horizontal lines in the plane of the inclined surface of the lithological boundary). In steep-sided valleys, boundaries show broad open 'V' shapes; the 'V' points in the direction of dip. The outcrop width of stratigraphical units depends on the dip: wider for low dips, narrower for high dips.

Fig. 2.4

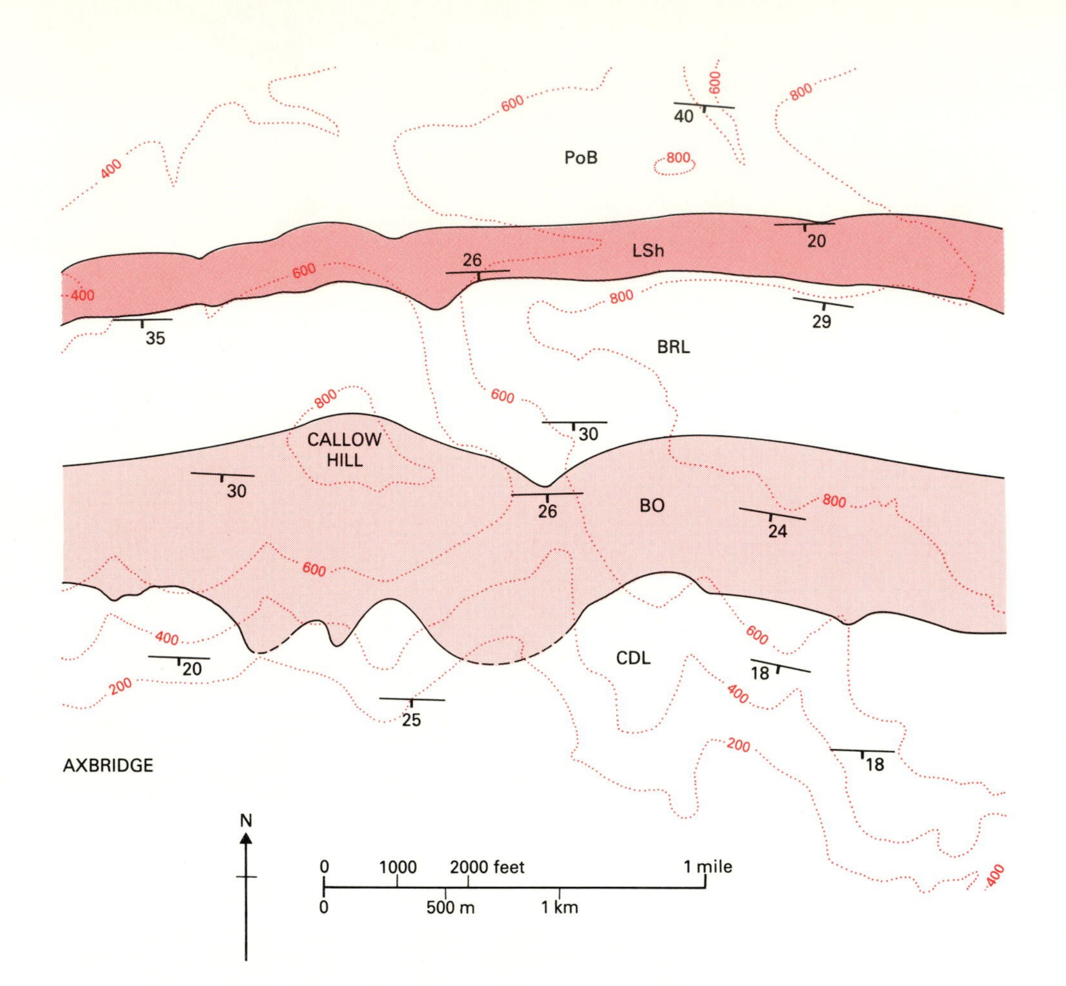

Fig. 2.4 Part of the central area of B.G.S. map ST45 of the Cheddar area, England, showing rocks of the Devonian and Carboniferous dipping at medium values to the south.

CDL	Clifton Down Limestone	Carboniferous
BO	Burrington Oolite	
BRL	Black Rock Limestone	
LSh	Lower Limestone Shale	
PoB	Portishead Beds	Devonian

Trias, Pleistocene, and Recent rocks omitted

The BRL/LSh boundary forms a strike line where it is parallel to the 800-foot contour for a distance of 1 km in the northeast corner of the map.

Figure 2.7 shows a block diagram of the same area. The eastern half of the area is the same as the southwestern part of Plate 1.

2.A.3 STEEP OR VERTICAL DIPS

Fig. 2.5

Geological boundaries are nearly or exactly parallel to each other. They show little or no deviation as they cross ridges and valleys.

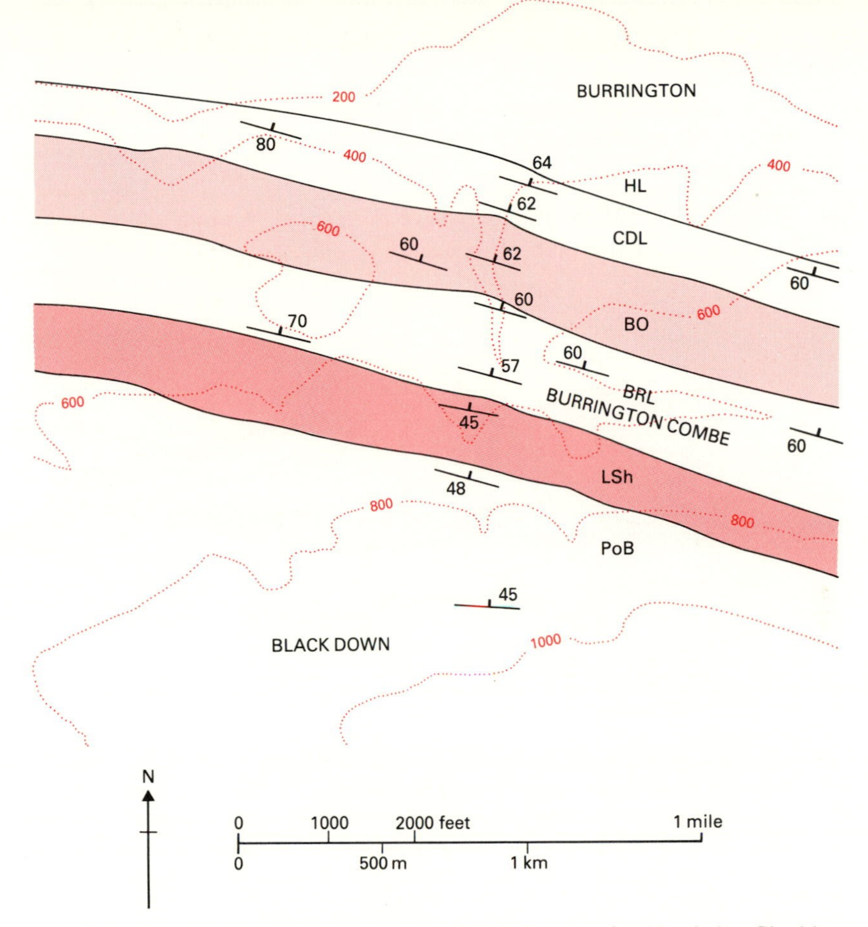

Fig. 2.5 Part of the northeast corner of B.G.S. map ST45 of the Cheddar area, England, showing rocks of the Devonian and Carboniferous dipping steeply to the north-northeast. Units labelled as in Fig. 2.4, and:

HL Hotwells Limestone Carboniferous
Trias, Pleistocene, and Recent rocks omitted.

The same area is shown in the northeastern part of Plate 1.

2.A.4 FLAT TOPOGRAPHY

Geological boundaries are strike lines (see Sect. 2.A.2). The amount of dip is not determinable unless the thickness of the formations is known (Sect. 2.C). Flat topography is commonly extensively covered by superficial deposits.

2.A.5 MOUNTAINOUS TOPOGRAPHY

Where the topographical slopes are appreciably steeper than the dip of the stratigraphical units, the outcrops resemble low-dip patterns even when the dips have medium or high values. Careful reading of the map is necessary to understand the intersections of the geological boundaries with the topographical surface.

2.B TO DETERMINE THE STRATIGRAPHICAL SUCCESSION

The stratigraphical succession is usually stated in the key to the map, but it is useful to be able to determine it by direct observation from the map.

2.B.1 NORMAL DIPS (the beds have been tilted through less than 90° from their orientation at the time of deposition)

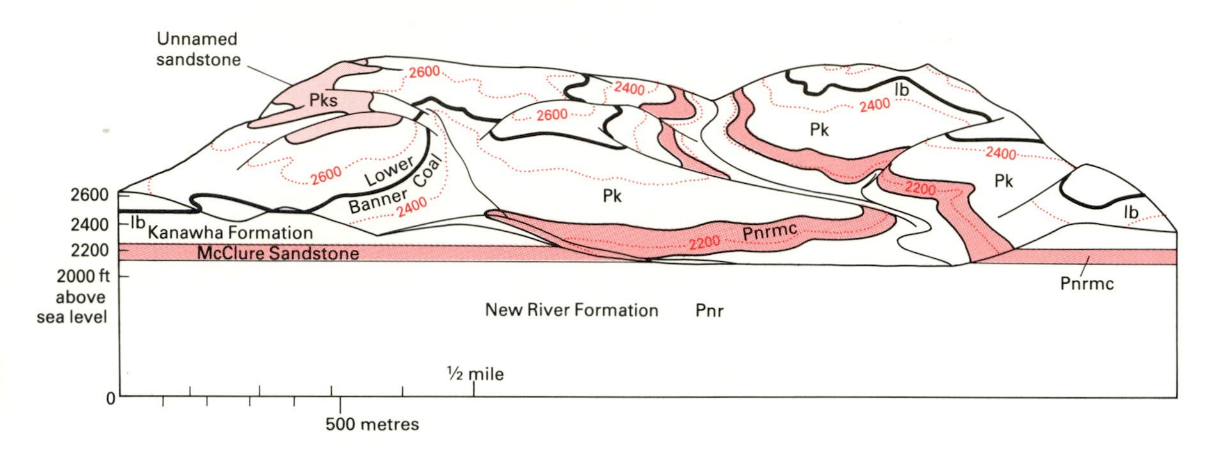

Fig. 2.6 Block diagram showing the area marked by a rectangle in Fig. 2.3. The stratigraphical succession is as set out in the caption of Fig. 2.3. (We have used block diagrams in this book as an aid to understanding the three-dimensional relationships of rock units. We do not intend that students should be able to construct block diagrams at this stage of learning map interpretation. The procedure for their construction is given in Badgley (1959) and Ragan (1985).)

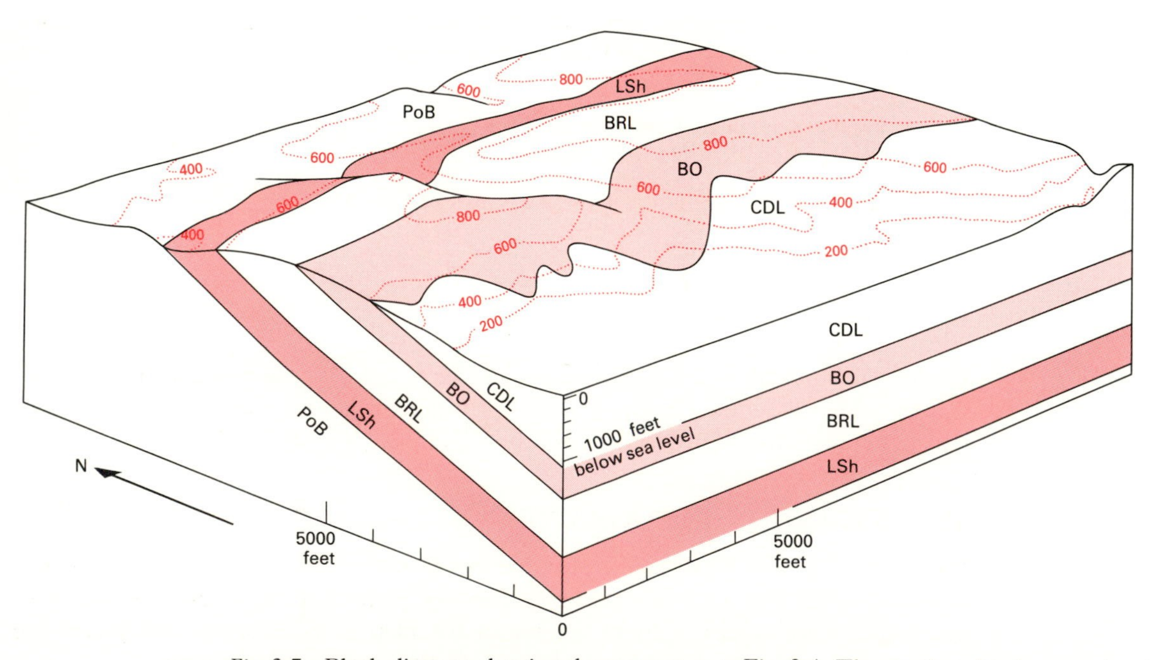

Fig. 2.7 Block diagram showing the same area as Fig. 2.4. The stratigraphical succession is as set out in the caption of Fig. 2.4.

Geologically younger beds lie above older beds. The beds become younger (higher up the stratigraphical succession) in the direction of dip (but see Section 2.B.2 for overturned dips).

If the lithological units are horizontal, the younger layers form hill tops and plateaus.

If the units are vertical, there is no direct means of determining the stratigraphical succession. Look for nearby non-vertical dips (but be careful to interpret overturned dips, Section 2.B.2, correctly) or use the general stratigraphical succession to determine the local succession.

Figs. 2.6, 2.7 If the stratigraphical succession is known, the direction of dip can be determined from the direction of younging of the units.

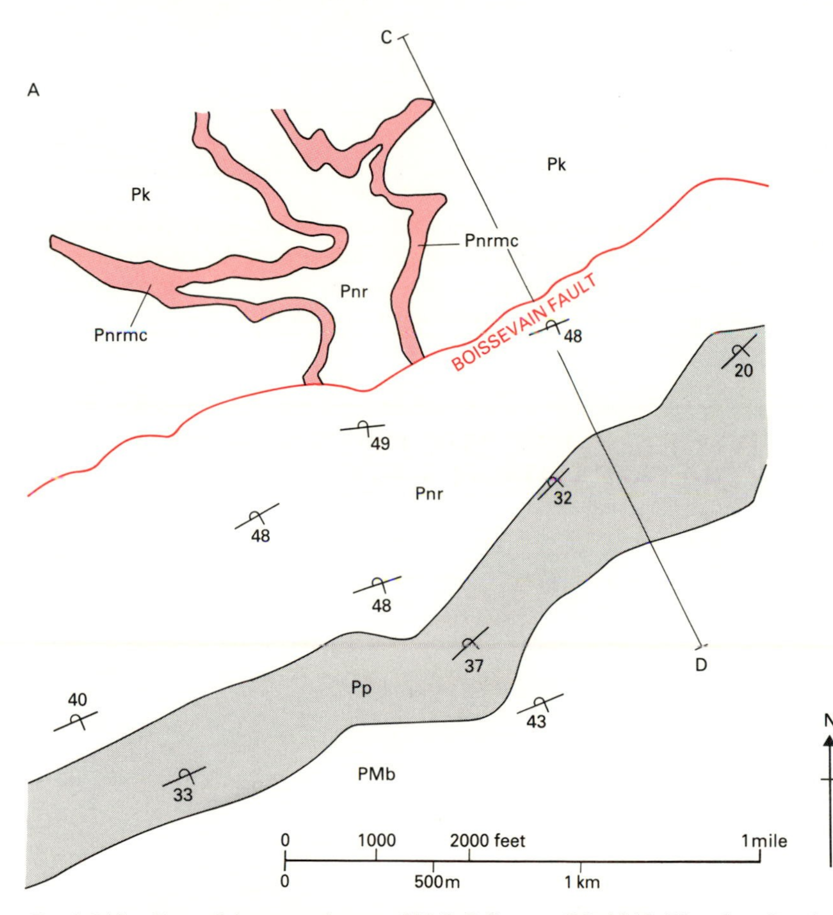

Fig. 2.8(A) Part of the central area of U.S.G.S. map GQ 1542 (Honaker Quadrangle, Virginia), slightly simplified, overlapping the southeast corner of Fig. 2.3, showing horizontal and overturned rocks of the Upper Mississippian and Lower Pennsylvanian. The direction of the (overturned) dip south of the Boissevain Fault is to the south, but the direction of younging is to the north. The stratigraphical succession is:

Pk	Kanawha Formation	Middle Pennsylvanian
Pnrmc	McClure Sandstone member of the New River Formation	Lower Pennsylvanian
Pnr	New River Formation	
Pp	Pocahontas Formation	
PMb	Bluestone Formation	Upper Mississippian – Lower Pennsylvanian

2.B.2 OVERTURNED DIPS (beds which were horizontal at the time of formation and which have since been tilted through more than 90°)

Figs. 2.8(A) and (B)

Dip measurements in such units are indicated by a special symbol – see the key to the map. Because of the extent of the deformation, younger rocks now lie *below* older rocks, and particular care must be taken in the determination of the stratigraphical succession.

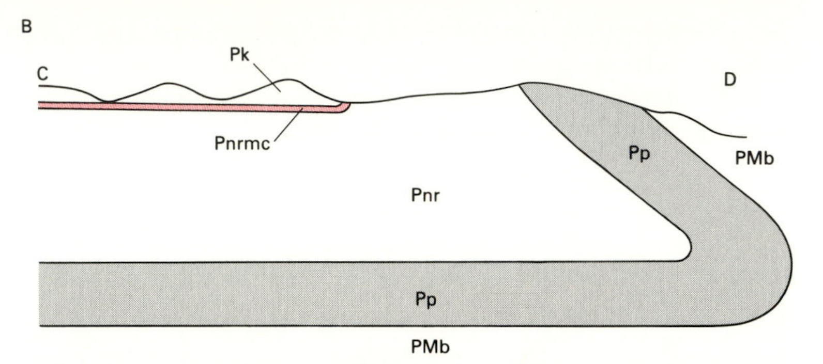

Fig. 2.8(B) Diagrammatic north-northwest – south-southeast cross-section along the line C–D in Fig. 2.8(A), showing the relation of the overturned to the horizontal rocks. The section has been simplified by omitting the effect of the Boissevain Fault, which has a displacement of only about 150 metres. Units labelled as in Fig. 2.8(A).

2.C TO DETERMINE THE THICKNESS OF A STRATIGRAPHICAL UNIT

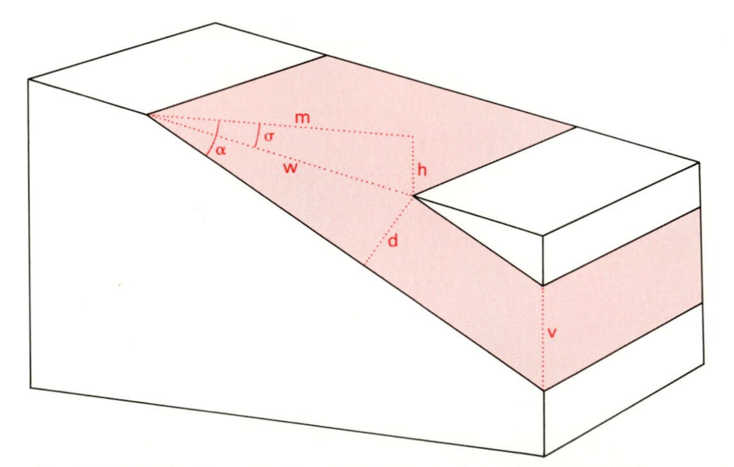

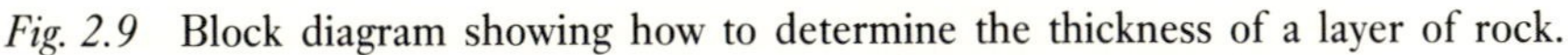

Fig. 2.9 Block diagram showing how to determine the thickness of a layer of rock.

m = horizontal distance between the outcrop of the base and the top of the layer (as measured on the map); the measurement must be made parallel to the direction of dip
h = vertical difference in height of the outcrop of the base and the top of the layer
σ = average angle of slope
w = true outcrop width on the ground
α = angle of dip
d = thickness of the unit
v = vertical depth of the unit

(*Caption of Fig. 2.9 cont.*)

The angle of the topographical slope is:

$$\sigma = \tan^{-1}\frac{h}{m}$$

and the true outcrop width on the ground is then:

$$w = \frac{m}{\cos \sigma}$$

The thickness of the layer is given by:

(i) *dip direction same as slope direction*

$$d = w \, . \, \sin\,(\alpha - \sigma)$$

(ii) *dip direction opposite to slope direction*

$$d = w \, . \, \sin\,(\alpha + \sigma)$$

In both cases $d = v \, . \, \cos \alpha$.

Use of these relationships for sloping topography can commonly be avoided by finding places where the top and base of a layer are at the same topographical level along a line parallel to the direction of dip (see Fig. 2.10).

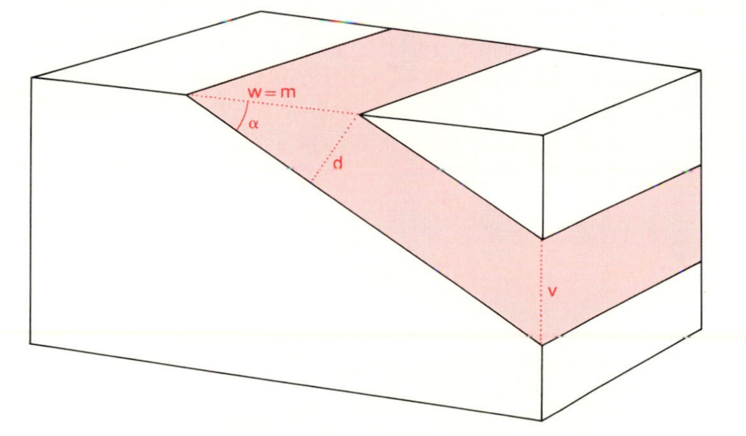

Fig. 2.10 Block diagram showing the outcrop of a layer on a horizontal topographical surface. Symbols as in Fig. 2.9. The thickness of the layer is $d = w \, . \, \sin \alpha$.

In selecting a position on the map to measure the outcrop width:

1. Check that there are no faults (displacements of outcrop – see Section 4.A) which would repeat or omit parts of the complete section.
2. Check that the dip is consistent both along and across the outcrop. In particular make sure that there are no folds (reversals of dip direction – see Section 5.A) which would repeat parts of the section.

   If the dips are variable, but in the same direction, determine the thickness by measurement of a succession of segments, assuming as a first approximation that the change of dip occurs at the mid-point between the dip observations.
3. Check that there are no intrusive igneous rocks later than the stratigraphical unit (cf. Sect. 2.I).

2.C.1 HORIZONTAL DIPS AND DISSECTED TOPOGRAPHY

Determine the topographical height of the base and the top of the unit. The difference is the stratigraphical thickness of the unit (d = v).

Fig. 2.11

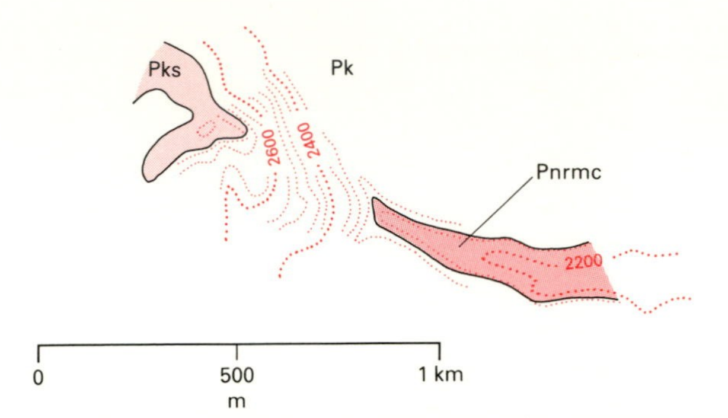

Fig. 2.11 Enlargement of part of Fig. 2.3 showing the base of the Kanawha Formation (Pk) and the base of the unnamed sandstone member of the same formation (Pks). Topographical contour interval 40 feet. The two horizons are at 2260 feet and 2720 feet respectively, giving the thickness of Pk up to the base of the sandstone as 460 feet or 140 metres.

2.C.2 MEDIUM DIPS

Measure the outcrop width on the map exactly parallel to the direction of dip (perpendicular to the direction of strike). Determine the difference (if any) in the topographical height of the top and base of the unit. Use the relationships above to calculate the stratigraphical thickness.

Fig. 2.12

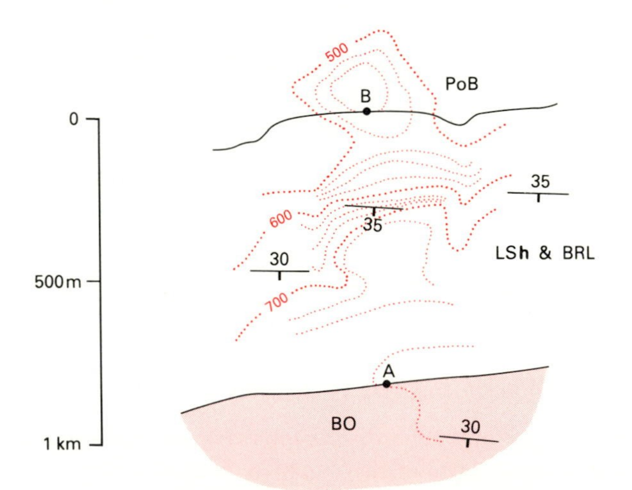

Fig. 2.12 Enlargement of part of Fig. 2.4 to show the boundaries of LSh and BRL (part of the Carboniferous) with adjacent units. Topographical contour interval 25 feet. The altitude of the top of BRL at point A is 775 feet, the base of LSh at point B is at 550 feet. The outcrop width on the map is 850 metres. The average dip is 33°.
From calculation:

h = 225 feet or 69 metres
σ = 4.6°
w = 853 metres
d = 520 metres

2.C.3 VARIABLE DIPS

Fig. 2.13

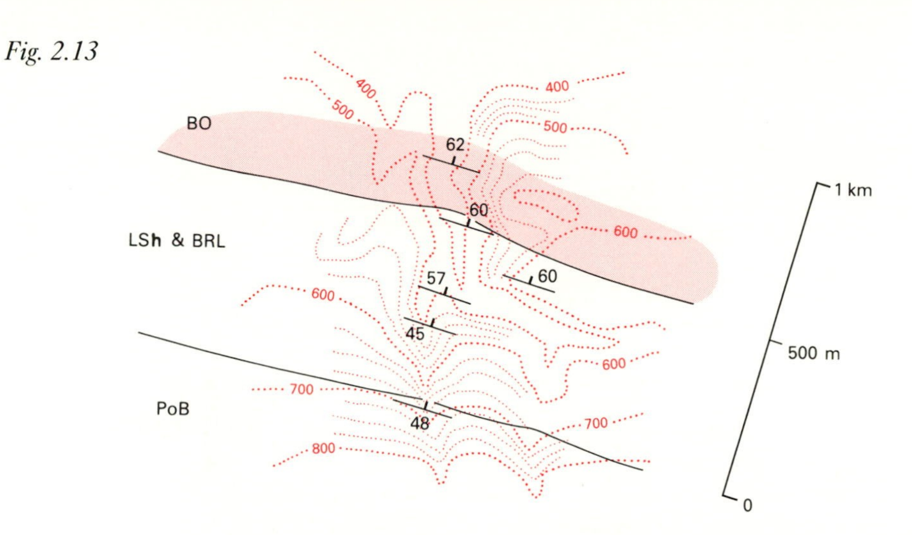

Fig. 2.13 Enlargement of part of Fig. 2.5 showing the boundaries of LSh and BRL (Carboniferous) with adjacent units, to illustrate the determination of thickness when the dip is variable. Topographical contour interval 25 feet and 100 feet. Measurements to determine the thickness of LSh + BRL are:

Segment	*h* (*metres*)	*m* (*metres*)	α
1	30	108	48°
2	15	150	45°
3	23	125	57°
4	15	188	60°

The calculated thickness of LSh + BRL is 410 metres.

2.C.4 VERTICAL DIPS

If the dip is vertical, the stratigraphical thickness is the same as the outcrop width on the map (d = w = m).

2.D TO CONSTRUCT A STRATIGRAPHICAL COLUMN

A stratigraphical column is in effect a representation of the rocks as though a borehole were drilled through them perpendicular to all the geological boundaries.

Select a suitable scale. Construct a vertical column to show the lithology, thickness, and names of the stratigraphical units. *Fig. 2.14*

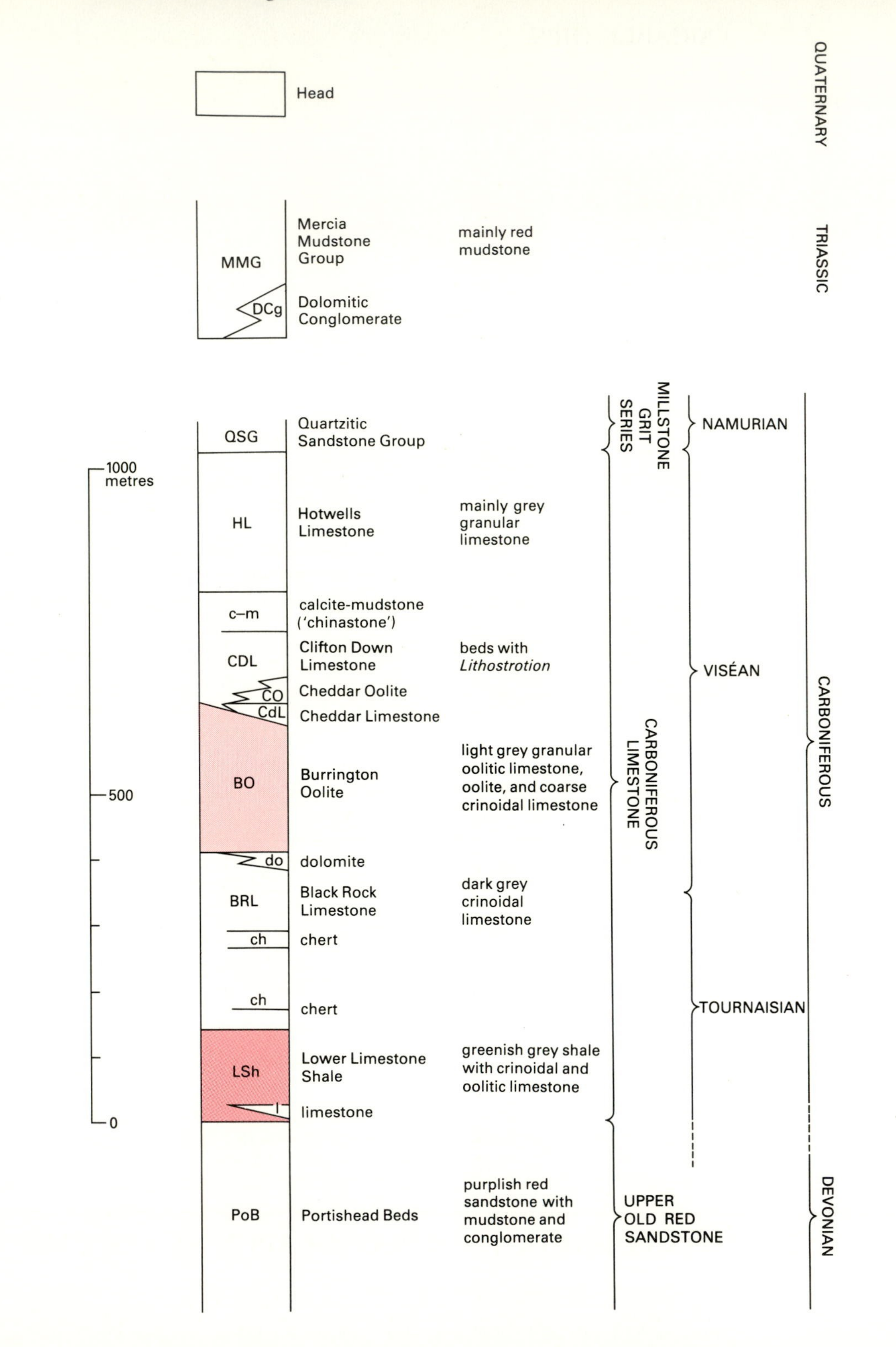

Fig. 2.14 Stratigraphical column simplified from the section published on B.G.S. map ST45 (Cheddar) and incorporating observations from Figs 2.4 and 2.5 and Plate 1.

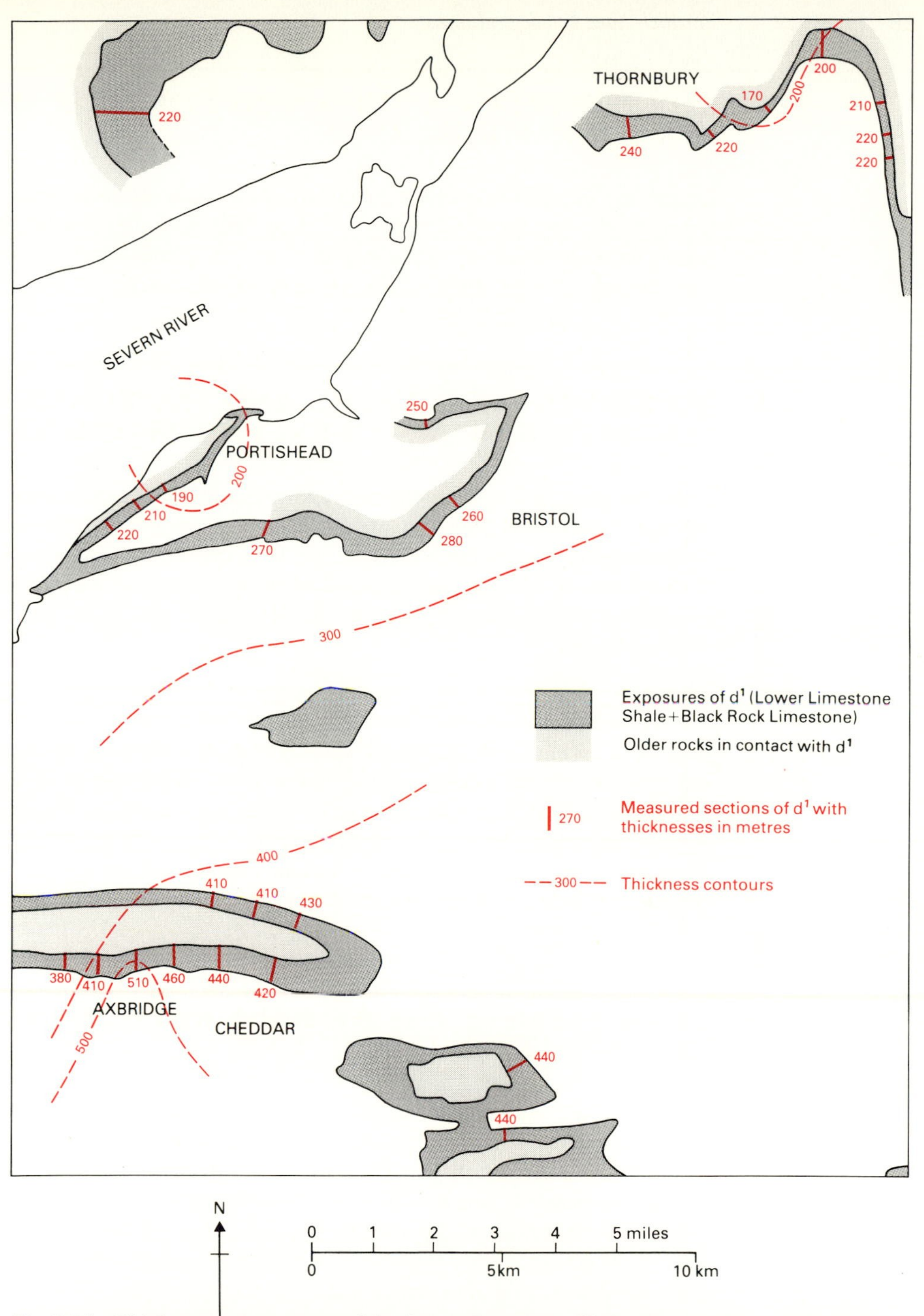

Fig. 2.15 Thickness contour map of the Lower Limestone Shale (d^{1a} or LSh) and Black Rock Limestone (d^{1b} or BRL) in the Bristol area, England (B.G.S. Special Sheet). The Cheddar area occupies the southwest corner of this map. Thicknesses of individual sections measured from the map are given in metres, and thickness contours have been approximately interpolated; because of the relatively sparse information, it is only possible to show thickness contours for small areas of the map. The greatest variation of thickness across the map is 300 metres in 40 km, or an angular divergence of the top and the base of the unit of 0.4°. Thus although the unit varies in thickness by a factor of $2\frac{1}{2}$, its top and base are essentially parallel; within a small area it shows a close approximation to the constant-thickness principle.

2.E TO CONSTRUCT A THICKNESS CONTOUR MAP OF A STRATIGRAPHICAL UNIT

Determine the thickness of the stratigraphical unit in different areas of the map, using the methods described in Section 2.C. Use bore-hole data if available. Plot the thicknesses on an outline copy of the map (reduced in scale if required). Interpolate contour lines between the locations of the measured thicknesses. *Fig. 2.15*

2.F TO IDENTIFY THE FACIES AND ENVIRONMENT OF FORMATION OF A SEQUENCE OF SEDIMENTARY ROCKS

Read the descriptions of the stratigraphical units on the published map and any additional descriptions that are available, noting particularly the rock types and any environmental interpretations. Make use of Table 2.1 or any other sources to identify the probable or possible environment of formation of the rock units, noting that only provisional identifications can be made unless the rock descriptions are very comprehensive.

Example The rock descriptions on Fig. 2.14 suggest that the Upper Old Red Sandstone corresponds to a continental (probably river) environment. The Carboniferous Limestone was formed in a coastal or shelf (probably carbonate shelf) environment. The sequence represents a marine transgression at the beginning of the Carboniferous, with a progressive decrease in the amount of terrigenous clastic sediment and the formation of pure limestones, followed by shallowing of the sea with the formation of oolites. The lithology of the Mercia Mudstone Group corresponds to several possible coastal or continental facies, and was in fact formed in an arid shoreline or desert environment.

2.G TO DETERMINE THE AVERAGE RATE OF DEPOSITION OF A STRATIGRAPHICAL UNIT

Determine the thickness of the unit (Sect. 2.C). Identify the stratigraphical age of the base and the top of the unit, using the legend of the map, supplemented if necessary by additional information (e.g. from the description or memoir accompanying the map sheet); correlate the stratigraphical age with the chronometric time-scale (Appendix 2). Divide the thickness of the unit by the age range of the unit to obtain the average rate of deposition. (This calculation necessarily includes the effects of compaction of the sedimentary pile after deposition.)

Note that sedimentation is normally an episodic process; the average rate of deposition is much slower than the rate of deposition of individual beds.

Example The Carboniferous Limestone of the Cheddar area (Fig. 2.14) has a total thickness of 1020 metres. The stratigraphical column shows that the base is near the beginning of the Tournaisian (355 Ma – Appendix 2), and the top is at the end of the Viséan (325 Ma), an interval of a little less than 30 million years. The average rate of deposition of the rocks was therefore approximately (1020 × 1000) millimetres/(30×10^6) years = 0.03 mm/y (cf. Table 1.2).

2.H TO CONSTRUCT A STRUCTURE CONTOUR MAP OF A GEOLOGICAL SURFACE

A structure contour line marks all the points on a surface which are at the same altitude and structure contours are used in the same way as topographical contours to reveal the three-dimensional shape of the surface. Structure contours are the same as strike lines (Sect. 2.A.2) – note that although in some synthetic map constructions strike lines are commonly shown as straight and parallel, in real geological situations they are generally curved and unequally spaced.

Mark all the points where the geological surface intersects a particular topographical contour line and connect the points with a smooth line; the structure contour lines must intersect or cross the topographical contour *only* at the marked points. Repeat for all available topographical contours and intersections.

The method can be used to construct a contour map of any geological surface – boundaries of lithological units (either sedimentary or igneous), faults (Sect. 4.C), and unconformities (Sect. 6.B). The construction of a vertical cross-section (Sect. 9.C) provides additional data for extending the structure contours far below the ground surface (Sect. 9.D).

The dip is perpendicular to the contours and in the direction of decreasing altitude; its value is given by:

$$\alpha = \tan^{-1}\frac{h}{m}$$

where h = contour interval and m = horizontal distance (measured on the map) between the contour lines. *Fig. 2.16*

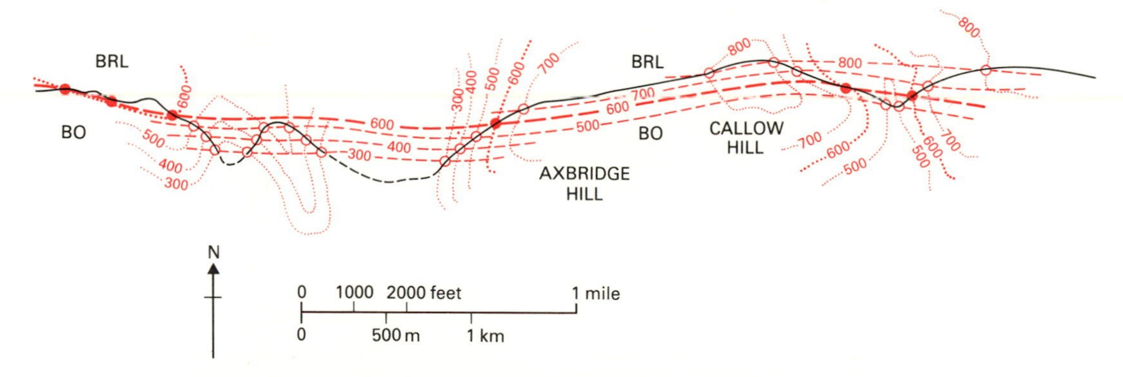

Fig. 2.16 Part of the west-central area of B.G.S. map ST45 of the Cheddar area, England (partly included in Plate 1), showing the boundary between the Black Rock Limestone (BRL) and the Burrington Oolite (BO) and some of the topographical contours. Red dots and circles mark the intersection of the boundary with the topographical contours. The 600-foot structure contour is shown by a long-dashed red line and the remaining structure contours by short-dashed red lines.

At Callow Hill the spacing of the 800- and 500-foot contours gives the dip as

$$\tan^{-1}\frac{(800-500)\text{ feet}}{625\text{ feet}} = 26° \quad \text{to the south.}$$

2.I PARALLEL-SIDED LITHOLOGICAL UNITS OTHER THAN SEDIMENTARY AND VOLCANIC ROCKS

Certain intrusive igneous rocks – sills and dykes (Sect. 3.D) – form parallel-sided rock units, the former predominantly parallel to adjacent sedimentary units and the latter at a high angle to stratigraphical boundaries. The direction and amount of dip and the thickness of sills and dykes can be determined as described in Sections 2.A, 2.C, and 2.H, but note that dykes may be mapped conventionally (e.g. by a ribbon of uniform width) because of their usually narrow outcrop width.

Intrusive rocks are necessarily younger in geological age than the stratigraphical units in which they occur. In estimating the thickness of stratigraphical units (Sect. 2.C), sills must be excluded from the measurements.

The method of distinguishing between a lava flow and a sill is described in Section 3.E.

3 LITHOLOGICAL UNITS WITH LESS REGULAR SHAPES (SEDIMENTARY, IGNEOUS, AND METAMORPHIC ROCKS AND MINERAL DEPOSITS)

The shape of a lithological unit on a map depends on its three-dimensional form and orientation and on the shape of the erosion surface on which it is displayed. Many rock units (especially those of sedimentary and volcanic origin, and intrusive sills and dykes) are parallel-sided, laterally extensive, and have essentially constant thickness (Ch. 2). In this chapter we describe less regular bodies of rock, including certain sedimentary rocks, the majority of igneous and metamorphic rocks, and many mineral deposits. The generalised shapes of rocks of these types are tabulated in Table 3.1 and illustrated in Figure 3.1; methods for recognising them from their outcrop patterns are set out in Section 3.A. Chapter 7 describes some present-day landforms and processes which are relevant for the reconstruction of ancient analogues as revealed on geological maps.

SEDIMENTARY ROCKS originally deposited in geographically restricted areas are laterally discontinuous and of variable thickness (Sect. 3.B). There may be no contiguous or nearby rocks of precisely the same age – for example, deposits in an isolated inland lake. Much more commonly, a particular lithology represents an environment or facies (Table 2.1) which changes laterally into another facies of the same age – as in a coastal or deltaic area where river, coastal, and marine rocks are all deposited contemporaneously in different parts of the area. In these situations the mapped boundaries between adjacent lithological units represent the geographical limits (usually gradational on a fine scale) between different environments. Such boundaries cut across time planes and are **diachronous** (Sect. 3.C).

IGNEOUS ROCKS originate as magmas at depths (typically between about 20 and 150 km below the surface of the Earth) where the temperature is high enough for rocks of the crust or mantle to become partially molten. The magma rises because of its buoyancy relative to solid rocks; the present shapes of igneous rocks reflect the processes by which the magma was emplaced within (**intrusive rocks**) or on (**extrusive rocks**) the surface of the Earth. Many volcanic piles and some individual units (especially flood basalts and ignimbrites) have parallel base and top as described in Chapter 2; individual lavas may have fan, shoestring, dome, or balloon shapes according to their rheology and the shape of the land surface onto which they were erupted (Table 3.1, Fig. 3.2, Fig. 7.1). Subaerial volcanoes are rapidly reduced by erosion, particularly if they include a large proportion of unconsolidated pyroclastic material; often an isolated vent or dyke is the only remaining evidence for a once substantial volcanic cone or fissure eruption. Fragments of the younger, upper components of old volcanic series may be preserved by down-faulting in calderas (Fig. 3.9).

Intrusive igneous rocks are classified as **concordant** or **discordant** according to their relationship to adjacent sedimentary or metamorphic rocks (Sect. 3.D). Concord-

(*Text continues on p. 37*)

Table 3.1 Typical shapes of some (non-parallel-sided) sedimentary, igneous, and metamorphic rock units

Idealised three-dimensional shape	*Typical idealised shape at outcrop*	*Typical shape in vertical cross-section*	*Illustration (Fig. 3.1)*	*Rock compositions and* *Sedimentary* *Composition*
Steep-sided cone	Circle, ellipse, broad-based triangle	Broad-based triangle	A	
Gentle-sided cone	Circle, ellipse, sub-parallel-sided band	Very broad-based triangle	A	
Steep-sided fan	Semicircle, wedge	Wedge	B	Conglomerates, sandstones
Gentle-sided fan	Low-angle wedge	Low-angle wedge	B	Sandstones, conglomerates
Lens	Circle, ellipse, double-tapering wedge	Double-tapering wedge	C	Limestones, dolomites
Sinuous elongate (shoestring)	Elongate (straight or meandering); short double-tapering wedge	Elongate; double-tapering wedge	D	Sandstones, mudstones
				Limestones, dolomites
Vertical sheet	Long, narrow, parallel-sided, cutting across other rock boundaries	Long, narrow, vertical parallel-sided	E	
Horizontal or low-dipping sheet (may be irregularly stepped)	Generally parallel to other rock boundaries, locally transgressive	Generally parallel to other rock boundaries, locally transgressive	F	

environments in which the shape may be developed				
	Igneous and metamorphic		*Notes*	*Examples*
Environment	*Composition*	*Environment*		
	Pyroclastic rocks, lavas of intermediate composition	Volcanic cone		Fig. 7.1
	Lavas of basaltic composition	Shield volcano		Plate 3
Alluvial fan	Individual agglomerate, lava flow (intermediate composition)	Volcano	Commonly accumulate to form composite piles which may have parallel top and base.	Plate 1, Fig. 7.3
Terrestrial, fluvial, and deltaic environments	Individual basaltic lava flow, ignimbrite, ash	Volcano	Margin of the pile may have tapering or saw-tooth profile.	Plate 4
Patch reef, reef knoll, Waulsortian mound	Intrusive igneous rocks	Sill, laccolith		Figs 3.4, 3.6
River channel deposit Beach deposit	Lava of basaltic composition	Lava flow in a valley		Figs 7.1, 7.5
Fringing reef				Fig. 3.6
	Intrusive igneous rock	Dyke	Often numerous: parallel or radiating pattern (dyke swarm)	Fig. 3.7, Plate 4
	Intrusive igneous rock	Sill	Commonly associated with dykes, vents, stocks	Fig. 3.7
	Layered basic and ultrabasic rock	Lopolith	Giant intrusion, thicker in the centre	

(*Table 3.1 cont.*)

Idealised three-dimensional stage	*Typical idealised shape at outcrop*	*Typical shape in vertical cross-section*	*Illustration (Fig. 3.1)*	*Rock compositions and*
				Sedimentary
				Composition
Vertical cylinder	Circle	Vertical, parallel-sided, cutting across other rock boundaries	G	
				Salt; clay, breccia
Balloon, pear	Circle	Ellipse, pear	H	Salt; clay, breccia
Dome	Circle	Semicircle	I	Limestone, dolomite
Hollow vertical cylinder	Hollow ring (may be incomplete or discontinuous); Ring intrusions are commonly coincident with a ring-shaped fault; younger rocks on the inside of the ring	Vertical, parallel-sided; Ring intrusions are commonly parallel to faults, with younger rocks inside	J	
	Hollow ring, parallel to an intrusive rock of vertical cylinder shape	Vertical, parallel to boundaries of igneous rock	J	
Steep-sided hollow cone	Hollow ring	Steeply-dipping parallel-sided converging or diverging downwards	K	
Irregular	Irregular	Irregular (shape in depth may not be determinable)	L	Any of the above rock unit

environments in which the shape may be developed				
	Igneous and metamorphic		*Notes*	*Examples*
Environment	*Composition*	*Environment*		
	Intrusive plutonic rock (diameter several kilometres)	Stock, pluton	May be composed of several successive intrusions	Fig. 3.8
Diapir	Intrusive hypabyssal rock (diameter a few hundred metres)	Vent, Stock	Shape may be adapted to pre-existing planes of weakness (faults, etc.)	Fig. 3.5, Plate 2
Diapir	Intrusive hypabyssal or plutonic rock (diameter several kilometres)	Diapir		
	Extrusive acid igneous rock (diameter a few hundred metres)	Dome		Fig. 7.1
Reef	Extrusive acid igneous rock (diameter of a few hundred metres)	Dome		Figs 3.6, 7.1
	Intrusive hypabyssal or plutonic rock	Ring intrusion (boundary of a caldera)		Fig. 3.9
	Metamorphic rocks	Zones of contact metamorphism around a cylindrical intrusive		Fig. 3.10, Plate 4
	Intrusive igneous rock	Ring intrusion (cone sheet, ring dyke)		Fig. 3.9
may develop irregular shape because of local environmental constraints				Fig. 3.5, Plate 2

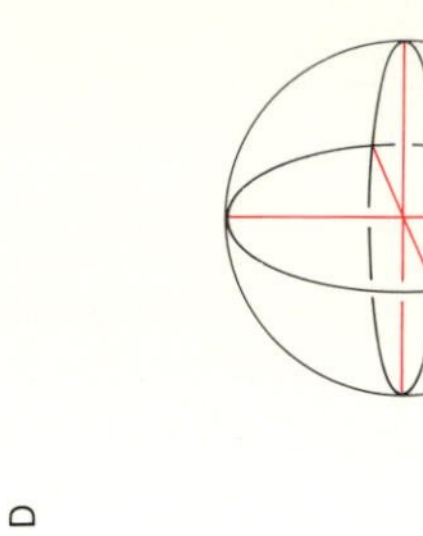
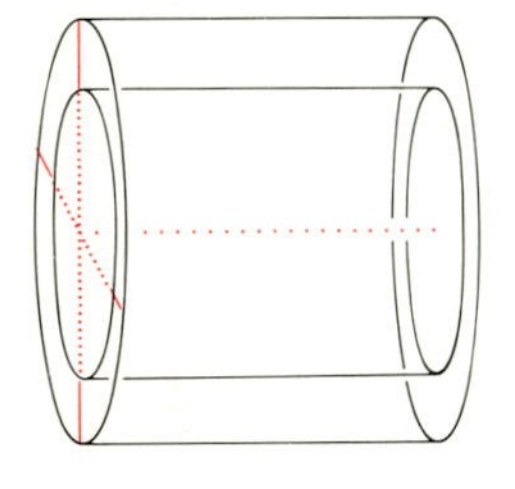
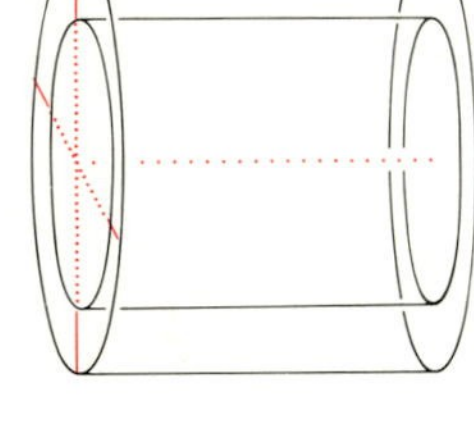
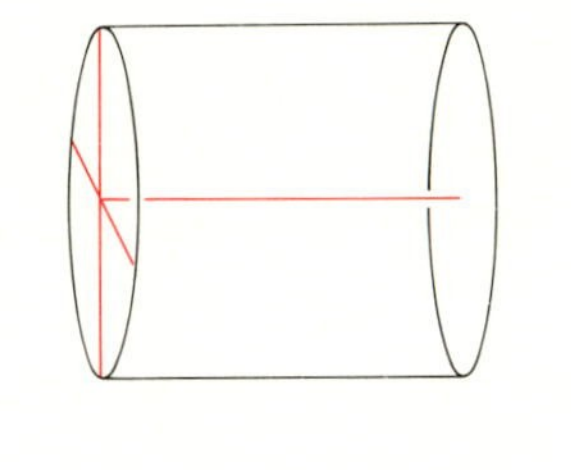
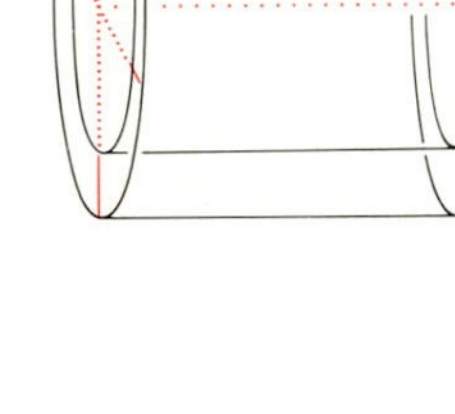
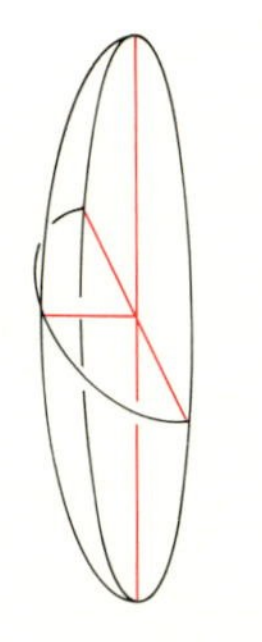
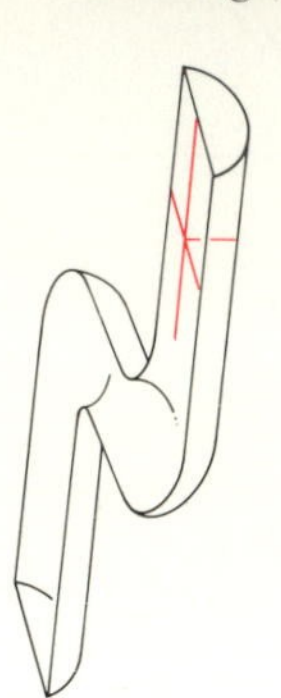
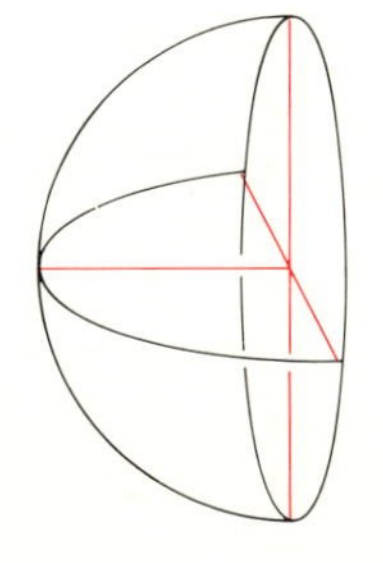

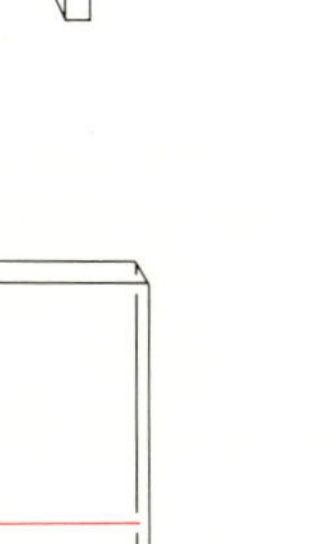
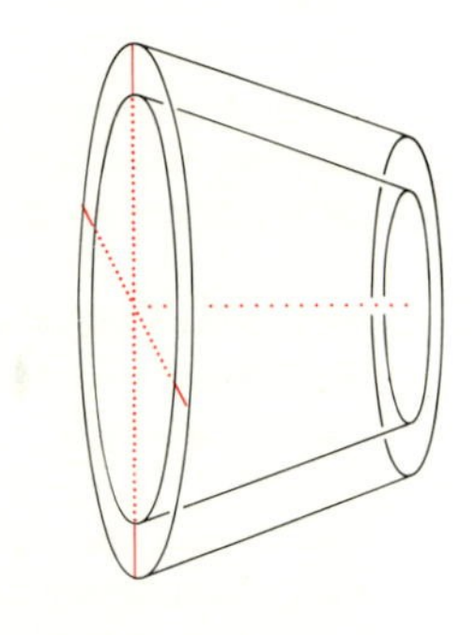
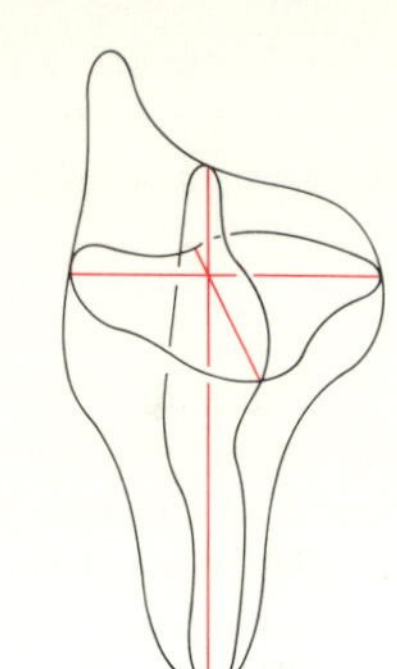

Fig. 3.1 Idealised shapes of some typical bodies of sedimentary, igneous, and metamorphic rocks (to be studied in conjunction with Table 3.1). A, cone; B, fan; C, lens; D, shoestring; E, vertical sheet; F, stepped horizontal sheet; G, vertical cylinder; H, balloon or pear; I, hemisphere (dome); J, hollow vertical cylinder; K, steep-sided hollow cone; L, irregular.

ant intrusives need to be distinguished from extrusive lava flows (Sect. 3.E). More precise details of the shapes and relationships of intrusive rocks provide information on their **mechanism of emplacement** (Sect. 3.F). Igneous rock-types that consistently occur together are grouped into **series** and these tend to be found in distinctive **plate-tectonic environments** (Table 3.2, Sect. 3.G).

Further details on igneous rocks are given in Williams & McBirney (1979) and Hyndman (1985).

Fig. 3.2 The volcano Teide (3734 m) on Tenerife in the Canary Islands. The summit cone of the volcano with its central crater, seen in profile, was built up within an older, larger crater, and has steeper slopes than the underlying structure. The slopes on the left are covered by a wide-spreading dark-coloured lava with steep sides and a lobate front which flowed across older, lighter-coloured flows on the lowest slopes of the volcano. In the immediate foreground are erosional relics of older flows and pyroclastic rocks. The slopes of the volcano have been eroded by deep gulleys which are relatively straight, typical of young erosion channels on a steep slope.

METAMORPHIC ROCKS are formed from pre-existing sedimentary, igneous, or older metamorphic rocks by recrystallisation in response to temperature, pressure, and stress. The increase of temperature in the rocks around a hot igneous intrusion gives rise to successive zones in a **metamorphic aureole**. The grade of regional metamorphism in rocks involved in mountain-building processes depends on the distribution of temperature and pressure in the orogenic belt, which in general increase systematically with depth. The **zones or facies of regional metamorphism** are layers approximately parallel to the Earth's surface as it was at the time of orogeny; these are now exposed on the present-day erosion surface as bands superimposed on (but often nearly parallel to) stratigraphical and structural boundaries in the metamorphosed rocks (Sect. 3.H). On geological maps, metamorphic rocks may be recorded on the simple basis of the structural features visible in the field (Table 3.3, first two columns). With further detail, the wide range of metamorphic conditions can be subdivided into **zones**

(*Text continues on p. 46*)

Table 3.2 Igneous rock types related to tectonic environment

Notes: This table is a simplified guide to the tectonic environment in which certain igneous rock types, as commonly recorded on published geological maps, probably formed. The correlations suggested are not definitive and much other information, for example chemical and isotopic features of the igneous rocks (not frequently supplied on geological maps) and the nature of contemporaneous interbedded sediments, is used by geologists making this kind of correlation.

Notice that the correlation is with the *site of formation*: subsequent tectonic movements may move igneous rock formations from one environment to another; for example, volcanic rocks formed at an oceanic spreading ridge may be obducted on to a continental edge.

			Divergent plate margins		
Rock series	*Volcanic rocks*	*Intrusive rocks*	*Oceanic spreading ridges*	*Continental passive margins*	*Continental rifts*
Tholeiitic series	picrite basalt	picrite	—	x	O
	tholeiite/tholeiitic basalt	gabbro, dolerite, diabase	X	X	O
	icelandite		—		O
	rhyolite	granite; granophyre; felsite; quartz porphyry	—	X	O
	obsidian, pitchstone		—	x	O
		plagiogranite	—		O
Calc-alkaline series	basalt	gabbro	O	O	O
	basaltic andesite		O	O	O
	boninite		O	O	O
	andesite	diorite	O	O	O
	dacite	granodiorite	O	O	O
	rhyodacite	monzogranite	O	O	O
	rhyolite	granite, porphyry, felsite, quartz porphyry	O	O	O

The following rock types recorded on geological maps are igneous rocks (name in brackets) metamorphosed under relatively low-temperature hydrous conditions. They are particularly associated with oceanic spreading ridges but can form in other settings if the metamorphic conditions are appropriate.

serpentinite (ultramafic rocks)
spilite (basic rocks)
keratophyre (intermediate and acid volcanics)

X distinctive rock type of the environment
x Rock type commonly present, occasionally abundant
— Rock type occasionally present
O Rock type unlikely in the environment
? Possible occurrence

A blank space indicates no equivalent rock type known or no information available.

Igneous rocks in back-arc or marginal basins have been omitted because of scarcity of information — some occurrences of tholeiitic and alkali basalts have been described; pyroclastic rocks are subordinate.

Convergent plate margins			*Transform plate margins*		*Intraplate environments*	
Island Arcs	*Active continental margins*	*Continental collision zones*	*Oceanic transform faults*	*Continental transform faults*	*Continental hot spots*	*Oceanic hot spots*
O	O	O		O	O	x
x	O	O		O	x	X
O	O	O		O	O	x
O	O	O		O	x	x
O	O	O		O	—	x
O	O	O		O	O	O
X	X	O	O	x	x	O
X	X	X				O
x	?	O	O			O
X	X	X	O	?	?	O
x	X	x	O	?		O
x	x	x	O			O
x	x	X	O			O

(*Table 3.2 cont.*)

Rock series	*Volcanic rocks*	*Intrusive rocks*	*Divergent plate margins*		
			Oceanic spreading ridges	*Continental passive margins*	*Continental rifts*
Alkalic series (1) mildly alkaline, non-feldspathoidal	alkali (olivine) basalt	alkali gabbro	These rock types are geographically associated with certain ridges but are parts of volcanic centres sited on hot spots	x	x
	trachybasalt	monzonite; syenodiorite		x	—
	trachyandesite	syenogabbro		x	—
	hawaiite			x	—
	mugearite			x	—
	benmoreite			x	—
	trachyte	syenite		x	x
	comendite			x	x
	pantellerite			—	x
	alkali rhyolite	alkali granite, granophyre		x	x
(2) mildly alkaline, feldspathoidal	basanite	teschenite		x	x
	tephrite	theralite		—	x
	phonolite	nepheline syenite tinguaite		—	x
(3) strongly alkaline, feldspathoidal; some ultramafic	nephelinite	ijolite	O	O	x
	melilite basalt		O	O	—
	melilitite		O	O	—
	limburgite		O		x
Potassic Series including feldspathoidal types	absarokite		O	O	—
	shoshonite		O	O	x
	latite	monzonite	O	O	O
	tristanite		see note above	O	O
	leucitophyre		O	O	O
	leucitite		O	O	—

Convergent plate margins			*Transform plate margins*		*Intraplate environments*	
Island Arcs	*Active continental margins*	*Continental collision zones*	*Oceanic transform faults*	*Continental transform faults*	*Continental hot spots*	*Oceanic hot spots*
O	O	O	x		x	X
O	O	O	x		x	x
O	O	O	?		x	x
O	O	O			x	x
O	O	O			x	x
O	O	O			x	x
O	O	O			x	x
O	O	O			x	x
O	O	O			x	
O	O	O	?		x	x
	O	O			—	x
	O	O			—	
	O	O			—	x
	O	O			—	x
	O	O				x
	O	O				x
	O	O				—
	O	—				
	x	—			—	
x	x	—			—	—
	O	O				
	?				—	
	?				—	

(*Table 3.2 cont.*)

			Divergent plate margins		
Rock series	*Volcanic rocks*	*Intrusive rocks*	*Oceanic spreading ridges*	*Continental passive margins*	*Continental rifts*
Komatiite Series (Precambrian only)	komatiite		?	?	
Miscellaneous rock types		dunite	x	x	—
		peridotite	x	x	—
		lherzolite	x	—	—
		harzburgite	x	x	—
		pyroxenite	x	x	—
		anorthosite		x	O
		lamprophyre	O	—	—
	carbonatite	carbonatite	O	O	x
	kimberlite	kimberlite	O	O	x
Characteristic volcanic features			fissure eruptions; pillow lavas; pyroclastics subordinate to absent	fissure eruptions; composite cones; shield volcanoes; pyroclastics subordinate	composite cones; shield volcanoes; fissure eruptions
Characteristic intrusive features			sheeted dykes; layered ultramafic masses	dyke swarms; layered intrusions; ring intrusions; sill complexes	stocks; ring intrusions; dykes
Present-day examples			Mid-Atlantic Ridge; East Pacific Rise	Margins of Red Sea; Eastern margins of North and South America	East African Rift; Rio Grande Rift, USA

Convergent plate margins			*Transform plate margins*		*Intraplate environments*	
Island Arcs	*Active continental margins*	*Continental collision zones*	*Oceanic transform faults*	*Continental transform faults*	*Continental hot spots*	*Oceanic hot spots*
?	O	O	O	O	?	O
	—	—				
	—	—				
	—	—				
	—	—				
	—	—				
	x	O		O	—	
	O	—				
	O	O			x	?
?	O	O	O	O	x	?
composite cones; calderas; domes; pyroclastics important to dominant	composite cones; calderas; ignimbrite eruptions; pyroclastics important to dominant	composite cones; calderas; ignimbrite eruptions; pyroclastics minor to important	pyroclastics subordinate	pyroclastics important	composite cones; domes; explosion craters; pyroclastics important, rarely dominant	composite cones; shield volcanoes; calderas; pyroclastics subordinate
dykes; stocks	dyke swarms; stocks; diapirs; batholiths	batholiths; dykes			dykes; sills; diatremes	dyke and sill complexes
Japan; Lesser Antilles; Mariana Islands	West coasts of North and South America	Himalayas; Turkey–Iran	Romanche Fracture Zone, Atlantic Ocean; Snaefellsness Peninsula, Iceland	Western USA; Angola? Western Australia? Western USA?	Jos Plateau, Nigeria; Transvaal, South Africa	Hawaii; Canary Is.

Table 3.3 Rock names and mineralogy of metamorphic rocks
Notes: This table provides a simplified guide to the names of metamorphic rocks and their mineral assemblages, as commonly recorded on geological maps, and their correlation with conditions of metamorphism.

There is no direct correlation of the generalised rock name based on the microstructure of the rock (in the 2nd column) with the more precise facies classification (in the 3rd and subsequent columns); if only a structural name is given, further information should be sought from rock descriptions in additional explanatory material for the map area before allocating a rock type to a particular set of metamorphic conditions.

Type of metamorphism	*Generalised rock-name (structural classification)*	*Facies*	*Characteristic mineral assemblage of metabasic rocks*
Contact (thermal) metamorphism	Hornfels	Hornblende hornfels	Oligoclase-andesine – hornblende
		Pyroxene hornfels	Labradorite – hypersthene – augite
Regional metamorphism (medium to high thermal gradient)	Relict igneous or sedimentary structures	Zeolite	Ca-zeolite – chlorite – albite
		Prehnite-pumpellyite	Prehnite – pumpellyite – chlorite – albite
	Slate	Greenschist	Albite – epidote – chlorite – actinolite
	Phyllite		
		Amphibolite	Albite-oligoclase – epidote – hornblende – almandine
	Schist		
	Gneiss	Granulite	Labradorite – hypersthene – augite – almandine – (hornblende)
	Migmatite		
Regional metamorphism (low thermal gradient)	Schist	Glaucophane schist (blueschist)	Glaucophane – lawsonite or epidote – almandine – chlorite
Lower crust and upper mantle	Eclogite	Eclogite	Jadeitic diopside (omphacite) – almandine-pyrope – kyanite
Cataclastic (dynamic) metamorphism	Breccia cataclasite mylonite pseudotachylite	–	The rocks are usually very fine grained, or with clasts of the adjacent rocks. The mineralogy is usually similar to that of the adjacent rocks.

Characteristic mineral assemblage of metapelitic rocks	*Other characteristic minerals*	*T(°C)*	*P*	*Depth*
Muscovite – biotite – andalusite – (chiastolite) – cordierite – quartz	Wollastonite (no calcite)	400–700	low	< 10 km
Orthoclase – hypersthene – sillimanite – cordierite	Wollastonite (no micas, amphiboles or calcite)	700–900	low	< 10 km
Muscovite – chlorite – albite – quartz	Zeolites, calcite	< 150	low	< 10 km
Muscovite – chlorite – albite – quartz	Stilpnomelane, calcite	150–250	low	< 15 km
Muscovite – biotite – chlorite – epidote – albite – quartz – (chloritoid)	Stilpnomelane, calcite	250–400	low/medium	5–20 km
Muscovite – biotite – almandine – staurolite or kyanite or sillimanite – quartz	Calcite (no chlorite)	400–700	medium	10–30 km
Orthoclase – plagioclase – almandine – sillimanite – cordierite – (biotite)	(No chlorite or muscovite)	700–900	high	15–40 km
Muscovite – chlorite – glaucophane – quartz	Jadeite, aragonite, stilpnomelane (no biotite)	< 400	high	15–30 km
(Metapelitic rocks are rarely found in eclogite facies conditions)		400–1000	> 10 kb	> 35 km
The rocks are usually very fine grained, or with clasts of the adjacent rocks. The mineralogy is usually similar to that of the adjacent rocks.		low	low to high	—

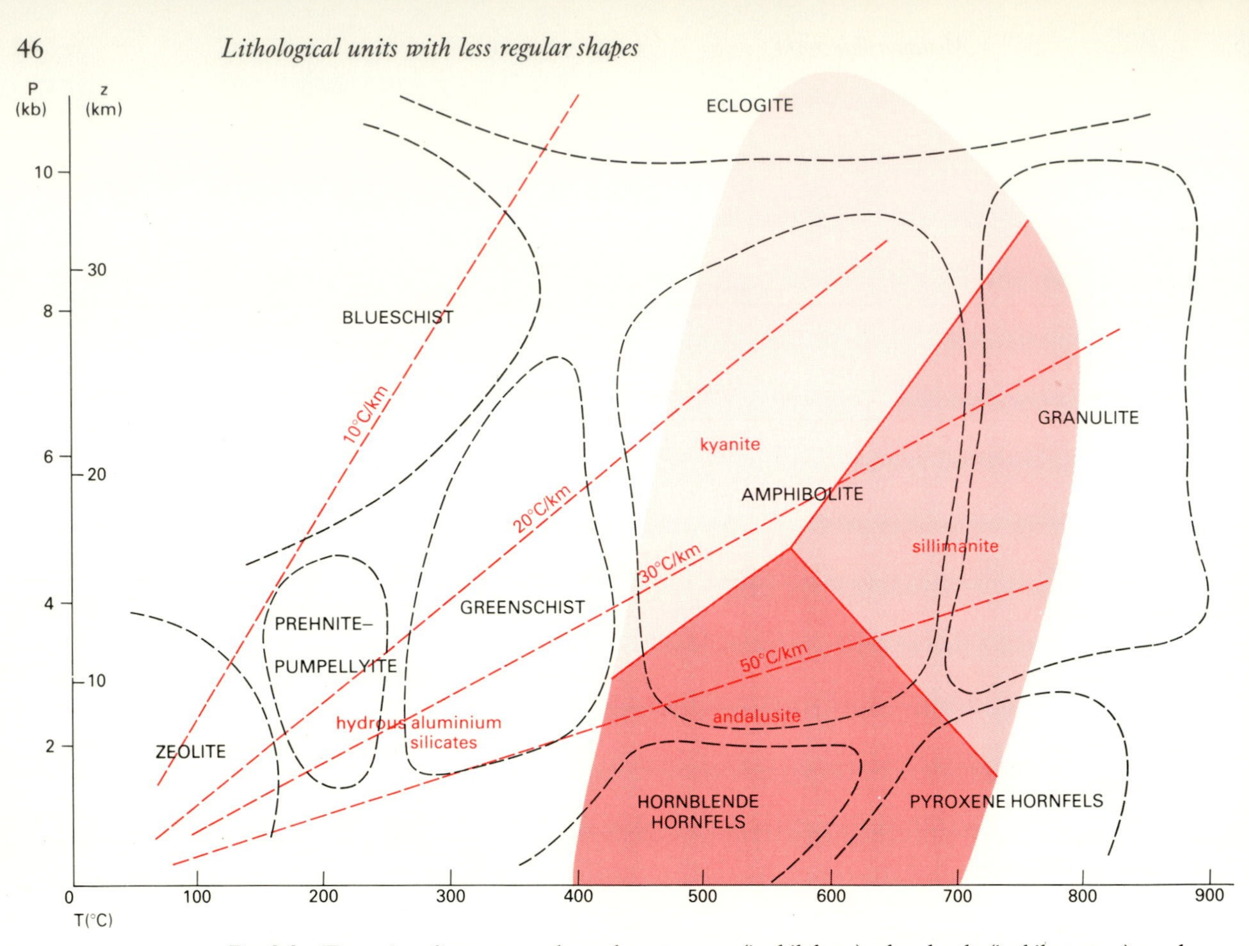

Fig. 3.3 Tentative diagram to show the pressure (in kilobars), the depth (in kilometres), and the temperature (in °C) of the metamorphic facies, adapted from Turner (1981) and Vernon (1976). The unlabelled areas represent mineral assemblages which are transitional between the adjacent facies. The stability fields of kyanite, andalusite, and sillimanite are from Greenwood (1976). Most regional metamorphism occurs in conditions represented by thermal gradients in the range 50 °C/km to 10 °C/km. Contact (thermal) metamorphism occurs mostly in the hornblende hornfels and pyroxene hornfels facies.

(defined by the presence of index minerals, and separated by **isograds**) or **facies** (defined by mineral assemblages, and separated by **facies boundaries**) – Table 3.3 and Figure 3.3. Provided that sufficient data on the mineral assemblages of the rocks are given in the explanation of the map, it is possible to identify the facies of the rocks, and so to obtain some limited information on the temperature, pressure, and depth of metamorphism (Sect. 3.I).

This very brief account of metamorphic rocks can be supplemented by reference to Miyashiro (1973), Vernon (1976), and Gillen (1982).

MINERAL DEPOSITS are largely formed by an extension of normal rock-forming processes, but leading to geochemical enrichment of less common chemical elements up to concentrations at which they are economically valuable. Certain types of mineral deposit occur in particular plate-tectonic environments (Table 10.1). **Syngenetic ores** are formed at the same time as, and have shapes compatible with, the sedimentary or igneous rocks in which they occur; **stratiform ores** form layers within normal sedimentary rock sequences and in layered intrusions. **Epigenetic ores** are formed later than the rocks in which they occur and in general have different shapes and discordant

Table 3.4 Typical shapes, compositions, and environments of mineral deposits (based on Hutchison 1983)

Typical shape	*Ore mineral or element*	*Associated rocks*	*Type of mineral deposit*	*Examples*
Magmatic ores				
Vertical pipes, dykes	Diamond	Kimberlite	Kimberlite pipe	South Africa; Siberia
Pods, lenses	Chromite	Peridotite, serpentinite	Alpine-type peridotite (ophiolite)	Philippines; Turkey
Stratiform layers in basic intrusion	Chromite, Pt	Basic and ultrabasic intrusive rocks	Layered intrusion	Bushveld complex, South Africa
Layers, lenses, veins	Ni-Cu	Basic volcanic and intrusive rocks	Layered intrusion	Sudbury, Ontario; Kambalda, W. Australia
Stocks, ring intrusions	Apatite, Nb, rare earths	Nepheline syenite, etc.; carbonatites	Alkaline igneous rock	Kola, USSR; Mountain Pass, California
Veins	Li, Cs, Sn, U, etc.; mica	Granites, regionally metamorphosed rocks	Pegmatite	Bikita, Zimbabwe
Hydrothermal ores				
Veins; disseminated	Cu, Mo	Porphyritic rocks of intermediate to acid composition	Porphyry copper, molybdenum	Western America; Bougainville, Papua New Guinea
Veins, irregular lodes	Sn, W, Cu, U	Granites or country rocks of granites	Hydrothermal vein	SE China; Southwest England; Massif Central, France
Veins, irregular lodes	Au, Ag, Sn, Sb	Volcanic rocks	Epithermal vein	American Cordillera
Veins; irregular (stockwork)	Cu, Zn, Pb	Volcanic rocks	Kuroko ore	Japan; Rio Tinto, Spain
			Cyprus-type	Cyprus
Stratabound veins, flats, irregular	Zn, Pb, fluorite, barite	Limestones, dolomites	Mississippi-valley-type	Central USA; Pine Point, Canada; Pennines, England
Exhalative ores (Hydrothermal fluids discharged into sedimentary depositional environment).				
Stratiform layers	Cu, Zn, Pb (W, Sn, Ag)	Marine sediments, volcanic rocks	Stratiform massive sulphide ore	Noranda, Canada; ?Broken Hill, NSW; (Red Sea)
			Kuroko ore	Japan
Metamorphic ores				
Irregular	W, Cu, Fe	Contact area of limestones and igneous rock	Skarn	SE China; Ely, Nevada

(*Table 3.4 cont.*)

Typical shape	*Ore mineral or element*	*Associated rocks*	*Type of mineral deposit*	*Examples*
Sedimentary ores				
Beds	Fe	Marine sediments	Sedimentary ironstone	European Mesozoic; Appalachians, USA
		Marine sediments (Precambrian)	Banded iron formation	Lake Superior, USA and Canada; Western Australia
Beds, lenses	U	Sandstones	Sandstone-uranium ore	Western USA
		Black shales	Uraniferous black shale	Sweden
Beds	Na, K, Mg; gypsum, anhydrite	Marine sediments, desert sediments	Evaporites	North Sea Basin
Lenses, layers	Au, Sn, Ti, U; diamond	Sandstones, conglomerates	Placers	Witwatersrand, South Africa; Southeast Asia
Ores formed by weathering				
Irregular (related to water table)	Fe, Cu	Any lower-grade metalliferous ore	Supergene enrichment	Venezuela
Irregular (soil)	Al, (Ni)	Any rock in wet tropical climate	Laterite, bauxite	Australia; Jamaica; (Ni – New Caledonia)

(cross-cutting) relations to the adjacent rocks; **stratabound ores** (Sect. 3.J) are epigenetic but tend to occur within particular stratigraphical units, as a result of selective replacement – for example, because of the easier solubility and higher chemical reactivity of a limestone within a sequence of sandstones and shales. Many epigenetic ores are deposited in pre-existing voids and so may occur in fault planes and in porous rocks such as breccias or in dolomitised limestones. Many minerals of economic value form only small ore bodies and on general geological maps may only be distinguished by the chemical symbols of the elements that are present.

Detailed accounts of mineral deposits can be found in Stanton (1972), Evans (1987), and Hutchison (1983).

The **relative ages** of all types of rocks and of structures are determined by application of logical principles based on the geometrical relationships observed on geological maps (Sect. 3.K). Radiometric ages for the crystallisation of igneous and metamorphic rocks are published on modern geological maps, and provide numerical dates in the geological history of an area. Correlation, on a world-wide basis, of the stratigraphical column with radiometric dates gives the **chronometric time-scale** (Appendix 2); this can be used to interpolate further dates for the stratigraphical and structural events of an area (Sect. 3.L).

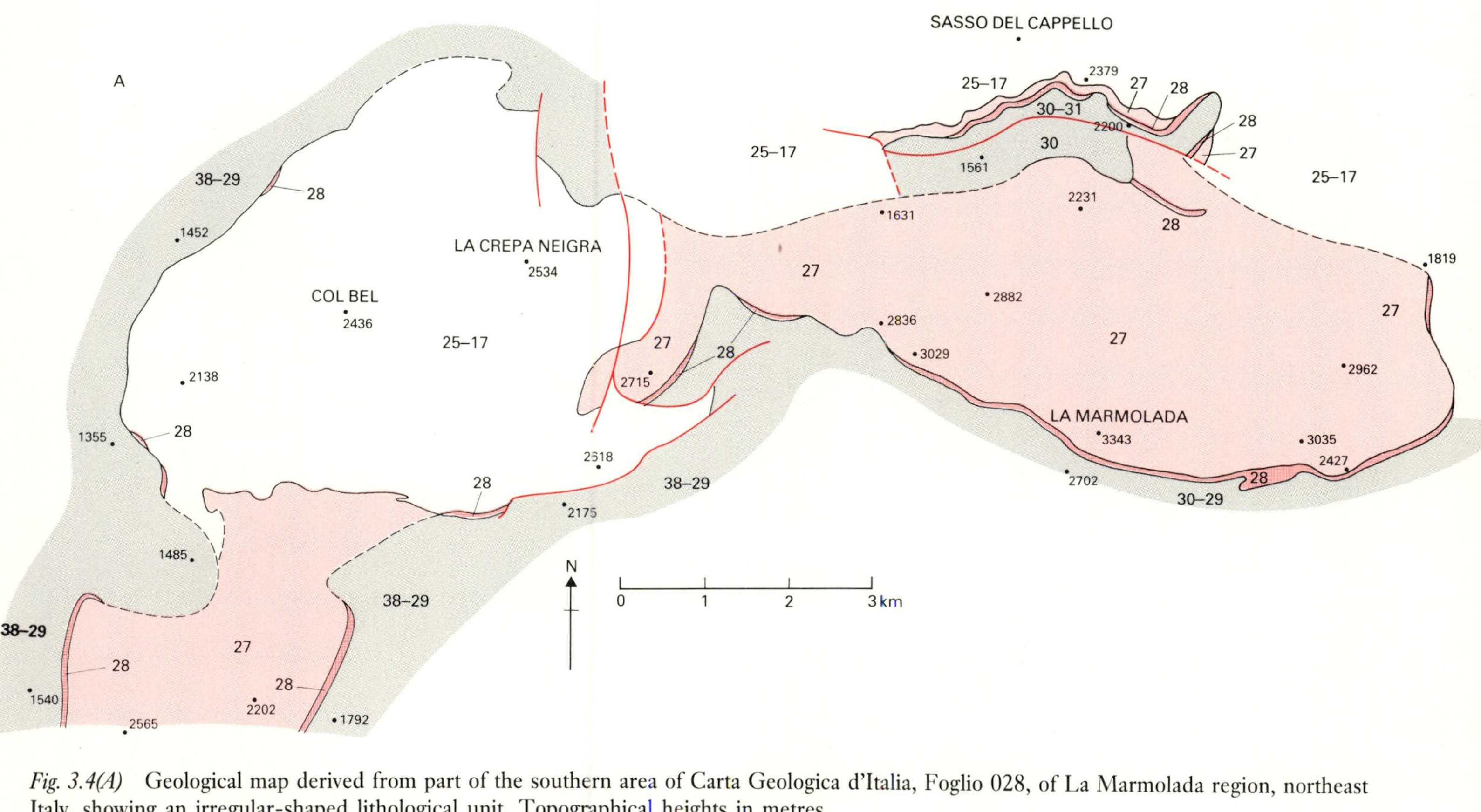

Fig. 3.4(A) Geological map derived from part of the southern area of Carta Geologica d'Italia, Foglio 028, of La Marmolada region, northeast Italy, showing an irregular-shaped lithological unit. Topographical heights in metres.

25–17	Units younger than Calcare della Marmolada (volcanic rocks with feeder dykes)	Upper Ladinian
27	Calcare della Marmolada (massive biogenic and detrital limestone)	
28	Formazione di Livinallongo (nodular and detrital limestones)	Lower Ladinian
38–29	Anisian, Scythian, and Permian	

3.A TO RECOGNISE IRREGULARLY-SHAPED LITHOLOGICAL UNITS

In comparison with parallel-sided lithological units (Ch. 2), which are laterally extensive and have upper and lower boundaries which maintain approximate parallelism over distances of several kilometres or more, bodies of irregular shape are laterally restricted, forming closed loops or more complex patterns. Particular shapes are distinctive of sedimentary and of igneous rocks (Table 3.1). Stratigraphical sections and vertical cross-sections on published maps may provide further information about the shapes of lithological units.

The following procedures, in appropriate conditions, provide clues to the shape of an irregular body of rock:

1. Determine whether the boundaries of irregular units diverge from or cut across parallel-sided units.
2. Use Table 3.1 as a guide to the shapes often adopted by particular compositions of rocks.
3. Determine the thickness of the unit at a number of points along its outcrop (Sect. 2.C).
4. From these thickness measurements construct a thickness-contour map (Sect. 2.E).
5. If the unit occurs in an area of strong topographical relief, construct a structure contour map (Sect. 2.H). If the unit is approximately equidimensional, construct a single map for its boundary; if it is elongate, construct a map showing both its longer boundaries – non-parallelism of the surfaces will be shown by unequal intervals between the contours for each surface.

Fig. 3.4(A) and (B)

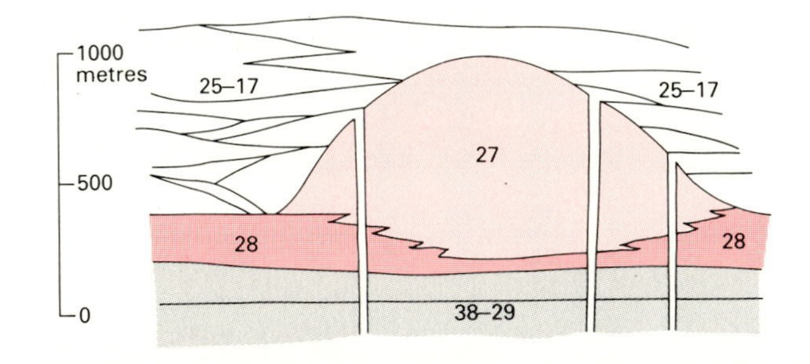

Fig. 3.4(B) Stratigraphical section derived from the section on the published map. The Calcare della Marmolada is up to 800 metres thick on La Marmolada, a few tens of metres thick south of Sasso del Cappello, and absent in the northwest around Col Bel and La Crepa Neigra. The stratigraphical section shows that it is lenticular in cross-section, corresponding to a reef limestone.

Further examples are shown in subsequent figures in this chapter.

3.B TO RECOGNISE LATERALLY DISCONTINUOUS STRATIGRAPHICAL UNITS

In addition to the procedures of Section 3.A, determine the stratigraphical succession at a number of points in the area (Sect. 2.B) and construct stratigraphical columns (Sect. 2.D) with correlation lines to show the relationships of units and their boundaries.

Fig. 3.5(A) and (B)

Laterally discontinuous units may result from deposition of a succession of layers at different times and covering different areas, with time intervals of erosion or of non-deposition in places where beds (present elsewhere) are absent. Such relationships are not necessarily diachronous (but see Sect. 3.C).

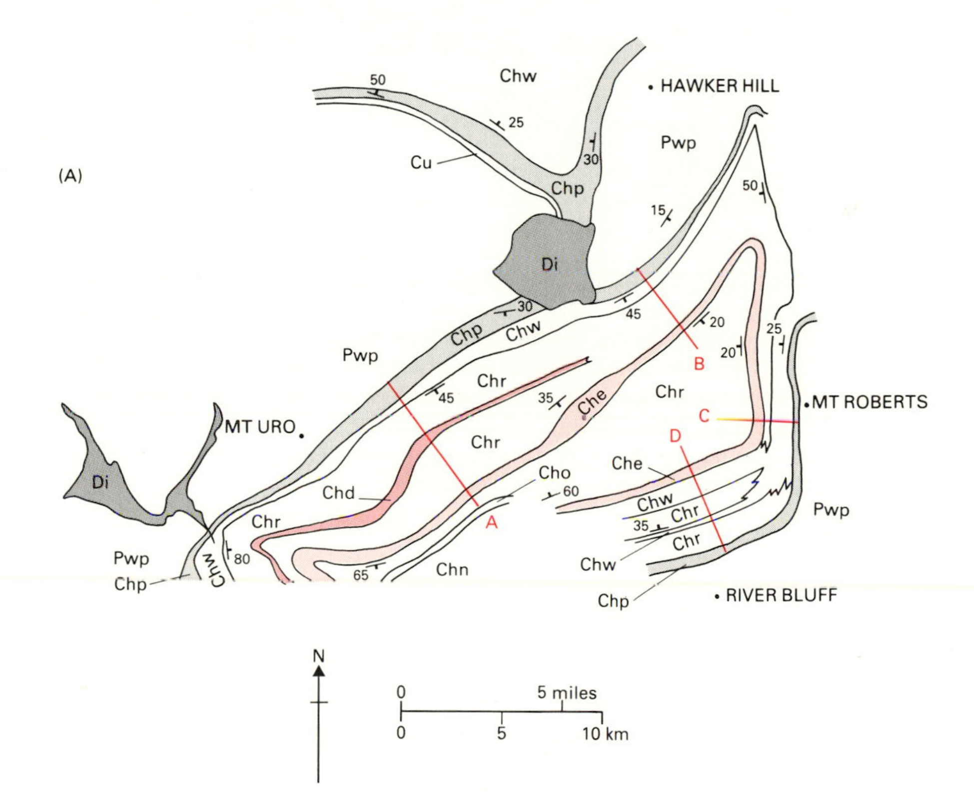

Fig. 3.5(A) Part of the southeastern corner of South Australia Geological Atlas Series sheet SH 54–9 (Copley) (see Plate 2), slightly simplified, showing stratigraphically discontinuous units.

Chn	Narina Greywacke	Lower Cambrian
Cho	Oraparinna Shale	
Chr	Parara Limestone	
Che	Nepabunna Siltstone	
Chd	Midwerta Shale	
Chw	Wilkawillina Limestone	
Chp	Parachilna Formation	
Cu	Uratanna Formation	
Pwp	Pound Quartzite	Proterozoic
Di	Diapiric breccia	

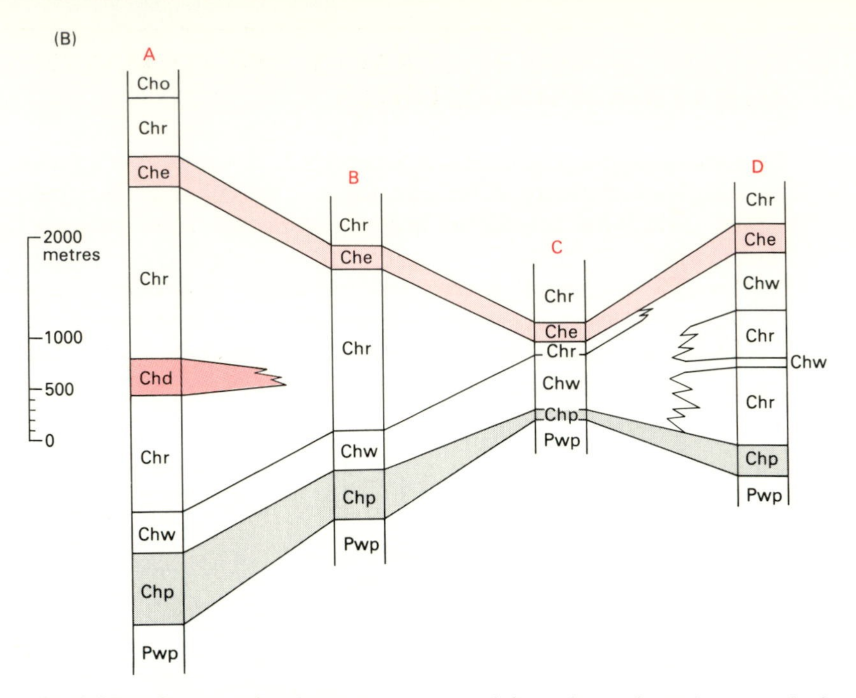

Fig. 3.5(B) Stratigraphical sections, measured from the geological map at the locations marked A to D. Correlation lines indicate the variation within the stratigraphical succession.

Chp and Che are continuous through the area shown. Between these two units, Chw, Chd, and Chr show lateral variation with changes in both thickness and stratigraphical sequence. The mapped formations represent changes of environment in both time and place and are probably diachronous (Sect. 3.C).

See Section 6.A.3 for a comment on the stratigraphical relationships of Cu, and Section 3.F for a comment on the diapiric breccias.

3.C TO RECOGNISE DIACHRONOUS STRATIGRAPHICAL UNITS

Diachronous stratigraphical units are deposited *at the same time* because of variation of facies within an area of deposition. Detailed fossil evidence may be necessary to prove diachronism. Lateral variation of lithology within a marine succession (where intervals of non-deposition are less likely than on land), and the presence of laterally continuous units above and below suggest diachronism.

Fig. 3.6(A) and (B)

Examine the stratigraphical succession (Sect. 3.B), both on the map and on stratigraphical columns and cross-sections, looking for evidence of contemporaneity of lithological units as detailed above.

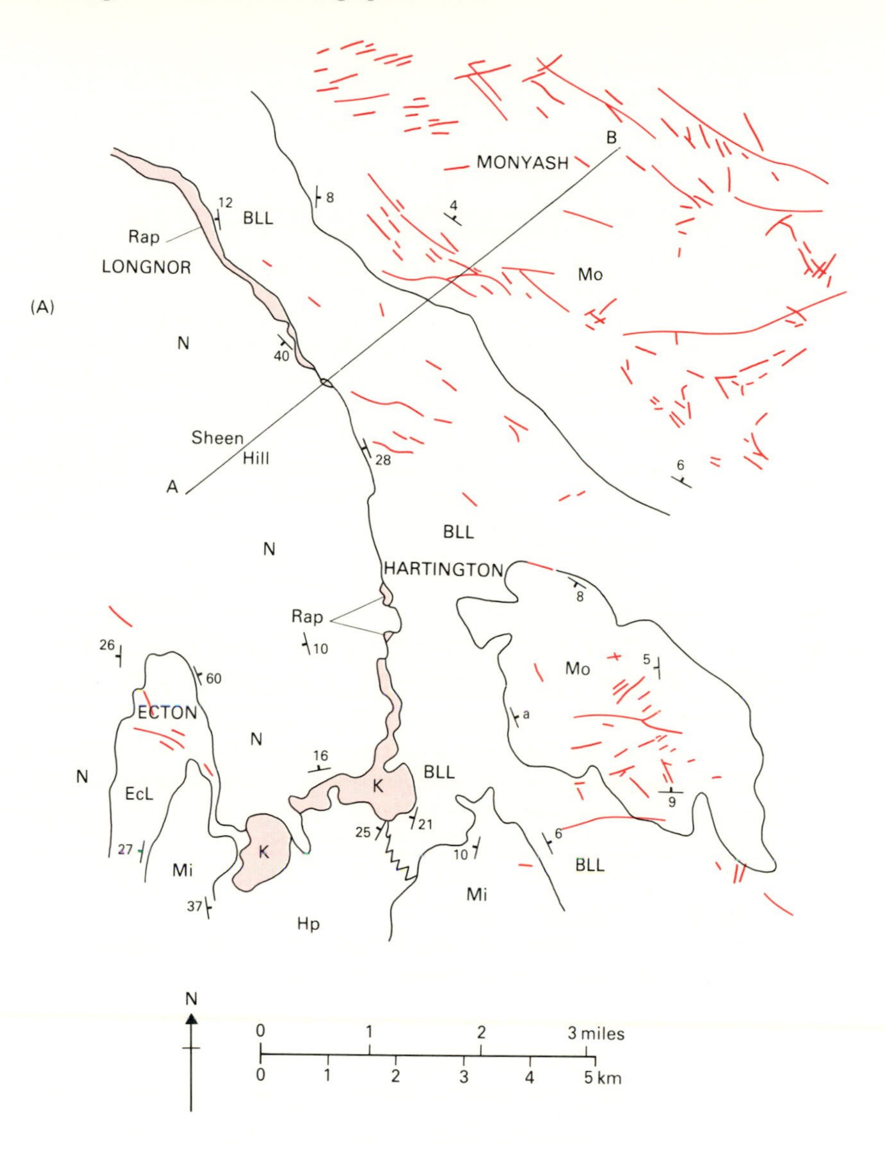

Fig. 3.6(A) Part of the south-central area of B.G.S. map 111 (Buxton), omitting faults and with other simplifications, showing diachronism within part of the Lower Carboniferous.

N	Millstone Grit Series (sandstones and mudstones) and Mixon Limestone-shales		Namurian and uppermost Viséan
Mo	Monsal Dale Limestones	D2 zone	Viséan
EcL	Ecton Limestones	D1 zone	
Hp	Hopedale Limestones		
BLL	Bee Low Limestones		
K	Knoll-reefs		
Rap	Apron reefs		
Mi	Milldale Limestones and Woo Dale Limestones	S2 and earlier zones	

Short red lines: mineral veins

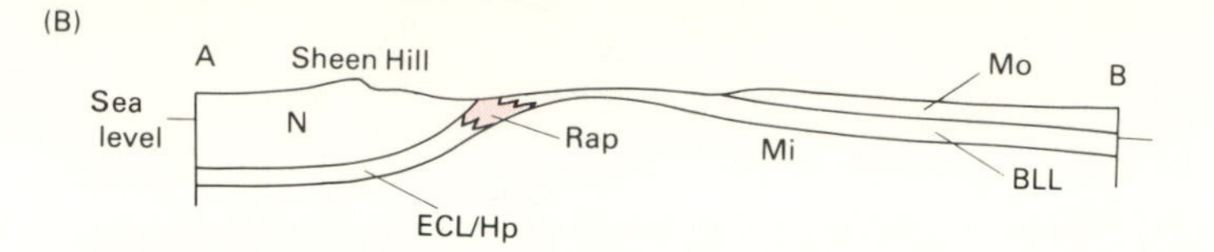

Fig. 3.6(B) Vertical cross-section along the line A – B in Figure 3.6(A), simplified from the section published on B.G.S. map 111, showing the stratigraphical relations of three of the facies of the D1 zone rocks.

The five different limestone facies of the rocks of the D1 zone correspond to a shallow shelf sea in the east (BLL), fringed by an apron reef (Rap) and knoll-reefs (K), and progressively deeper water to the west (Hp and EcL). The change of facies from shallow to deep water probably marks a pre-Carboniferous structural boundary and has been re-activated to form a post-Carboniferous monoclinal fold.

(Several other examples of diachronous units are shown on the published map.)

See Section 3.J for a comment on the mineral veins.

3.D TO DISTINGUISH BETWEEN CONCORDANT AND DISCORDANT INTRUSIVE IGNEOUS ROCKS

High-level minor intrusives a few tens or hundreds of metres thick are emplaced along planes of weakness in the adjacent country rocks – these tend to be either gently-dipping bedding planes in sedimentary units or steeply-dipping joints or pre-existing fault planes.

At a deeper crustal level the same principles apply, but other processes may operate in addition (Sect. 3.F).

Examine the outcrop of the intrusive to determine whether it is predominantly parallel to the boundaries of adjacent lithological units (*concordant*) or in cross-cutting relationship (*discordant*). *Fig. 3.7*

3.E TO DISTINGUISH BETWEEN A LAVA FLOW AND A SILL

Certain igneous rocks (especially basalt) can occur either as an extrusive (volcanic) rock interlayered with sediments in stratigraphical succession, or as an intrusive rock emplaced into older rocks. Because an intrusive sill is predominantly concordant (Sect. 3.D) it may be particularly difficult to distinguish from a lava flow. Apply the following criteria:

1. Examine the boundaries of the igneous rock in relation to adjacent lithological units – if at any point they transgress boundaries of adjacent rocks (but take care with unconformable relationships – Ch. 6) it is likely to be a sill; if it maintains a constant stratigraphical position it may be a sill or a lava flow.

2. If the relative or chronometric ages of the rocks are stated in the legend of the map or are otherwise known, a younger igneous rock in contact with older rocks *above and below* is a sill; if the ages of the igneous and sedimentary rocks are not known or are very similar the igneous rock may be either a sill or a lava flow.

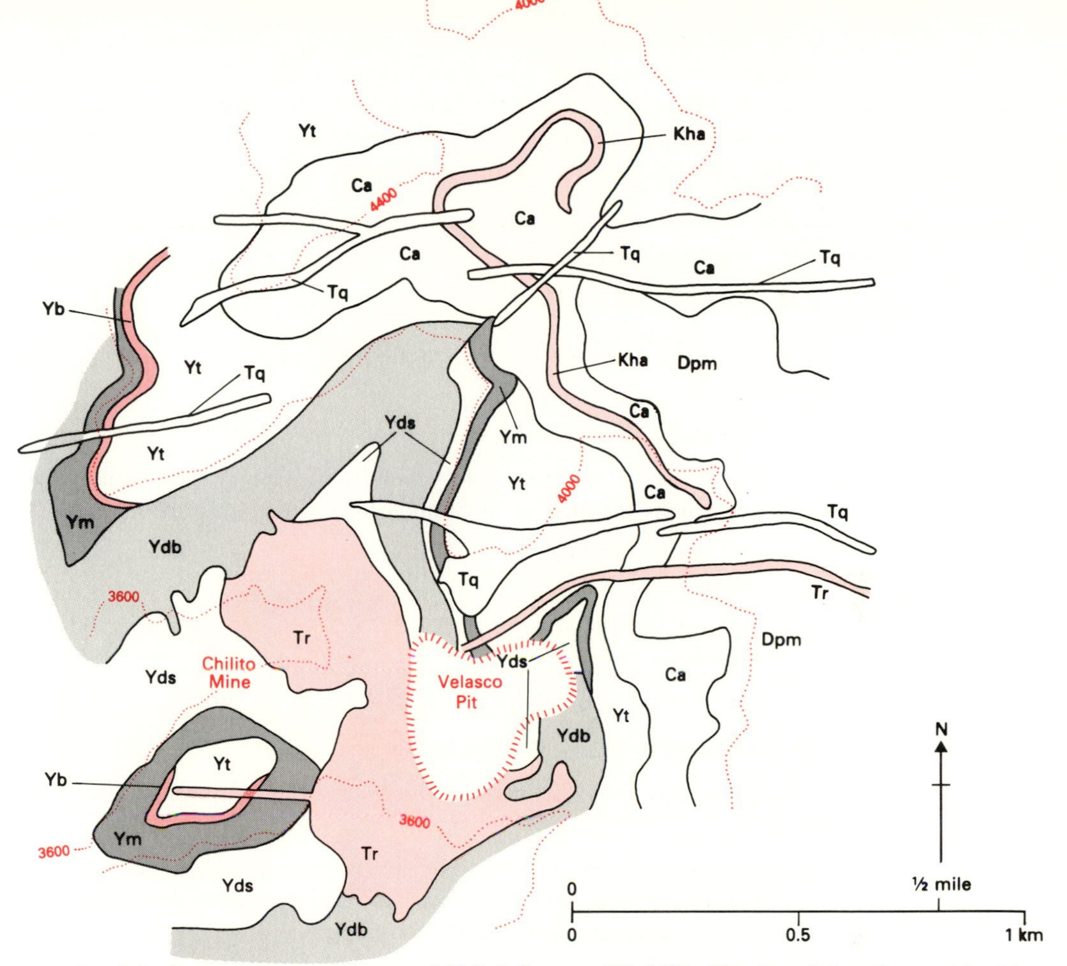

Fig. 3.7 Part of the central area of U.S.G.S. map GQ-1391 (Hayden, Arizona), considerably simplified and omitting faults and Quaternary sediments, showing the field relationships of some intrusive and extrusive igneous rocks. In time sequence (oldest at the base):

Tq	Quartz latite porphyry	Paleocene
Tr	Rhyodacite porphyry	
Kha	Hornblende andesite porphyry	Cretaceous
Dpm	Percha Shale and Martin Formation	Upper Devonian
Ca	Abrigo Formation	Upper and Middle Cambrian
Ydb	Diabase	
Yt	Troy Quartzite	
Yb	Basalt (one or more flows)	Precambrian
Ym	Mescal Limestone	
Yds	Dripping Spring Quartzite	

The rocks comprise a nearly horizontal Precambrian stratigraphical succession (Yds to Yt) intruded by a diabase (Ydb), and followed by Cambrian and Devonian sediments (Ca and Dpm). The rocks are intruded by a Cretaceous sill (Kha) and by Paleocene dykes (Tq) and an irregular-shaped rhyodacite porphyry (Tr), which has associated copper mineralisation.

Tq and Tr cut across stratigraphical boundaries and are discordant; Kha is concordant; Ydb is in part concordant (top left, centre, and bottom left), and in part discordant (transgressive) where it changes level from below Ym (at the top left) to within Yds (at the top centre).

Example In Figure 3.7 the hornblende andesite porphyry (Kha) can be positively identified as a sill because older rocks occur *above* as well as below, and it must therefore be a younger intrusive rock. The basalt (Yb) occurs at a constant stratigraphical level between Ym and Yt (and at the same level in nearby maps up to at least 20 miles away – this strongly suggests (but does not prove) that it is a lava flow, and this is confirmed by the statement in the legend of the map.

3.F TO RECOGNISE THE MECHANISM OF EMPLACEMENT OF AN INTRUSION

Examine all the information presented on the map and any accompanying data, noting the following:

1. The shape of the intrusion.
2. The presence or absence of xenoliths of country rock either large enough to be individually mapped or mentioned in information provided with the map.
3. Structures internal to the intrusion such as foliation, layering, and crystal orientation.
4. Structures in the country rock – dip and strike of sediments, foliation and lineation of metamorphic rocks.
5. The location and orientation of faults (Ch. 4) and folds (Ch. 5).

The following features are significant:

Forceful emplacement (the magma forces entry by deforming and displacing country rocks).

1. The intrusion has a rounded or elongate shape with smooth, often steep or vertical margins, concordant or discordant, lacks country rock xenoliths, and displays internal foliation parallel with the margin.
2. Older structures in the country rock become re-oriented parallel with the margin.
3. New structures develop in the country rock such as arcuate faults and folds and cleavage.

Permissive emplacement (the country rock, under stress, dilates or moves apart as magma wells into the space).

1. The intrusion is sheet-like (either gently or steeply dipping) or conical (Fig. 3.1) with sharp margins; it contains few, if any, country-rock xenoliths, and generally lacks internal foliation.
2. Older country rock structures may be cut across by the intrusion but are not significantly deformed or re-oriented; certain structures may control the emplacement, e.g. sills intruding bedding planes and dykes infilling faults.

Stoping (rising magma detaches blocks of country rock, which sink into it)

1. The intrusion may be any shape but characteristically has many apophyses projecting into the country rock and contains many disoriented, angular xenoliths of country rock, especially in the marginal and roof zones.
2. The contacts are discordant towards country rock structures, but these are not significantly deformed and no new structures are produced.

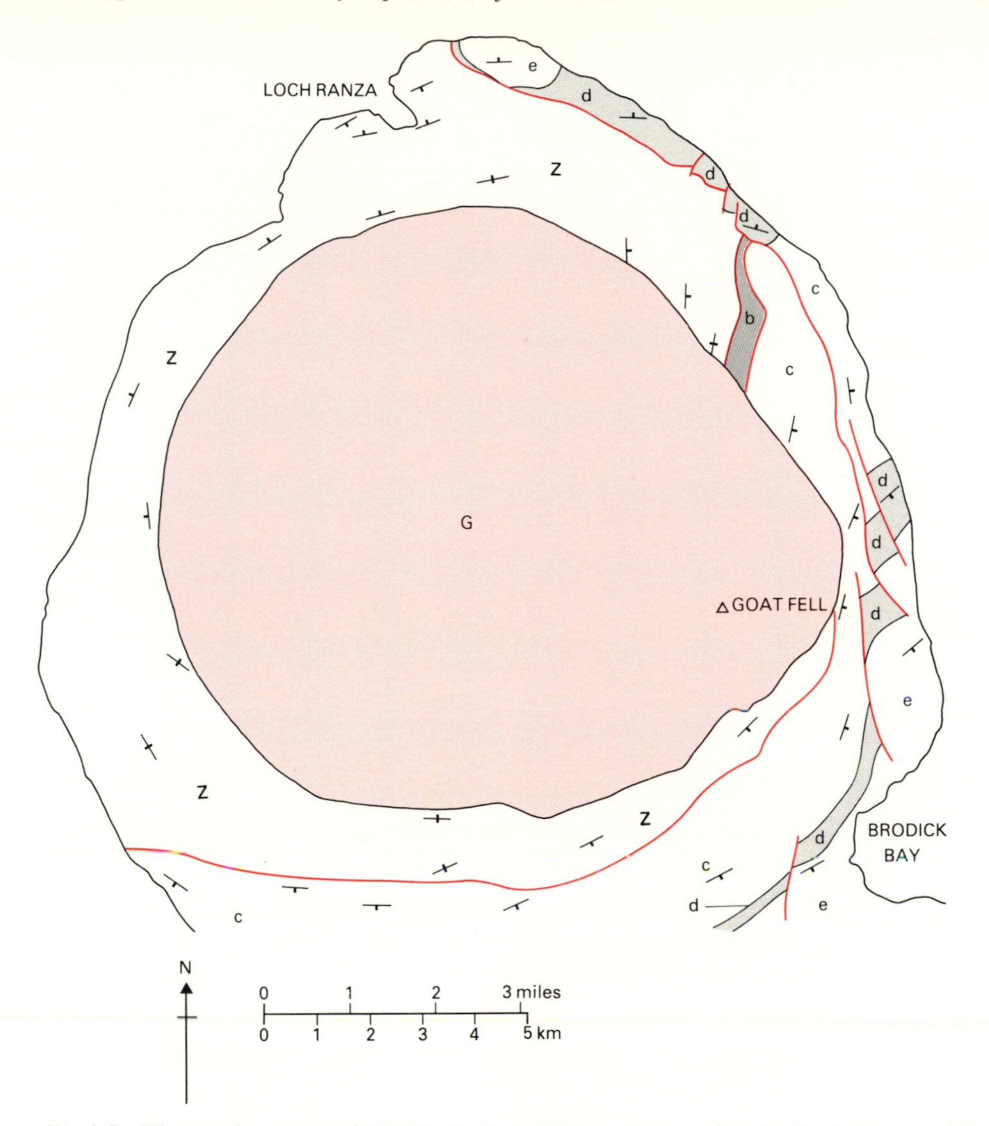

Fig. 3.8 The northern part of B.G.S. Geological Map of Arran, Scotland, slightly simplified, showing the North Arran Granite.

G Granite (Tertiary)
e Permian sediments and volcanics
d Carboniferous sediments and volcanics
c Old Red Sandstone (Devonian)
b Ordovician sediments and volcanics
Z Dalradian metasediments

The granite outcrop has a rounded shape and a discordant, vertical contact. Country rocks generally dip steeply away from the granite, with arcuate folds (in the north) and faults. Xenoliths are absent; there is no evidence of internal foliation in the granite. The evidence indicates that the North Arran Granite is a forceful intrusion.

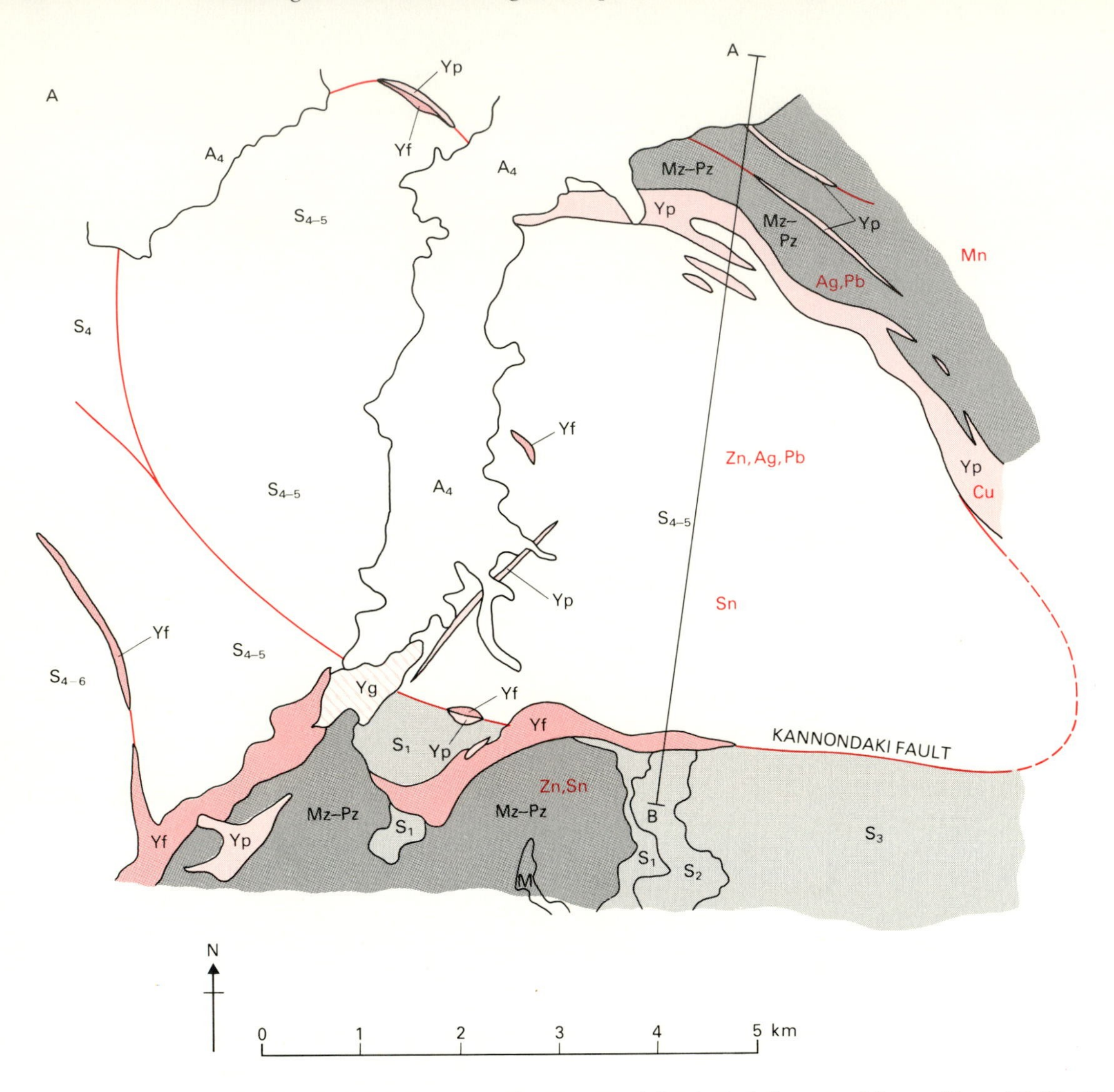

Fig. 3.9(A) Part of the southeast area of Geological Survey of Japan Sheet 15–23 (Taketa), showing intrusions emplaced along fault lines.

A_4	Pyroclastic flow deposits (rhyolite tuffs, welded tuffs, ashes, and pumice, from Aso volcano)	Late Pleistocene
Yg	Biotite Granite	Miocene
Yp	Granite porphyry and quartz porphyry	
Yf	Felsite	
S_{4-6}	Late Sobosan Volcanic Rocks (andesite and dacite lavas, tuffs, and tuff breccias)	
S_{1-3}	Early Sobosan Volcanic Rocks (rhyolite and dacite lavas, tuffs, tuff breccias, and welded tuffs)	
M	Mitate Formation (conglomerate and sandstone)	?Paleogene
Mz-Pz	Igneous, metamorphic and sedimentary formations	Palaeozoic and Mesozoic

Mineralization indicated by chemical symbols (see Sect.3.J).
Dashed red line shows continuation of Kannondaki Fault in map area to the east.

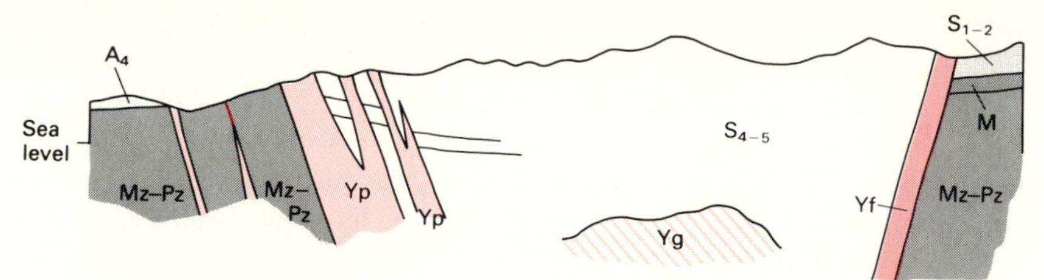

Fig. 3.9(B) Vertical cross-section along the line A – B on Figure 3.9(A) adapted from the section on the published map.

This caldera structure consists of an oval-shaped area of younger lavas and pyroclastics (S_{4-6}) of the Miocene Sobosan Volcanic Rocks downfaulted within a curved fault system (partly obscured by Late Pleistocene ashes and tuffs from the Aso volcano) against older volcanics of the Sobosan series and Mesozoic and Palaeozoic rocks. Arcuate intrusions were emplaced permissively along the fault line as magma rose into place during the formation of the caldera.

Melting and assimilation (magma ingests fragments of country rock by melting and chemical reaction).

1. The intrusion is usually irregular in shape, sometimes elongated, has diffuse margins and contains many country rock xenoliths which show all stages of rounding and digestion but are not significantly disoriented.
2. The intrusion is generally concordant with country rock structures and no significant new structures are produced.

Figs. 3.8 and 3.9

Further examples Figure 3.7: The majority of the dykes (Tq and Tr) and the sill (Kha) are parallel-sided, dilatational, permissive intrusions.

Figure 3.11: The shapes of the granite bodies in the Mount Painter Complex suggest permissive emplacement or stoping.

Figure 3.5: The diapiric breccias were intruded permissively along fault planes and structural highs in the overlying sediments.

3.G TO CHARACTERISE AN IGNEOUS ROCK ASSOCIATION

Examine all the information presented on the map, noting the following:

1. The range of rock types and any information on the chemical compositions of the rocks.
2. The type of igneous activity, e.g. extrusive and/or intrusive; explosive or effusive volcanism; types of intrusion etc.
3. The time range of the igneous activity, using ages relative to stratigraphical units and/or radiometric data; note any change of activity with time.
4. The geographical limits of the igneous activity.
5. The associated rock types including the contemporaneous sediments, metamorphism, mineral deposits, and the structural setting.

Use Table 3.2 to determine the petrological series to which the igneous rock association belongs.

Example In Figure 3.9 volcanic rocks ranging in age from Miocene to Holocene comprise lavas and pyroclastic flow deposits, including welded and non-welded tuffs, assigned to several successive stages in the evolution of calderas. The intrusive rock types include granodiorite and granite. The extrusive rock types are andesite, dacite and rhyolite. Table 3.2 identifies these rock types and this style of explosive volcanic activity with the calc-alkaline igneous association, produced at a convergent plate margin (actually an island-arc environment).

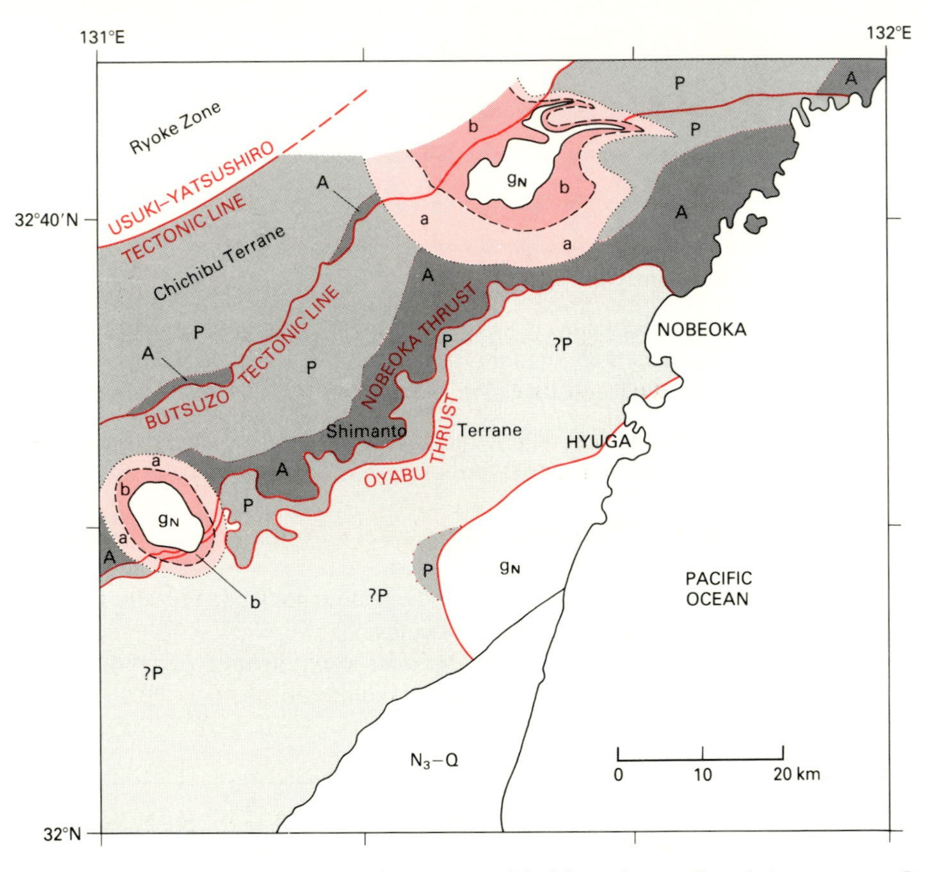

Fig. 3.10 Structural and metamorphic map simplified from the small-scale inset map on Geological Survey of Japan Sheet NI–52–6 (Nobeoka), showing zones of regional and contact metamorphism.

N_3–Q Late Miocene to early Pleistocene (not metamorphosed)
g_N Granites (Miocene)

Zones of metamorphism:

	Regional		*Contact*
?P	?Prehnite-pumpellyite zone	a	Actinolite zone
P	Prehnite-pumpellyite zone	b	Biotite zone
A	Actinolite zone		

The grade of regional metamorphism increases southeastwards within the Chichibu and Shimanto terranes towards the Butsuzo and Nobeoka thrusts respectively, while the general grade increases towards the northwest. During the Miocene the thrust sheets were intruded by granites (g_N), which are surrounded by metamorphic aureoles with zones increasing in grade towards the granite contacts.

3.H TO RECOGNISE AND DISTINGUISH BETWEEN CONTACT, REGIONAL, AND CATACLASTIC METAMORPHISM

Zones of contact metamorphism are concentric to the boundaries of intrusive igneous rocks and are superimposed on older structures (including older metamorphic zonation). Regional metamorphism is laterally extensive, on the scale of the orogenic belt in which it was formed. Cataclastic metamorphism is marked by a zone of breccia, cataclasite, or mylonite within a few tens of metres of a major thrust or fault. (For an example of the latter, see Fig. 4.5.)

Fig. 3.10

3.I TO DETERMINE THE CONDITIONS OF METAMORPHISM

The amount of information that can be obtained is usually very limited. Identify facies boundaries or isograds (e.g. the boundary between kyanite and sillimanite zones). Correlate these with Figure 3.3; this will usually indicate limiting conditions of temperature, pressure, and thermal gradient.

Example Table 3.3 indicates that the actinolite zone of regional metamorphism on Figure 3.10 is likely to be equivalent to the greenschist facies. Figure 3.3 shows that the temperature of the prehnite-pumpellyite / greenschist facies boundary is at approximately 250 °C; the depth is not determinable, but the thermal gradient in the area was in the range 15 °C/km to 50 °C/km.

3.J TO DETERMINE THE POSSIBLE METHOD OF FORMATION OF A MINERAL DEPOSIT

On most general geological maps the information relating to mineral deposits is sparse, and may be inadequate for characterisation. Determine as much as can be learned from the map and any other available information about the shape, size, and composition of the ore body and its associated rocks. Use Table 3.4 as a guide to the possible type of mineral deposit and method of formation. See Section 9.E for techniques of assessment of mineral resources.

Fig. 3.11

Further examples Figure 3.6: The mineral veins contain galena, sphalerite, fluorite, barite, and calcite, and are post-Carboniferous, probably of Permian or Triassic age. They have cross-cutting relationships to the adjacent rocks, are particularly abundant in the Monsal Dale Limestone (Mo), and are virtually absent from the sandstones and mudstones of the Upper Carboniferous – they are thus *epigenetic*, *stratabound* and probably of *Mississippi-Valley-type*.

Figure 3.7: A small porphyry copper mineral deposit associated with the Paleocene igneous activity.

Figure 3.9: The base metal and silver deposits occur in veins and skarns associated with the Miocene acid intrusives.

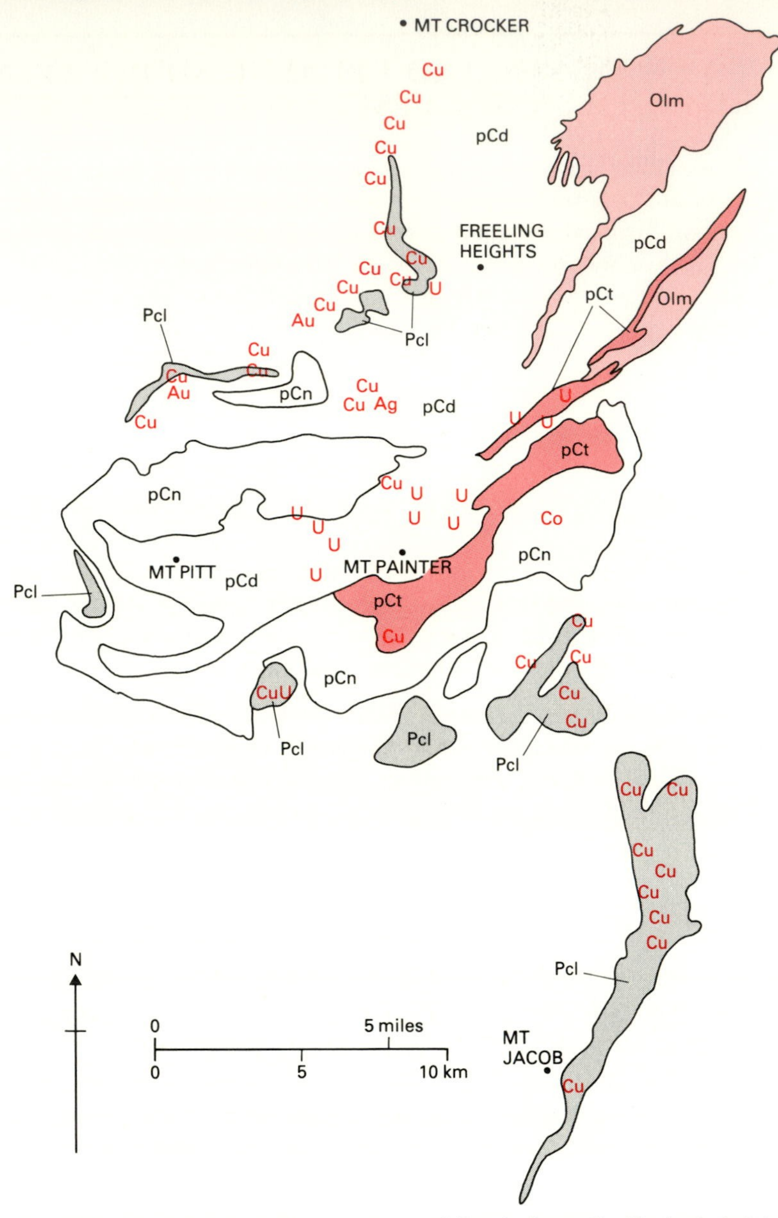

Fig. 3.11 Part of the northeast corner of South Australia Geological Atlas Series sheet SH 54–9 (Copley), showing mineralisation in the Mount Painter Complex. Only the following units are shown:

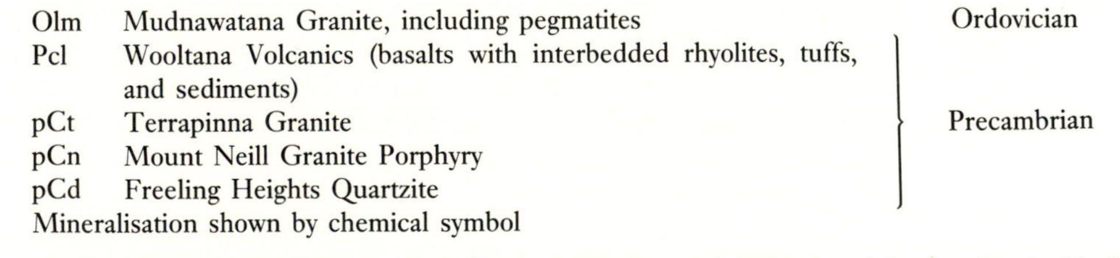

Olm	Mudnawatana Granite, including pegmatites	Ordovician
Pcl	Wooltana Volcanics (basalts with interbedded rhyolites, tuffs, and sediments)	Precambrian
pCt	Terrapinna Granite	
pCn	Mount Neill Granite Porphyry	
pCd	Freeling Heights Quartzite	

Mineralisation shown by chemical symbol

The location of the copper mineralisation suggests a volcanogenic origin associated with the Wooltana Volcanics. The uranium may be of hydrothermal origin related to the Terrapinna Granite.

3.K TO DETERMINE THE RELATIVE AGE OF A LITHOLOGICAL UNIT

The fundamental logic of the determination of relative age is that a rock or structure is older than rocks or structures that are superimposed upon or that cut it, and younger than rocks or structures on which it is superimposed or which it cuts.

So, for example:

1. Stratigraphically younger rocks rest upon older rocks (Sect. 2.B).
2. The relative date of an extrusive igneous rock is given by the position it occupies in the stratigraphical succession (Sect. 3.E, Fig. 3.7).
3. An intrusive igneous rock is younger than the country rock (including any other igneous rocks) which it cuts (Sect. 3.E, Fig. 3.7).
4. Metamorphism is younger than the protolith (the rock as it existed before metamorphism) (Fig. 3.10).
5. The date of contact, regional, or cataclastic metamorphism is contemporaneous with the igneous intrusion, orogenic process, or tectonic movement (respectively) that caused the metamorphism (Fig. 3.10, Fig. 4.5).
6. A mineral vein is younger than the country rock into which it is emplaced; a mineral vein in the plane of a fault is generally younger than the fault. A mineral vein may be contemporaneous with nearby igneous rocks (Sect. 3.J, Fig. 3.6).

Similar principles apply to the determination of the ages of faults (Ch. 4), folds (Ch. 5), and the processes associated with the formation of an unconformity (Ch. 6).

3.L TO DETERMINE THE CHRONOMETRIC DATE OF A LITHOLOGICAL UNIT

Use relative dates (Sect. 3.K) in conjunction with the chronometric time-scale (Appendix 2), together with radiometric dates stated on the map or published elsewhere, to evaluate the limiting dates (or range of dates) for the formation of the lithological unit.

Example On Figure 3.9 the date of the biotite granite, Yg, is stated in the explanation that accompanies the map to be (probably) 21 Ma. The granite intrudes Sobosan Volcanic Rocks S_{1-6} of Miocene age. The beginning of the Miocene was at 23.7 Ma (Appendix 1). The Sobosan Volcanics are therefore constrained to a small range of dates between 23.7 and 21 Ma in the early Miocene. (See Sect. 4.J for an extension of this argument.)

4 FAULTS

The response of rocks to stress by brittle fracturing and relative movement of the blocks of rock each side of the fracture produces a surface of discontinuity or fault (Fig. 4.1). Faults are recognised on geological maps by the displacement of outcrops of rocks which are otherwise continuous and by the juxtaposition of rocks of different ages each side of the fault (Sect. 4.A).

The fracturing of rocks adjacent to a fault plane makes them less resistant to weathering so that the outcrop of a fault is commonly marked by an erosional feature – a stream, valley, lake, escarpment, or coast line. Large faults which juxtapose rocks of different lithology may be further marked by a contrast between areas of differing topography, vegetation, and land use.

The structural significance of faults lies in the information that can be derived from them about the stress situation in which they were formed (Hobbs, Means, & Williams 1976; Park 1983). The geometry of fault displacement is shown in Fig. 4.2, and the principal types of fault in relation to the stress environment and direction of movement in Fig. 4.3.

Fig. 4.1 An area near Sitia, east Crete, showing the outcrop of a fault as a mappable feature extending for several kilometres across the hillsides. The fault cuts Tethyan (pre-Neogene) limestones; its continued activity to the present day maintains it as a topographical feature. Photograph: Dr Carol Lister.

The procedure for the characterisation of different types of faults is set out in Table 4.1, and described in Sections 4.B to 4.G.

Although it is rare to find features that allow the true displacement of a fault (Fig. 4.2(A)) to be determined, the magnitude of a fault can be indicated by the distance of separation of originally continuous lithological layers or other planar structures (Fig. 4.2(B), Sect. 4.F, 4.H).

The determination of the **relative age** of a fault (Sect. 4.I) is important for the geological history of an area; evaluation of the **chronometric time-interval** of the fault movement allows determination of the minimum rate of movement of the fault during its formation (Sect. 4.J).

Once a fault has been initiated as a surface of weakness and easy movement, further strain in the region may be preferentially taken up by further movement on that surface, rather than by folding or by formation of additional faults. Consequently the amount of movement on a fault plane may be large compared with the amount of deformation shown by adjacent contemporaneous structures. However, though localised on particular surfaces, faults are the product of a **regional stress system**, and need to

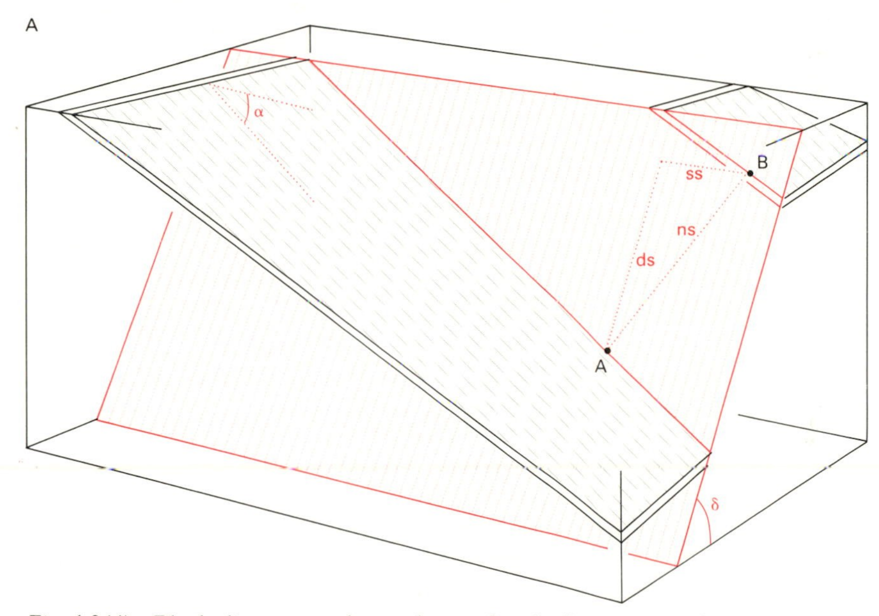

Fig. 4.2(A) Block diagram to show a layer of rock dipping at angle α towards the front right lower corner, displaced by a fault dipping at angle δ towards the front of the block. The front right upper corner of the block has been omitted. Points A and B were originally in contact.

ns **net slip**, the true displacement of the fault
ds **dip slip**, the component of the true displacement parallel to the dip of the fault plane
ss **strike slip**, the component of the true displacement parallel to the strike of the fault plane.

The true displacement can only be determined if points such as A and B can be identified, and this is uncommon in practice.

The front half of the block has moved down (with a small additional component of strike-slip movement) relative to the rear half and is referred to as the **downthrow** side.

The following terms are also used in the description of faults:

hanging wall: the block of rock that lies on the upper side of the fault plane
footwall: the block of rock that lies on the lower side of the fault plane
hade: 90°-angle of dip.

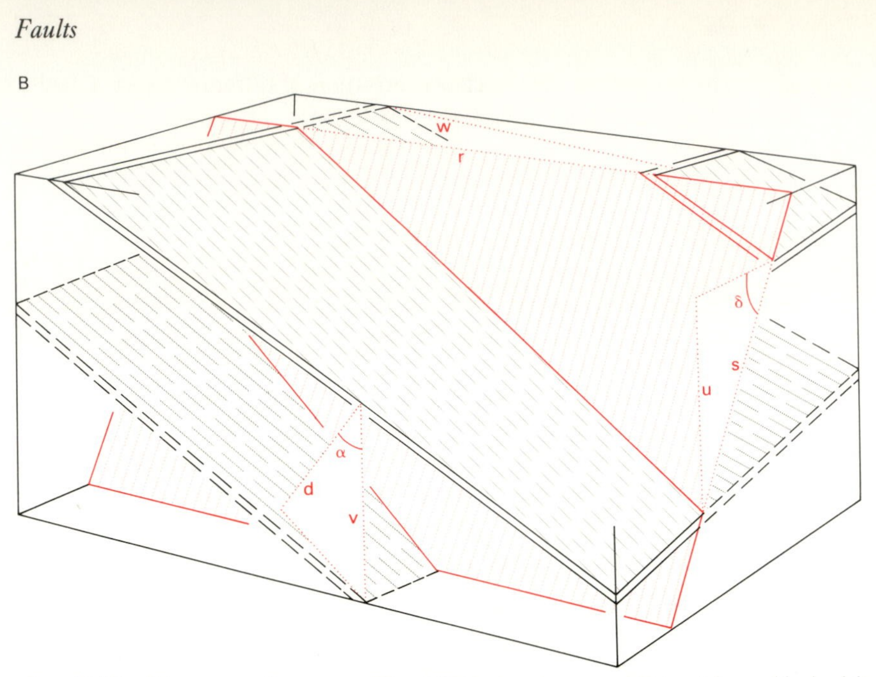

Fig. 4.2(B) The same diagram as Fig. 4.2(A) showing an additional layer (dashed lines), which has been brought into contact with the first layer by the fault movement. When the true displacement of the fault cannot be determined, any of the following measurements may be used to provide an estimate of the magnitude of the fault movement:

r **strike separation** of the outcrops of a displaced layer on a horizontal surface
s **dip separation** of a displaced layer measured along the dip of the fault plane
d **stratigraphical displacement**, the stratigraphical distance between two layers now in contact across the fault plane
u **throw** of the fault, the vertical component of the dip separation.

The **outcrop width** (unfaulted) between the two layers brought into contact by the fault is w, and the vertical depth between the two displaced layers is v (Fig. 2.9).

In the most general case, illustrated in Fig. 4.2(B), where the strike directions of the rock units and the fault are oblique, the three-dimensional geometry is complex and the displacement of the fault is most easily determined by graphical methods. In the simpler, and geologically common, situation of the strike of the fault being perpendicular to the strike of the displaced layers, the following relationships apply:

$w = r$, and
$u = v = r.\tan\alpha = s.\sin\delta = d/\cos\alpha$

Further simplifications emerge if either the fault is vertical ($\delta = 90°$) or the layers are horizontal ($\alpha = 0°$). The relationships become more complex and not readily usable if (a) the topographical surface is sloping or irregular, (b) the dip of the units is different each side of the fault.

be understood in relation to other structures of an area; Section 4.K illustrates some distinctive regional fault patterns. On a global scale, faults reveal the relative motion of regions of the crust as a result of **plate-tectonic movements**.

Faults may continue to be planes of weakness for millions of years after their active movement. Consequently they may be reactivated in later periods of earth movement, controlling subsidence of basins and deposition of sediments ('growth faults'), allowing the emplacement of igneous intrusions, acting as sites of mineralisation, or taking part in further episodes of deformation. Movement on ancient fault planes produces many present-day earthquakes.

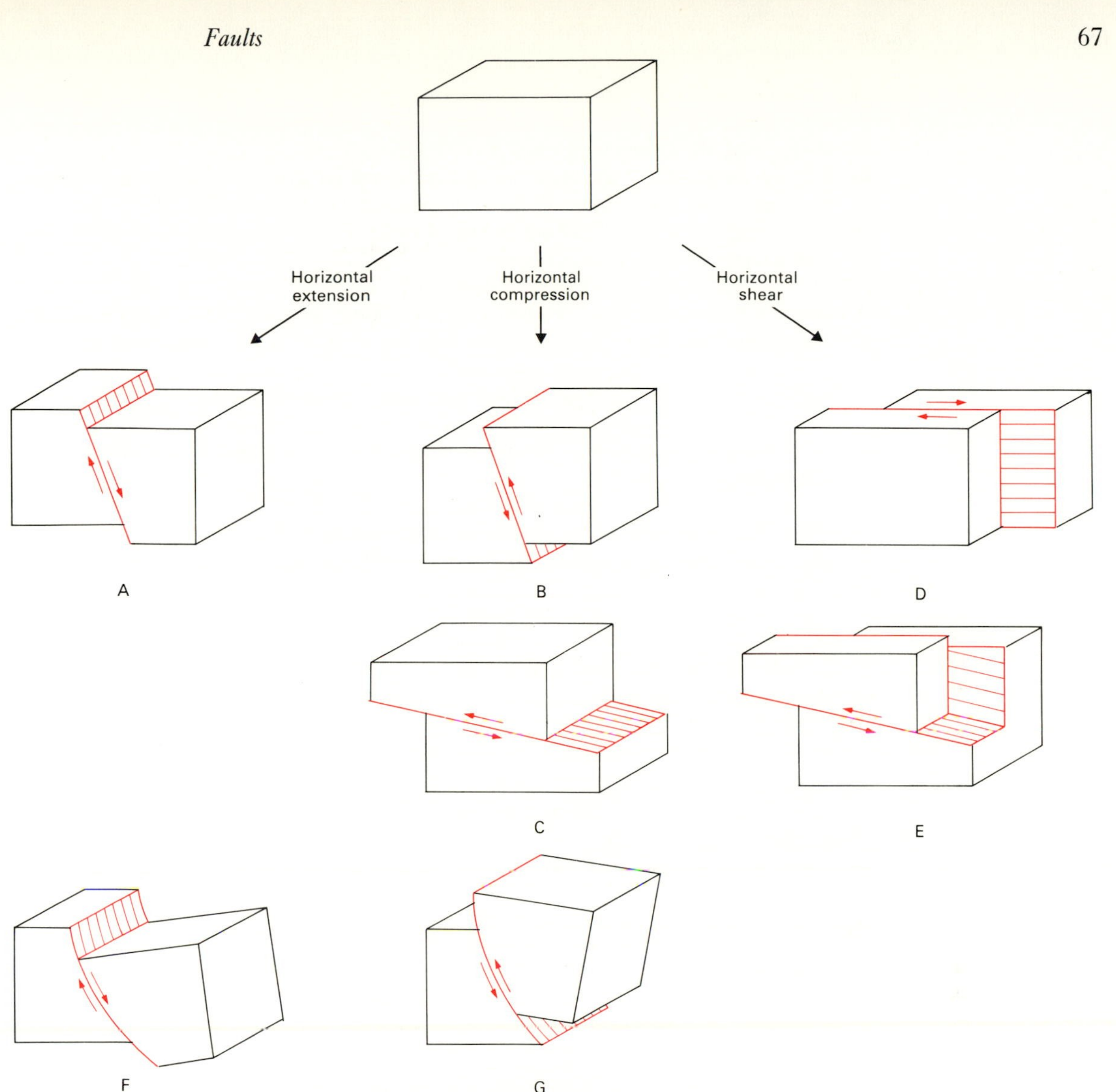

Fig. 4.3 Schematic representation of the principal types of fault.

(A) Normal fault – horizontal **extension** with slip parallel to the dip of the fault surface.

(B) Reverse fault – horizontal **compression** with slip parallel to the dip of the fault surface (greater than 45°).

(C) Thrust – horizontal **compression** with slip parallel to the dip of the fault surface (less than 45°).

(D) Strike-slip fault – horizontal **shear** with slip parallel to the strike of the fault surface.

(E) A combination of thrust and strike-slip fault. Slip parallel to the line of intersection of the two parts of the fault surface.

(F) Listric normal fault – horizontal **extension** with rotational movement of one or both blocks of rock.

(G) Listric reverse fault – horizontal **compression** with rotational movement. The steeply-dipping region is equivalent to a reverse fault and the low-dipping region is equivalent to a thrust.

Table 4.1 Sequence of observations for characterising a fault

1. Assess the kind of topography (Sect. 2.A, 2.H, 4.C):
   A. Rugged topography – use measured dips or construct structure contours to determine the geometrical relationships of the fault and adjacent boundaries
   Proceed to Step 4 or 7.
   B. Flat topography or small-scale maps
   Proceed to Step 2.
2. Determine the direction of dip of lithological boundaries adjacent to the fault (Sect. 2.B):
   A. Boundaries adjacent to the fault dip at low angles or all in the same direction

   Type of fault may not be determinable

   B. Boundaries adjacent to the fault dip at moderate to steep angles and in different directions
   Proceed to Step 3.
3. Determine the direction(s) of offset of lithological boundaries (Sect. 4.B):
   A. Boundaries dipping in opposite directions are offset in opposite directions

   Dip-slip fault

   Proceed to Step 4.
   B. Boundaries dipping in opposite directions are offset in the same direction

   Strike-slip fault

   Proceed to Step 7.
4. *Dip-slip faults*:
   Determine the direction of downthrow of the lithological units adjacent to the fault (Sect. 4.C.2):
   Stratigraphically *younger* rocks occur on the *downthrow* side of the fault
   Proceed to Step 5.
5. *Dip-slip faults*:
   Determine the direction of dip of the fault (Sect. 4.C.1):
   A. Fault vertical or direction of dip not determinable

   Type of dip-slip fault may not be determinable.

   B. Fault dips towards the downthrow side

   NORMAL FAULT

   C. Fault dips away from the downthrow side
   Proceed to Step 6.
6. *Dip-slip faults*:
   Determine the angle of dip of the fault plane (Sect. 4.C.1):
   A. Dip of the fault plane $>$ 45°

   REVERSE FAULT

   B. Dip of fault plane $<$ 45°

Table 4.1(cont.)

7. *Strike-slip faults*:
   Determine the direction of relative movement of lithological boundaries adjacent to the fault (Sect. 4.G):
   A. Boundaries off-set to the left on the far side of the fault

LEFT-LATERAL STRIKE-SLIP FAULT

   B. Boundaries off-set to the right on the far side of the fault

RIGHT-LATERAL STRIKE-SLIP FAULT

4.A TO RECOGNISE A FAULT

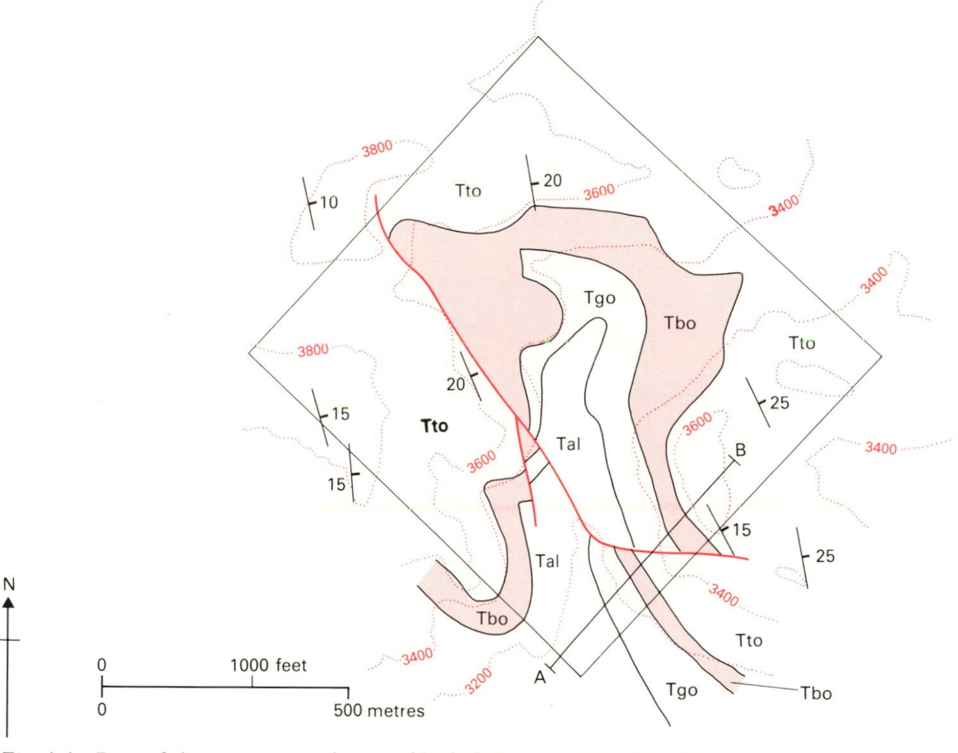

Fig. 4.4 Part of the west–central area of U.S.G.S. map GQ 1559 (Teapot Mountain, Arizona) showing Miocene rocks cut by two faults. Lithological units (in stratigraphical sequence):

Tto	Older tuff	(youngest)	Miocene
Tbo	Basalt		
Tgo	Older gravel		
Tal	Apache Leap tuff	(oldest)	

The outcrop pattern of Tbo has been slightly simplified. Tgo is not mappable in the western outcrops southwest of the faults.

Both faults displace lithological boundaries and bring units of different ages into contact.

The rectangle marks the area of the block diagram shown in Fig. 4.10. (See also Figs 4.9 to 4.12 and 4.14.)

On large-scale maps, faults and thrusts are usually shown with distinctive line symbols different from all other geological boundaries, and in addition may be individually named; the legend of the map should be inspected to identify the conventions used. Small-scale maps may show all boundaries the same.

Faults displace outcrops and boundaries that are otherwise continuous, and bring lithological units of different age into contact. *Fig. 4.4*

The displacement on a fault may be so large that the rocks on each side are all of different age and origin, and no units can be matched from one side of the fault to the other.

Examine the rock-types and structures to see whether the boundary in question can be explained by a fault (or by an unconformity – younger rocks on the upper side of the boundary – see Chapter 6, or by an igneous intrusion – igneous rocks on one side of the boundary – see Section 3.D). *Fig. 4.5*

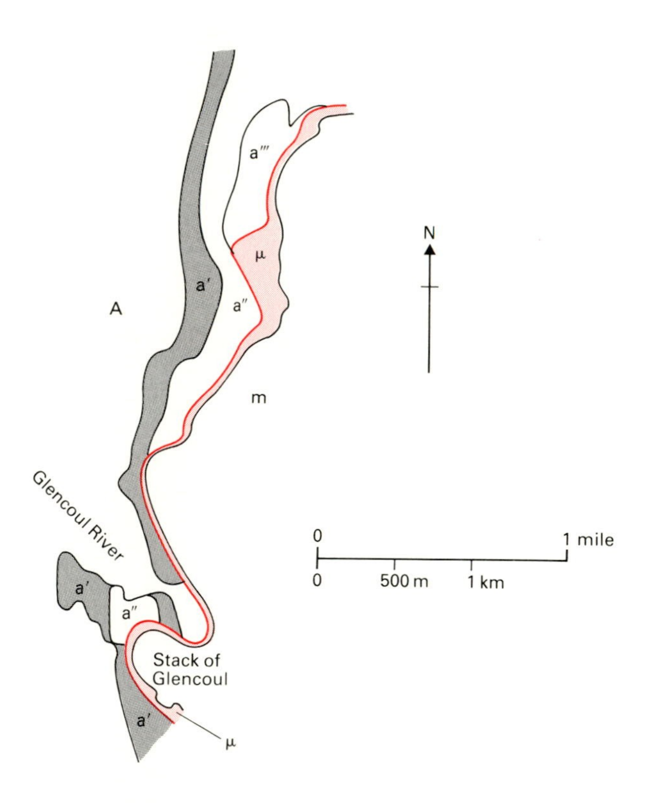

Fig. 4.5 Part of the northeast area of the B.G.S. Special Sheet of Assynt, Scotland, showing the Moine Thrust cutting across Precambrian gneisses and Cambrian sediments and bringing Moine Schists (older than the Cambrian) into contact with them. A zone of mylonite (recrystallised crush rock – Table 3.2) occurs on the fault plane.

m	Moine Schists	Late Precambrian
μ	Mylonite	
a'''	Fucoid Beds	Cambrian
a''	Pipe Rock	Cambrian
a'	Basal Quartzite	Cambrian
A	Lewisian gneisses	Early Precambrian

4.B TO DISTINGUISH DIP-SLIP AND STRIKE-SLIP FAULTS

Different types of fault may be distinguished by symbols – see the legend of the map.

Observe the direction of dip of lithological units and the direction of offset of their outcrops at as many points as possible along the line of the fault.

Positive identification of dip-slip and strike-slip faults may not always be possible – for example, if the units on both sides of the fault all dip in the same direction. Caution is particularly necessary in identifying strike-slip faults, which may produce a similar outcrop pattern to dip-slip faults cutting unidirectionally dipping units; strike-slip faults are much less common than dip-slip faults (see Table 4.1).

4.B.1 DIP-SLIP FAULTS

Fig. 4.6 The outcrops of lithological units dipping in opposite directions are offset in opposite directions by the fault.

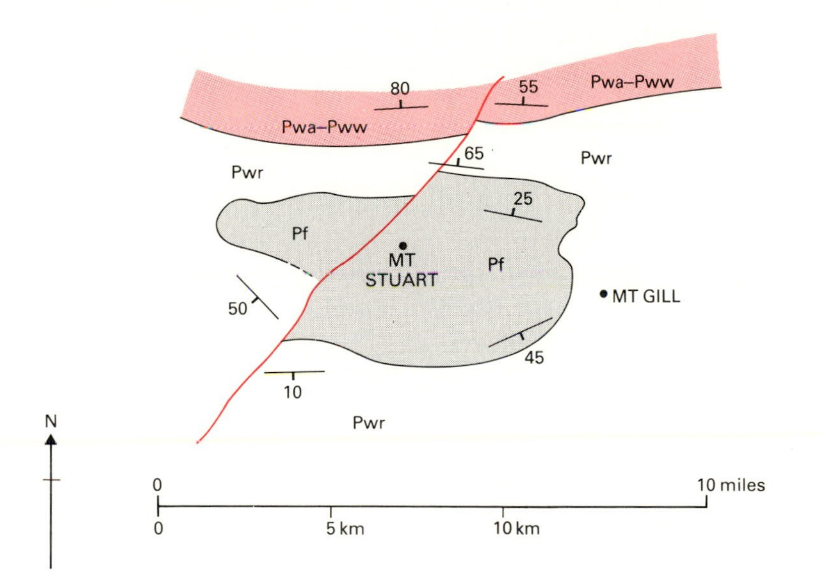

Fig. 4.6 Part of the south–central area of South Australia Geological Atlas Sheet SH54–9 (Copley) showing Precambrian rocks with variable dip direction displaced by a dip-slip fault.

Pwa-Pww	Units of Wilpena Group	Precambrian
Pwr	Brachina Formation	
Pf	Farina Sub-Group	

Minor faults, diapiric rocks, and Pleistocene omitted.

Fig. 4.7 Dip-slip faults cause little offsetting of the outcrops of steeply-dipping or vertical features (e.g. stratigraphical units, igneous dykes, margins of vents and plutons, many fold axial planes) whose strike is perpendicular to the strike of the fault.

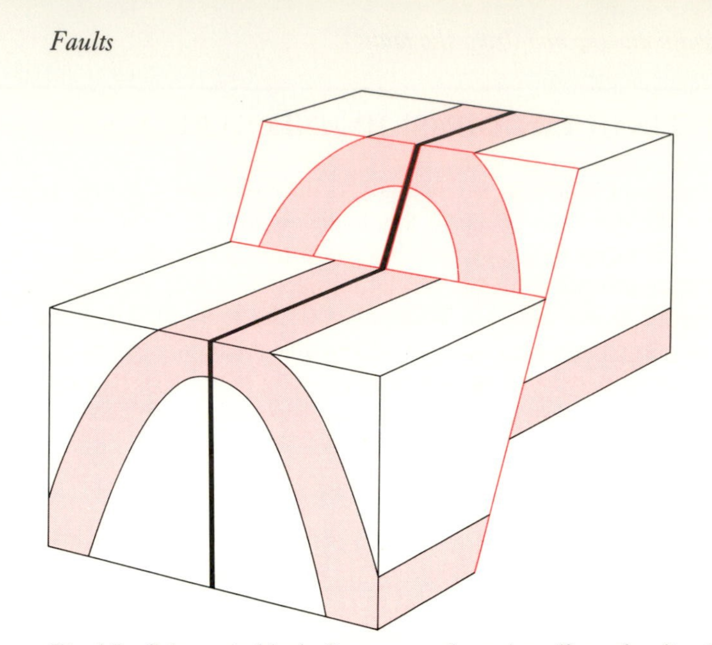

Fig. 4.7 Schematic block diagram to show the effect of a dip-slip fault on a vertical feature. An upright fold (cf. Fig. 5.3) with a vertical igneous dyke (solid black) marking its axial plane is displaced by a dip-slip fault (red) whose strike is perpendicular to the trend of the dyke. The outcrops of dipping units on the limbs of the fold are offset by the fault (cf. Fig. 4.6), but the vertical dyke is not offset.

4.B.2 STRIKE-SLIP FAULTS

The outcrops of moderately to steeply-dipping lithological units and structural features are all offset in the same direction. (Gently-dipping or horizontal units may show an inconsistent direction of offsetting.)

Fig. 4.8

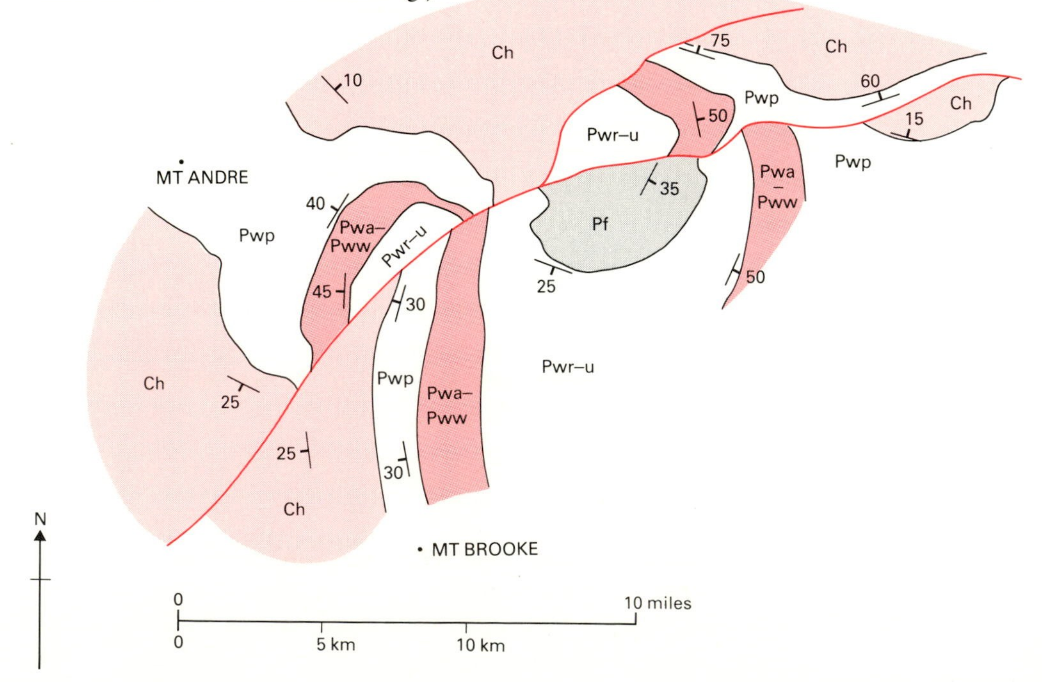

4.C TO DETERMINE THE ESSENTIAL INFORMATION FOR CHARACTERISING A DIP-SLIP FAULT

4.C.1 TO DETERMINE THE DIRECTION AND AMOUNT OF DIP OF A FAULT PLANE

The outcrop pattern of a fault is influenced by the topography in the same way as lithological boundaries (Sect. 2.A). But since fault surfaces are commonly both steeply inclined and gently curved, detailed inspection of the outcrop shape in relation to the topographical surface is necessary in order to determine whether deflections of outcrop are due to the intersection of a nearly planar surface with the topography or due to actual curvature of the surface.

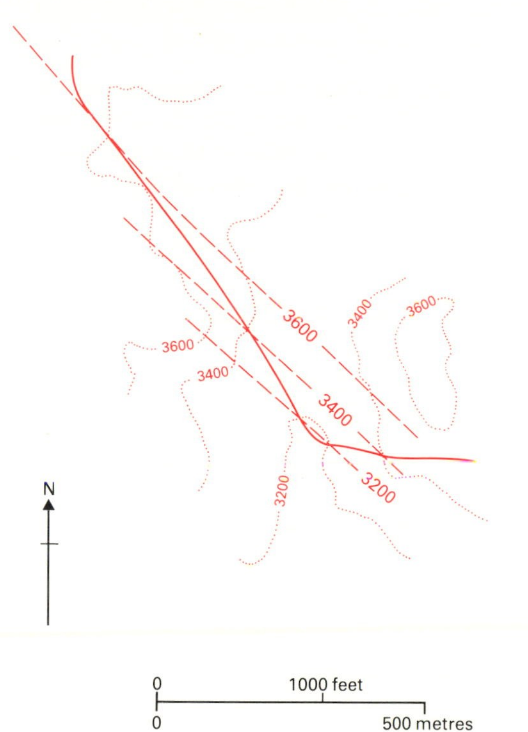

Fig. 4.9 The main fault and topographical contours from Fig. 4.4, with structure contours added. The dip of the fault is given by $\delta = \tan^{-1} 200 \text{ feet}/225 \text{ feet} = 42°$ (to the southwest).

Fig. 4.8 Part of the south–central area of South Australia Geological Atlas Sheet SH54-9 (Copley), showing Precambrian and Cambrian rocks cut by strike-slip faults.

Ch	Hawker Group		Cambrian
Pwp	Pound Quartzite	Wilpena Group	Precambrian
Pwa-Pww	other units of the Wilpena Group		
Pwr-u	Brachina Formation		
Pf	Umberatana Group		

Minor faults, diapiric rocks, and Pleistocene omitted.

The relative movement on the faults is such that all geological boundaries, irrespective of the direction of dip, are offset towards the northeast on the southeast sides of the faults.

The area is also shown in Plate 2.

The direction of dip of a fault (thrust) dipping at medium or low angles can be assessed in the same way as dipping lithological units (Sect. 2.A.1, 2.A.2).

The direction and amount of dip can be accurately determined provided that:

(a) The topography of the area is irregular, with the fault outcropping over a range of heights.
(b) Topographical contours are legible on the map or can be derived from a topographical map of the same area.

The procedure for constructing structure contours (Sect. 2.H) should be followed. The strike of the fault is given by the direction of the structure contours, and the dip, δ, is given by:

$$\tan \delta = \frac{\text{vertical interval between the structure contours}}{\text{horizontal distance between the structure contours}}$$

If the above conditions are not found (for instance where the topography is flat or only gently sloping, or on small-scale maps), the dip of the fault may be indicated in a cross-section in the margin of the map or in an accompanying description of the area.

Fig. 4.9 If even this information is lacking, the direction of dip may be indeterminable.

4.C.2 TO DETERMINE THE DIRECTION OF DOWNTHROW OF A DIP-SLIP FAULT

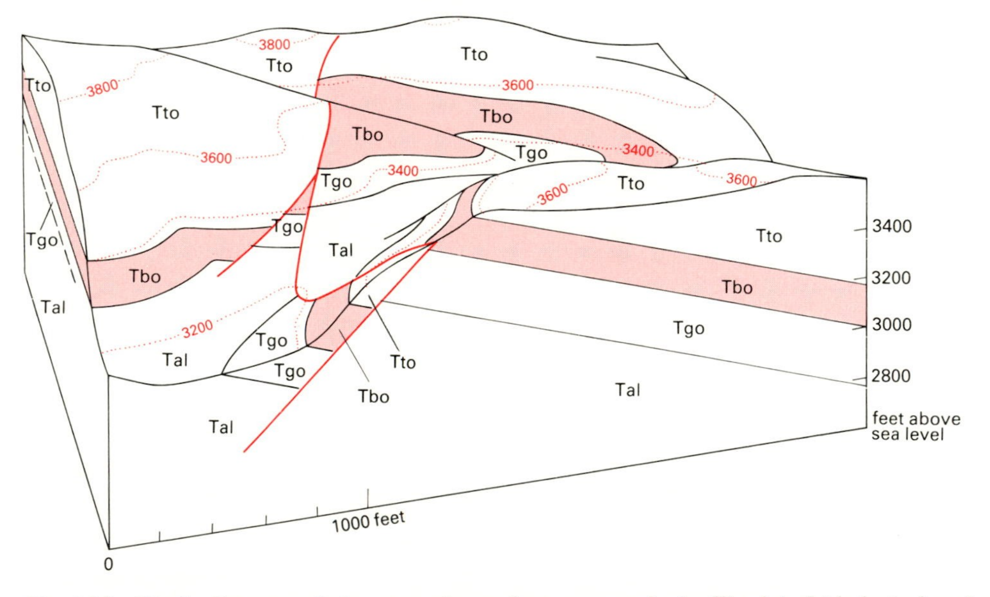

Fig. 4.10 Block diagram of the area shown by a rectangle in Fig. 4.4. Lithological units labelled as in Fig. 4.4.

Both the main fault and the small fault (on the far side of the valley in the foreground) bring Tbo, Basalt, on the southwest side into contact with older rocks (Tgo, Older gravel, and Tal, Apache Leap tuff) on the northeast side. The downthrow side of both faults is therefore on the southwest.

The direction of relative displacement (see Fig. 4.2(A)) is commonly indicated by a symbol such as a short tick on the downthrow side of a fault or by toothed marks on the upper plate of a thrust (see the legend of the map for the conventions used). It must be emphasised that the direction of movement is *relative*; the actual direction of movement cannot usually be determined, nor can the direction of true displacement (cf. Fig. 4.2 caption).

To identify the downthrow side of a fault from observations on the map, determine the stratigraphical succession in the area adjacent to the fault. Check that dips in the area are normal. (If the stratigraphical succession is inverted, as shown by overturned dips – Section 2.B.2 – determine the local sequence of lithological units from structurally lowest to highest (the reverse of the stratigraphical succession) and use this to identify the relative up and down movements each side of the fault.)

The effect of dip-slip movement is to bring structurally *higher* (with normal dips, stratigraphically *younger* rocks) on the *downthrow* side into contact with structurally *higher* rocks on the other (upthrow) side.

If there is strong topographical relief, a lithological boundary runs into the fault at a lower altitude on the downthrow side than on the upthrow side.

The direction of relative movement should be determined at a number of points, since it is possible for the direction to change (as in a scissors movement) along the length of the fault.

Fig. 4.10

4.D TO CONSTRUCT A VERTICAL CROSS-SECTION THROUGH A FAULT

A vertical cross-section shows the way the rocks would appear on a vertical slice through the area. The vertical and horizontal scales should be the same, so that observed or calculated dip values can be plotted directly on the section. The construction of an extended cross-section through an area is described in Section 9.C.

1. Select a position of interest and draw a line perpendicular to the strike of the fault plane. (To avoid marking the map, use a tracing paper or film overlay.)
2. Using a sheet of graph paper, construct the topographical profile along the line (see Appendix 3).
3. Mark the outcrop of the fault on the topographical profile, and construct a line to represent the fault plane dipping at the calculated angle (Sect. 4.C.1). If the dip is not known, draw a line at a reasonable angle to match what is known or inferred about the fault.
4. Mark the outcrops of lithological boundaries on the profile and construct lines at the known dip values to represent the inclination of the boundaries. Extend the lines from each side of the fault only up to the fault plane.
5. Make use of information on stratigraphical thicknesses of the units (Sect. 2.C) to ensure that the cross-section is self-consistent and geologically realistic.

Fig. 4.11

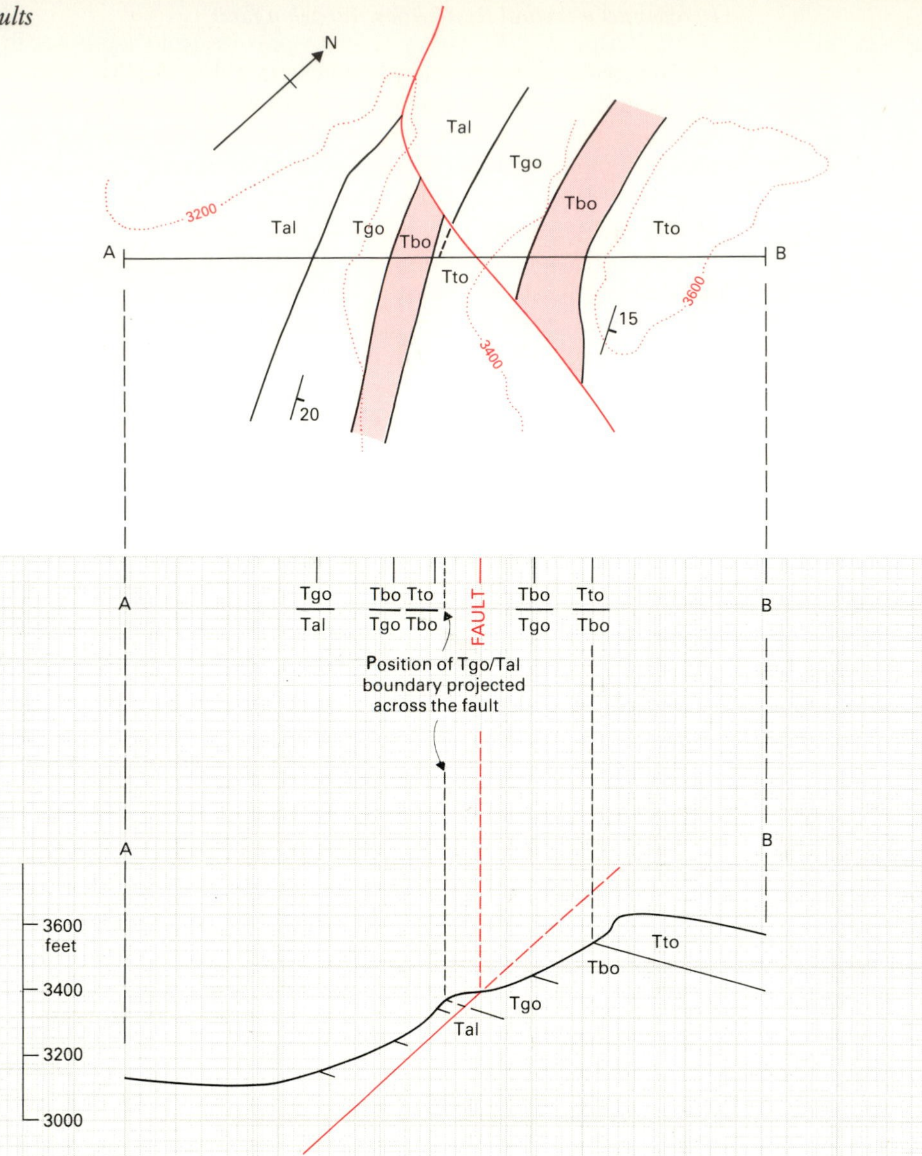

Fig. 4.11 The southeast part of Fig. 4.9, enlarged, and a partly completed vertical cross-section along the line A–B. The dashed line on the map is the geometrical extension of the Tgo/Tal boundary on the northeast side of the fault, to find where the boundary would have intersected the line of section if the fault had not been there.

The positions of topographical contours and of geological boundaries have been marked on the edge of the graph paper. These have been used to construct, in the following order:

(i) the topographical profile
(ii) the outcrop of the fault and a line to represent its appearance in cross-section, using the calculated dip of 42°
(iii) the Tto/Tbo boundary on the northeast side of the fault, using the dip angle of 15° shown on the map
(iv) the Tgo/Tal boundary on the northeast side of the fault has been approximately located from its projected position (dashed lines) on the southwest side of the fault
(v) the rest of the stratigraphical boundaries have been marked on the topographical profile ready for completion of the section.

The completed section is shown in Fig. 4.12.

4.E TO IDENTIFY NORMAL, REVERSE, AND THRUST FAULTS

Identify the direction and amount of dip of the fault and the direction of downthrow (Sect. 4.C). Then apply the criteria of Table 4.1, Steps 5 and 6.

Figs. 4.12 and 4.13 See Fig. 4.3 for the relationship between the directions of movement and of dip of the fault and the stress environment of each type of fault.

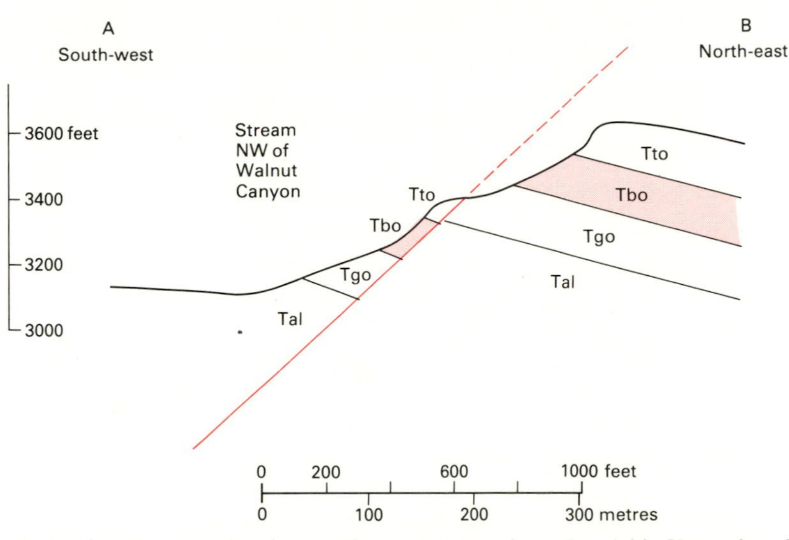

Fig. 4.12 The completed vertical cross-section from Fig. 4.11. Vertical and horizontal scales equal. The fault dips to the southwest and the downthrow side is on the southwest – the fault is therefore a *normal* fault, produced by horizontal extension.

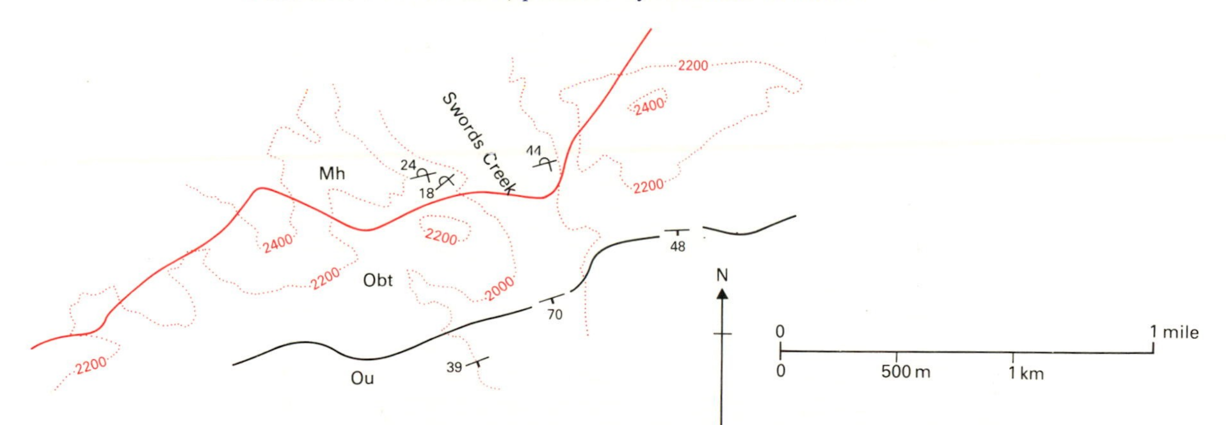

Fig. 4.13 Part of the central area of U.S.G.S. Map GQ 1542 (Honaker Quadrangle, Virginia), showing the Pine Mountain Fault cutting Ordovician and Carboniferous rocks in an area of irregular topography. Topographical contour interval 200 feet.

Mhr	Hinton Formation	U. Mississippian
Ou	Limestone	early M. Ordovician
Obt	Beekmantown Dolomite	L. Ordovician

Structure contours can readily be constructed on the fault surface and show that it dips at 25–35° between SSE and SE. The downthrow side is determined from the relative ages of the rocks each side of the fault and is on the north (the overturned dips in the Mississippian do not affect this argument). The fault is therefore a *thrust*, produced by horizontal compression.

4.F TO DETERMINE THE AMOUNT OF DISPLACEMENT OF A DIP-SLIP FAULT

Either the dip separation, s, or the stratigraphical displacement, d, or the throw, u, may be used to indicate the magnitude of the fault movement (see Fig. 4.2).

The displacement of a dip-slip fault can be simply determined provided that:

(a) The same lithological boundary can be recognised on each side of the fault plane, or two different boundaries can be recognised which are a known stratigraphical distance apart.
(b) The dip of the lithological units is the same each side of the fault.
(c) The fault is planar.

In other instances (very large faults which bring wholly different rock units into contact, or faults which cut folded rocks, or in which there is rotation in addition to dip-slip movement, or which have a history of successive movements), displacement of the fault must be determined from regional structural considerations (cf. Fig. 4.15; footnotes 3, 4, and 5 of Table 10.3).

Provided the above criteria are satisfied the displacement of a fault can be determined by calculation from measurement of the strike separation or from the outcrop width of displaced lithological boundaries (Fig.4.2(B)). In many cases, however, it is more convenient to measure the displacement from a vertical cross-section.

The displacement of a fault varies along its length from zero at the ends of the fault to a maximum value. The place where the displacement is measured should be stated; a complete description of a fault should include determination of displacement at as many points as possible.

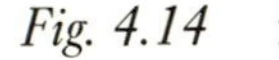
Fig. 4.14

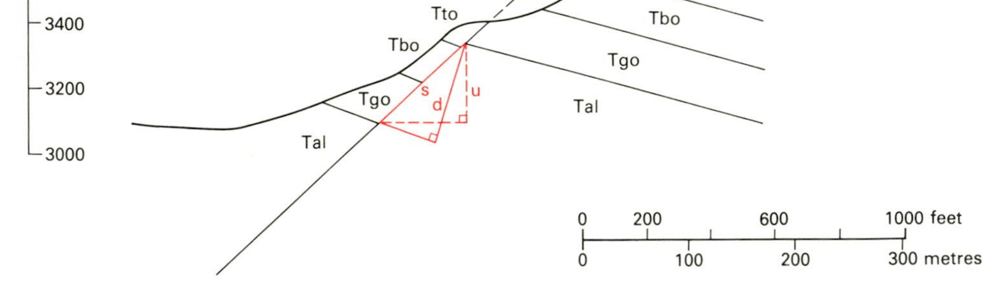

Fig. 4.14 The same vertical cross-section as Fig. 4.12, showing the measurements to indicate the displacement of the fault.

Dip separation	= s	=	109 metres (360 feet)
Throw	= u	=	75 metres (245 feet)
Stratigraphical displacement	= d	=	96 metres (315 feet)

The stratigraphical displacement of the fault can also be approximately determined from the map, Fig. 4.9, if the thickness of the lithological units is known. The top of Tbo on the southwest side of the fault is just below the base of Tgo on the northeast side. The thickness of Tbo + Tgo (measured from a section such as Fig. 4.12, or calculated by the method of Section 2.C.2) is 86 metres (282 feet). The stratigraphical displacement, d, of the fault is therefore a little more than 86 metres.

4.G TO DETERMINE THE DIRECTION OF RELATIVE MOVEMENT OF A STRIKE-SLIP FAULT

(Strike-slip fault surfaces are commonly steeply inclined or vertical. Commonly too they cause erosional features, so that information on their dip is hard to obtain. They can usually be assumed to be vertical unless there is evidence to the contrary.)

The direction of relative movement is horizontal (or predominantly so). Find stratigraphical or structural features that can be identified on both sides of the fault. Then apply the following criteria:

Picture yourself standing on a rock unit on one side of the fault:

If the continuation of the rock unit on the far side is to the right:
Right-lateral (or dextral) strike-slip fault.

If the continuation of the rock unit on the far side is to the left:
Left-lateral (or sinistral) strike-slip fault.

Examples The strike-slip fault shown in Fig. 4.8 was produced by left-lateral movement.

The northwest–southeast fault in the northwest corner of Plate 2 is a right-lateral strike-slip fault.

4.H TO DETERMINE THE DISPLACEMENT OF A STRIKE-SLIP FAULT

When a vertical feature (a lithological boundary or the axial plane of an upright fold – Fig. 5.3) is cut by a vertical strike-slip fault intersecting it at an angle, the displacement of the outcrop is the same as the strike separation, r, and is equal to the strike slip, ss (Fig. 4.2). Lacking such ideal relationships, the amount of movement can be determined approximately from the displacement of outcrops of boundaries with variable dip. More precise determination may be possible by construction or by calculation, using the geometry of Fig. 4.2 as a basis for relating strike separation of outcrops to the amount of strike-slip movement.

Identify corresponding boundaries on each side of the fault. Check that the dips of the units are moderate to steep and are variable in direction and amount. Measure the distance between corresponding boundaries each side of the fault. Similarity of distance or a systematic variation of distances between successive boundaries implies a consistent amount of strike-slip movement.

Example In Fig. 4.8 the strike separation of the Cambrian and Precambrian outcrops along the fault (including its northern branch) varies between 5 km in the south and 8 km in the north.

4.I TO DETERMINE THE TIME INTERVAL DURING WHICH A FAULT MOVED

The *older limit of age* of a fault is given by the youngest rock or structure that is cut by the fault. A fault is clearly younger than the stratigraphical or chronometric age of the rocks it cuts, but it is commonly possible to establish the age of the fault more precisely relative to later episodes of tilting, folding, or metamorphism that have affected those rocks.

Care is necessary in the determination of the *younger limit of age* of a fault because the end of a fault as shown on the map may represent:

(a) Movement on the fault became too small to be mappable.
(b) The forces that produced the fault by brittle fracture in one set of lithological units produced a different style of deformation (e.g. attenuation, folding) in adjacent, overlying, or underlying units.
(c) Economic priorities determined the detail of mapping. Faults of both large and small displacement are mapped in rocks of economic interest, but only the larger faults are mapped in older and younger lithological units.

Consequently the clearest evidence for determining the younger limit of age of a fault comes from unconformities (Ch. 6) or igneous intrusions that lie across or intersect the fault at a point where it has a *large* displacement in rocks older than the unconformity or igneous intrusion.

Intersections of faults may arise from the contemporaneous relative movement of three or more blocks of rock (analogous to paving stones moving on the joints between them); the contemporaneity (or otherwise) of such faults is usually difficult to determine. However, a fault which cuts and displaces a thrust can usually be recognised as later than the thrust.

Because a fault plane persists as a plane of weakness for millions of years after its initiation, several episodes of movement may be identifiable from extended study of the geological map.

To determine the time interval (or intervals) during which a fault moved, determine the age (stratigraphical or chronometric) of the youngest rock or structure that is cut by the fault, and determine the age of the oldest rock or structure that lies across, intersects, or deforms the fault.

Example On Fig. 3.9, the Kannondaki Fault cuts units S_1 to S_{4-5} of the Sobosan Volcanic Rocks, and is therefore younger than these. It is cut by the granite porphyry, Yf, and the intrusion follows and widens the fault plane. The fault is also cut by the biotite granite, Yg. (The Pleistocene Aso tuffs, A_4, lie across both the fault plane and the granite.) The earlier Sobosan Volcanics and the later intrusive rocks are Miocene, so the fault is also of Miocene date.

4.J TO DETERMINE THE RATE OF MOVEMENT OF A FAULT

Determine the true displacement of the fault (or any other measurement (see Fig. 4.2) that provides a good estimate of the relative movement of the rocks each side of the fault plane), and determine the time interval during which the fault moved (Sect. 4.I). Divide distance by time to give the rate of movement. (The estimate is likely to give a minimum value because the period of movement is usually less than the determinable time interval.)

Most faults have a small throw and give a very small minimum rate of movement, but major faults related to global crustal movement have rates of movement related to those of plate-tectonic processes (Table 1.2). *Fig. 4.15*

Example In Fig. 3.9(B) the throw of the Kannondaki Fault is at least the height of the section shown, i.e. a minimum of 2000 m. The fault is younger than (or possibly contemporaneous with) the Sobosan Volcanics and is older than the biotite granite, Yg (Fig. 3.9(A)); see the examples in Sections 3.L and 4.I. The duration of movement of

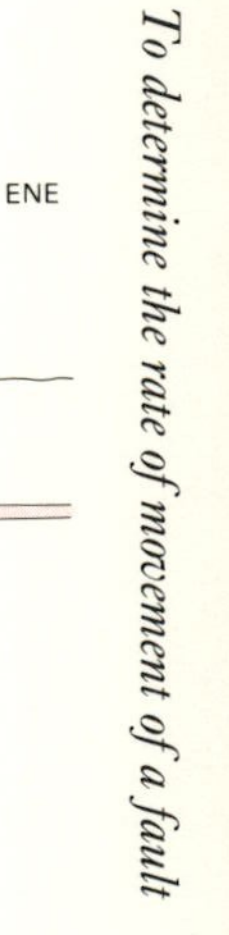

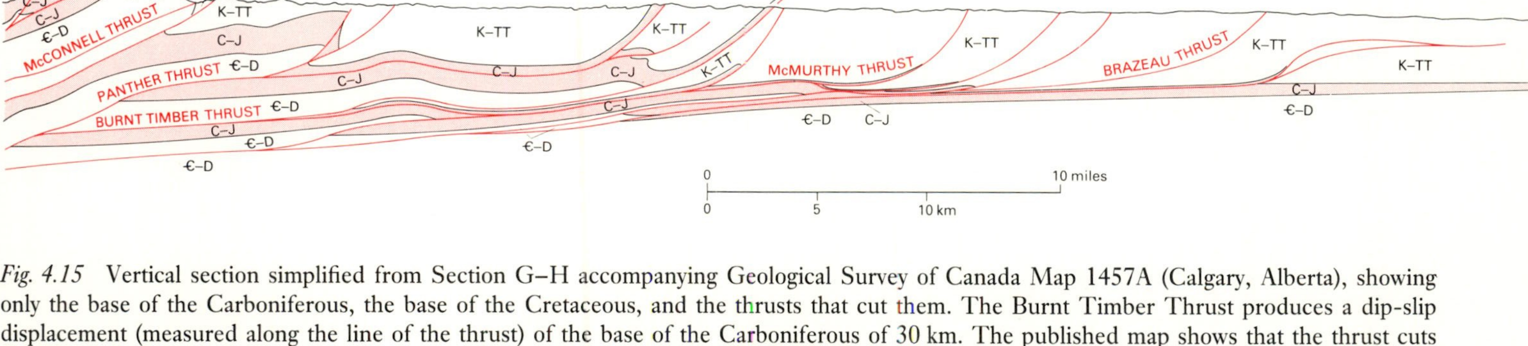

Fig. 4.15 Vertical section simplified from Section G–H accompanying Geological Survey of Canada Map 1457A (Calgary, Alberta), showing only the base of the Carboniferous, the base of the Cretaceous, and the thrusts that cut them. The Burnt Timber Thrust produces a dip-slip displacement (measured along the line of the thrust) of the base of the Carboniferous of 30 km. The published map shows that the thrust cuts Paleocene rocks, 65 to 55 Ma old; its younger limit of age cannot be determined from the map. The *minimum* rate of movement of the Burnt Timber Thrust is therefore 30 km/65 m.y. = 0.5 mm/y. The total displacement on all the thrusts below and to the east of the McConnell Thrust can be similarly measured from either the base of the Carboniferous or the base of the Cretaceous as 52–56 km, giving a minimum net rate of movement of 0.8 mm/y. (In fact the peak of the Rocky Mountain orogeny occurred in the mid-Eocene, when the rate of thrust movement would have been much faster than the estimated minimum rate–cf. Table 1.2.)

the fault is therefore confined to a very short interval early in the Miocene, not more than 2.7 m.y. and probably much less. The average rate of movement of the fault is therefore at least 2000×1000 mm/2.7×10^6 y = 0.75 mm/y and probably much greater (cf. the rapid rates of movement in present-day volcanic structures – Table 1.2).

4.K SOME EXAMPLES OF FAULT PATTERNS

Faults are the product of a regional stress system, and the type and relationships of faults reflect that system. Determination of the stress pattern may require more information than is usually available on a geological map. However, distinctive patterns are observable, of which some are illustrated below.

Note that the orientations of all the faults and their directions of movement are significant in determining their relationships.

4.K.1 FAULTS IN EXTENSIONAL STRESS SYSTEMS

Regional extensional strain uniformly distributed over an area results in a large number of small normal faults with a range of orientations and directions of movement. A preponderance of two particular directions may indicate a symmetrical relationship of the faults to the stress system. *Fig. 4.16*

A linear zone of crustal extension produces a graben or rift, with geologically younger rocks exposed in a down-faulted block between older rocks each side. The rift is frequently a topographical valley, and may be occupied by recent sediments or volcanic rocks and present-day rivers or lakes. *Fig. 4.17*

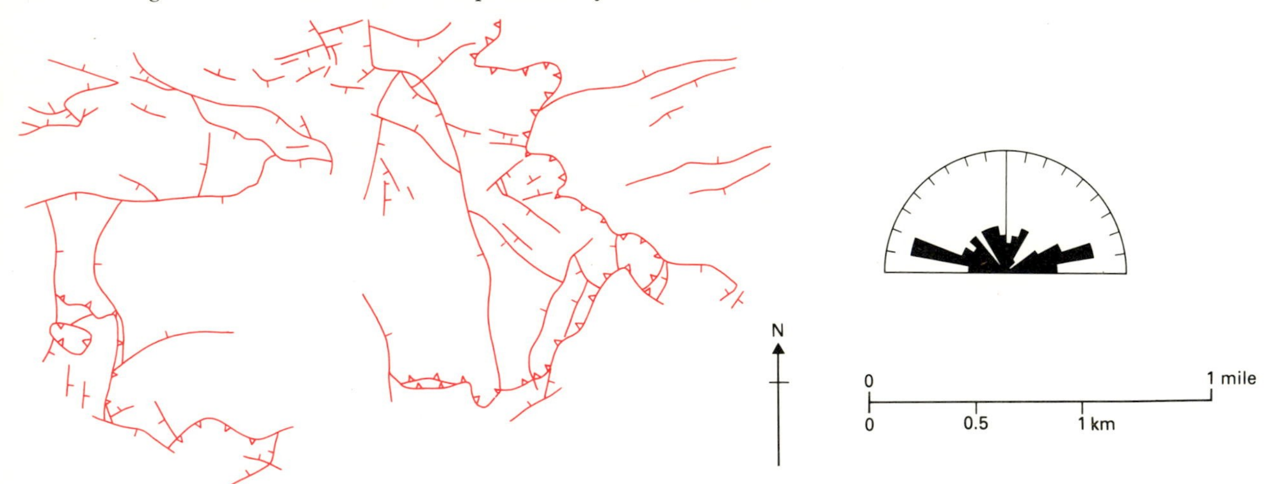

Fig. 4.16 Part of the north-central area of U.S.G.S. map GQ 1559 (Teapot Mountain, Arizona), showing only the faults and some of the thrusts. The area is within the Basin and Range Province of strong crustal extension. The numerous small faults have a range of orientations shown by the frequency diagram (inset), with maxima which are symmetrically related to the east–west direction of regional crustal extension. Detailed structural analysis of the area shows that the faults are initiated as steep-angle listric normal faults (Fig. 4.3(F)); as extension proceeds older faults are rotated to become low-angle gravity slides (thrusts – saw-tooth marks on upper plate), and new generations of normal faults (ticks on downthrow side) are produced. *Inset*: Frequency diagram of fault directions, constructed from segments of outcrops, 250 metres in length, of all the faults (excluding low-angle slides).

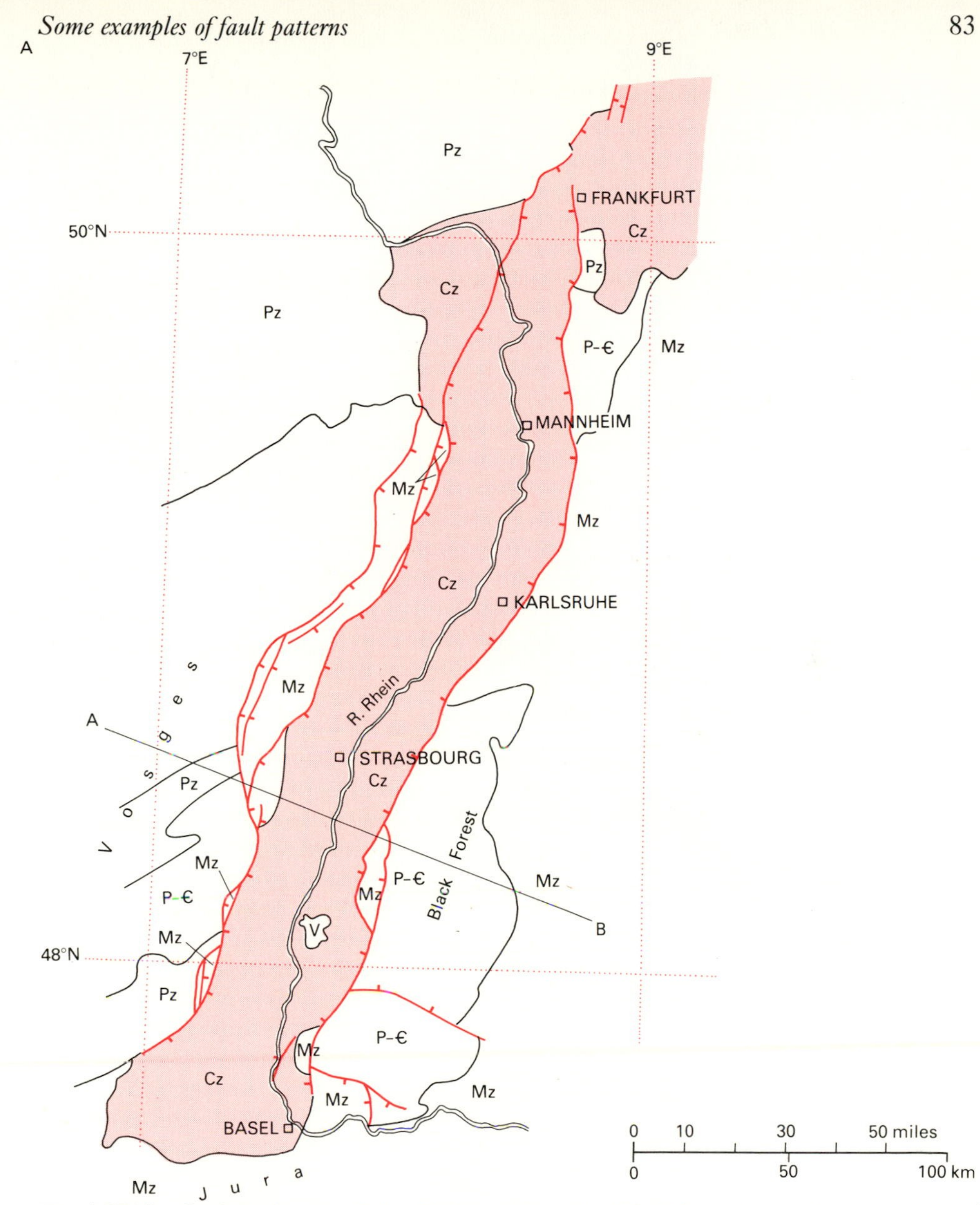

Fig. 4.17(A) Geological map of the Rhein graben, simplified from the southwest corner of Geologische Karte der Bundesrepublik Deutschland (1 : 1 000 000, 1973). Cz, Cenozoic; V, igneous rocks of the Kaiserstuhl volcano (Cenozoic); Mz, Mesozoic; Pz, Palaeozoic; P–C, Precambrian. The downthrow sides of faults are marked with a short tick. (See Fig. 1.1).

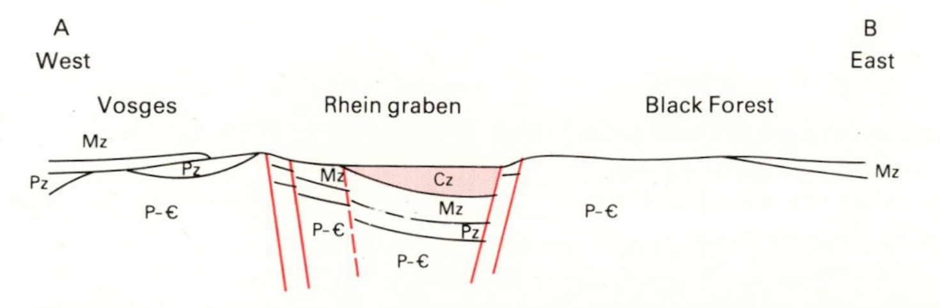

Fig. 4.17(B) Simplified cross-section along line A–B on Fig. 4.17(A). Cenozoic rocks are faulted down in the Rhein graben between Precambrian to Mesozoic rocks of the Vosges and the Black Forest.

4.K.2 FAULTS (THRUSTS) IN COMPRESSIONAL STRESS SYSTEMS

The boundary between an orogenic belt and relatively undeformed rocks of a stable block is commonly marked by a **thrust** – a laterally extensive plane of large-scale compression with movement of deformed over less deformed rocks.

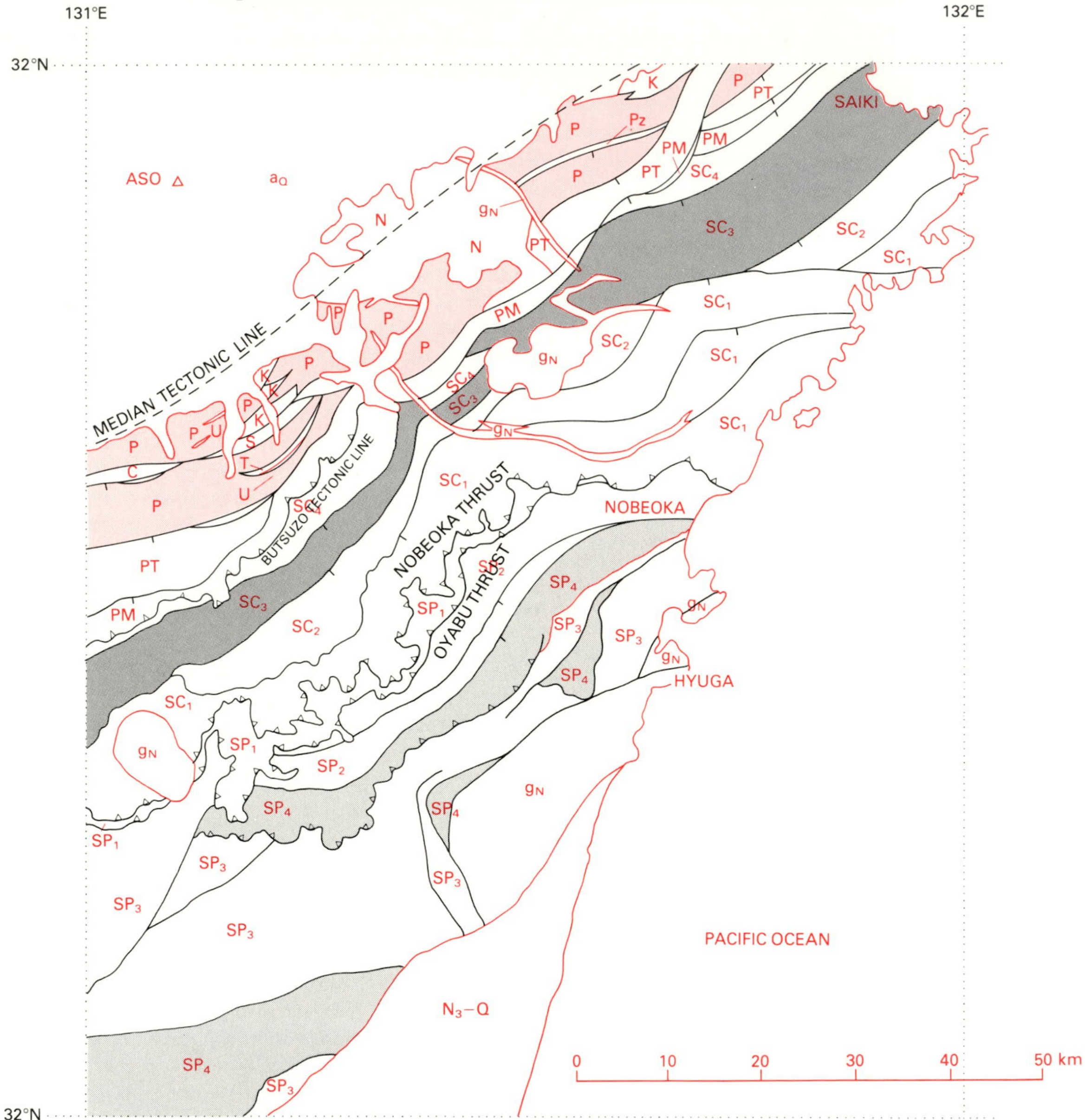

Fig. 4.18 Part of the northeast corner of Geological Survey of Japan Sheet 15 (Kagoshima), considerably simplified, showing thrusting in the east-central part of Kyushu. U peridotite and serpentinite, Pz Palaeozoic (phyllite), S Silurian and pre-Silurian, C Carboniferous, P Permian, PT Permian to Triassic, PM Permian to early Mesozoic, T Triassic, K Cretaceous, SC_1 Early Cretaceous, SC_2–SC_4 Late Cretaceous, SP_1–SP_2 Paleocene? to Eocene, SP_3–SP_4 Oligocene to early Miocene, (SC_1 to SP_4 Shimanto Supergroup), N Miocene, g_N granite (Miocene), N_3–Q late Miocene to Quaternary, a_Q volcanics from the Aso volcano (Quaternary). Downthrow sides of faults marked with a tick, upper side of low-angle faults (thrusts) shown with saw-tooth marks. The ages of the thrust sheets become generally younger towards the Pacific Ocean, and the Shimanto Supergroup represents a possible accretionary prism. Younger igneous intrusions (cf. Fig. 3.9) and the products of the active Aso volcano occur in the northeast of the area. Figure 3.10 shows the metamorphic zonation of the area.

Examples

Figure 4.5 shows the Moine Thrust between Cambrian sediments and regionally metamorphosed Moine Schists, forming the northwest boundary of the Caledonide mountains of the Highlands of Scotland.

Figure 4.15 shows the succession of thrusts that form the eastern margin of the Rocky Mountains in Canada, with flat-lying sediments of the foreland to the east.

At a convergent plate boundary (see Ch. 8), thrust slices of sediments formed in the oceanic trench at the continental margin or scraped off the subducting oceanic plate may form an **accretionary prism**; within the thrust slices individual sedimentary units young towards the continent, while the thrust slices themselves become younger towards the oceanic side. *Fig. 4.18*

4.K.3 FAULTS IN SYSTEMS WITH HORIZONTAL SHEAR MOVEMENT

Relative horizontal movement of blocks of rock at transform plate boundaries (Sect. 8.B.2) produces a strike-slip fault or system of faults, usually of very large displacement. On a smaller scale, strike-slip faults may be associated with folds or with thrusts and are the product of differential horizontal movement resulting from differing amounts of crustal shortening in adjacent blocks of rock (see Fig. 4.8, Plate 2). *Fig. 4.19*

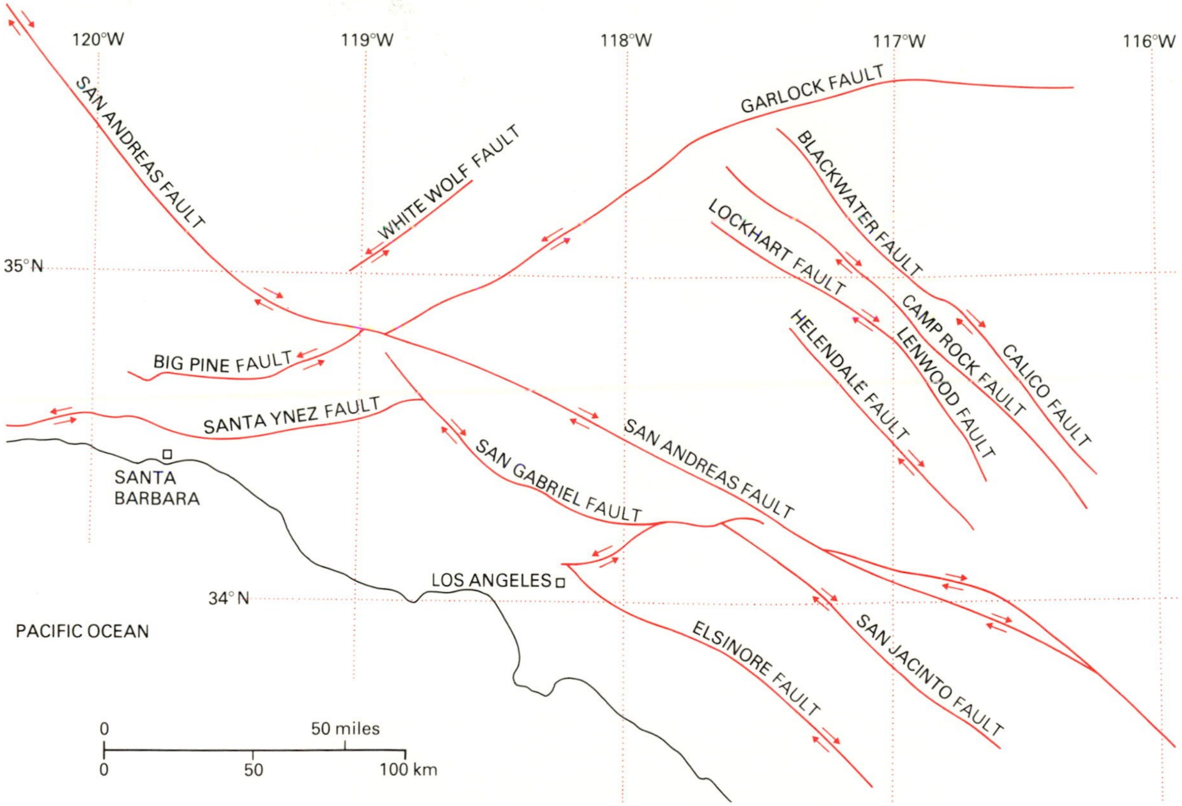

Fig. 4.19 Part of the south-central area of the 1 : 750 000 Geologic Map of California, showing only the major strike-slip faults and their directions of relative movement. NW–SE faults are right-lateral with large displacement (the San Andreas Fault is 1100 km long, with an estimated displacement at its north-western end of about 600 km); the E–W and NE–SW faults are left-lateral and smaller.

5 FOLDS

The response of rocks to stress by ductile deformation is for originally planar surfaces to become non-planar, that is, bent or buckled into folds. The fact that folds are found in such diverse materials as glacier ice, soft sediment, high-grade metamorphic rocks

Fig. 5.1 A minor fold in shales and gypsum beds of Eocene age, Sulaiman Range, Pakistan. Photograph by Professor W.D. Gill. The fold is a recent superficial structure produced by gravitational sliding (from left to right) of easily deformed sediments, but it illustrates characteristics of folding at a high level in the crust as found in many larger structures of tectonic origin:

(i) Lithological units maintain constant thickness around the fold.
(ii) Some units (for example in the core of the fold) respond to the stress by faulting rather than folding.
(iii) The amplitude of the fold is approximately equal to the thickness of the lithological units that are folded. Geometrical extrapolation of the fold structure is limited to the thickness of rocks involved; below this limit there must be some other structure (in this example a plane of sliding).

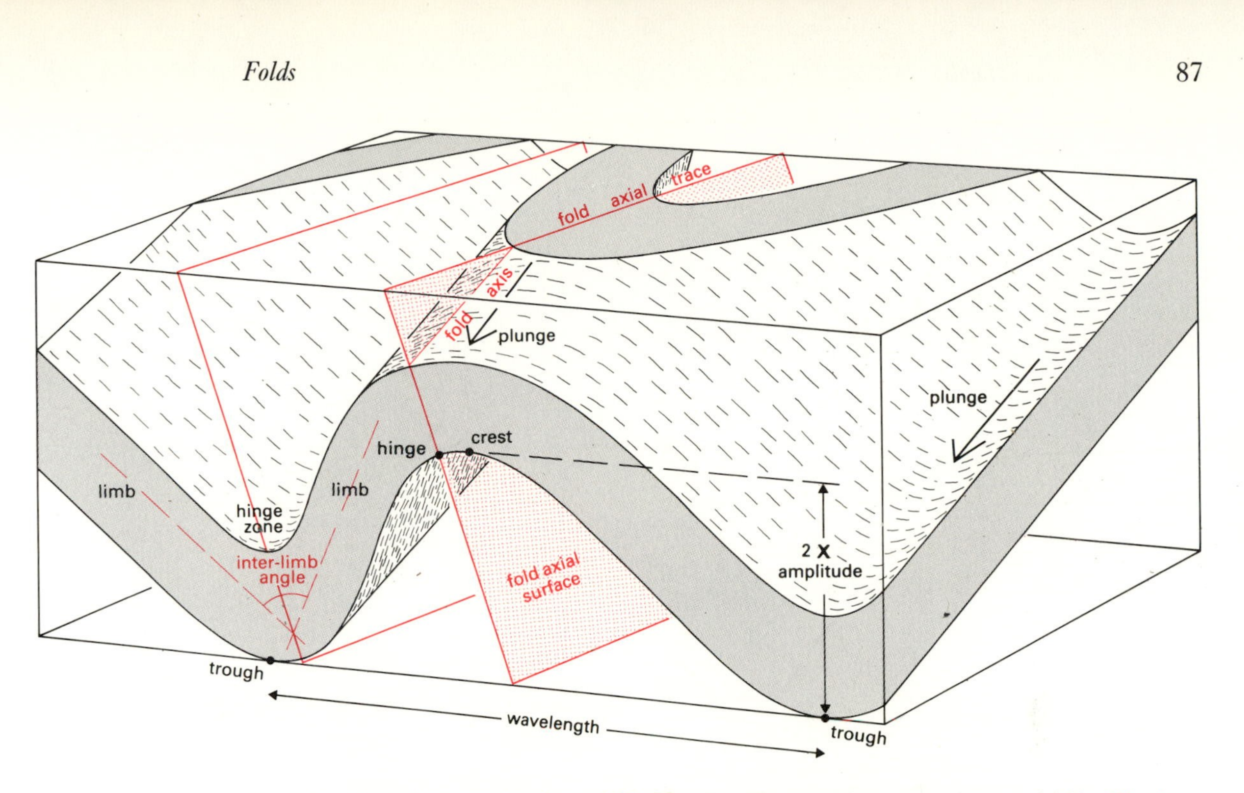

Fig. 5.2 Block diagram showing a folded layer to illustrate the nomenclature of folds. The layer is bounded by two folded surfaces (lithological boundaries or bedding planes). The **hinge** is the point of maximum curvature in any surface, and the **fold axial surface** (two shown, one dotted) passes through the hinge points of all the surfaces. The outcrop of the fold axial surface on the ground is the **fold axial trace**. The **fold axis** is defined by the intersection of the fold axial surface with any of the folded surfaces. **Crests** and **troughs** of surfaces do not necessarily coincide with hinges. The **plunge** of a fold is the direction and amount of dip of the fold axis, or of the crest (or trough) of the folded surface.

and as primary flow structures in igneous rocks, and that they range in amplitude from millimetres to kilometres, shows that the conditions of stress leading to fold formation are extremely varied.

On geological maps repeated 'V'-shaped or looped outcrop patterns suggest the presence of folds, but other information such as dip and strike directions, or the relative age of rock units is needed to confirm the structure (Sect. 5.A). Essential geometrical features of folds and their orientation are shown in Fig. 5.2. The shape of a fold in two dimensions is conveniently shown by means of **vertical cross-sections** (Sect. 5.B.)

Folds can be described and classified according to their geometrical shapes (Sect. 5.C, 5.D and 5.E).

The location of a fold on a map can be shown by its **fold axial trace** which marks the intersection of the fold axial surface with the ground surface (Sect. 5.C.1). The fold axis itself is horizontal in **horizontal folds** and inclined in **plunging folds** (Sect. 5.C.3). The principal directions and relative intensity of the compressional forces producing folds can be gauged from the inclination of the fold axial surface and the **tightness** of the fold profile measured as the **interlimb** angle (Sect. 5.C.5).

A fold is identified as an **antiform** if it closes upwards and a **synform** if it closes downwards (Sect. 5.C.6). If the stratigraphical succession is known, or can be deduced, the fold is an **anticline** if older rocks form the core, and a **syncline** if

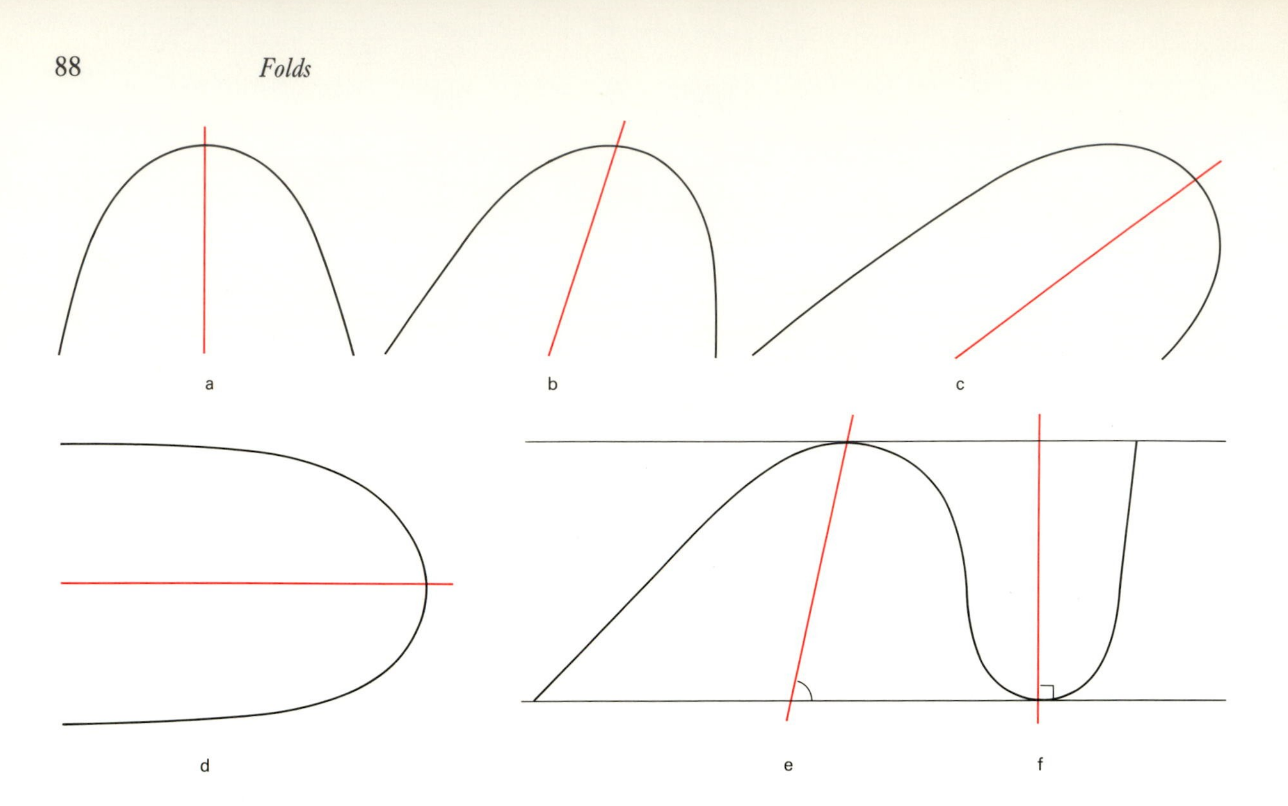

Fig. 5.3 Inclination of the axial plane as seen in profile, and description of folds. (a) **upright fold** (b) **inclined fold** (c) **overfold** (one limb inverted) (d) **recumbent fold** (e) **asymmetrical fold**: axial plane not perpendicular to enveloping surfaces; one limb shorter than the other (f) **symmetrical fold**: axial plane perpendicular to enveloping surfaces; limbs of equal length.

younger rocks form the core (Sect. 5.D). Special cases of these structures are **peri-clines, domes** and **basins** (Sect. 5.E).

Folding episodes may be dated in the same way as other geological events, that is, by determining the age of the youngest rocks or structures involved in the folding, and the oldest ones unaffected by the movements (Sect. 5.F). The effects of folding can be subdivided into a horizontal component of movement (layer shortening) and a vertical component (uplift). If the time interval of the fold formation is known, these measurements can be used to determine the **rates** of horizontal and vertical movement associated with the folding (Sect. 5.H).

The regional setting of folding and fold systems is related to the compressional forces that created the deformation; on the largest scale these can be related to the plate-tectonic environment (Sect. 5.I).

Detailed accounts of folds and the processes which create them are given by Park (1983) and Hobbs, Means, & Williams (1976).

5.A TO RECOGNISE FOLDS

On many geological maps the occurrence of folds is indicated by distinctive lines or other symbols marking the positions of the fold axial traces (Sect. 5.C.1). The legend of the map should be inspected to identify the conventions used.

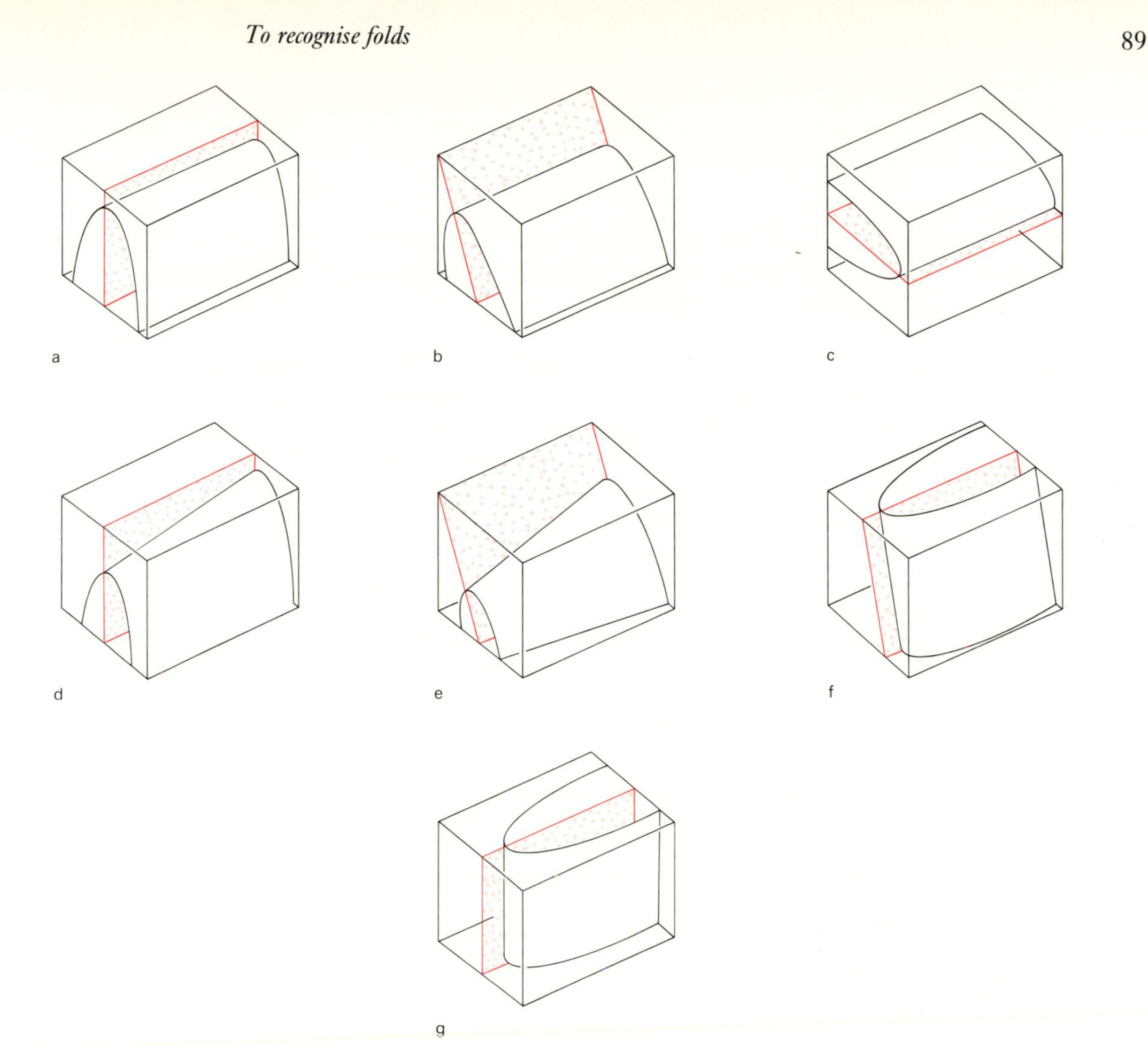

Fig. 5.4 Classification of folds based on orientation of the hinge line (or fold axis) and axial plane (or axial surface). The orientation of the hinge line is given first and that of the axial plane second in naming the fold.

Horizontal folds: (a) **Horizontal normal** (≡horizontal vertical) (b) **Horizontal inclined** (c) **Recumbent**

Plunging folds: (d) **Plunging normal** (≡ plunging vertical) (e) **Plunging inclined:** strike of axial plane oblique to trend of fold axis (f) **Reclined:** strike of axial plane perpendicular to trend of fold axis (g) **Vertical**

The outcrop pattern characteristic of eroded simple folds has V-shaped or looped outlines (on a scale larger than that caused by topographical effects – Sect. 2.A). Folds produce symmetrical repetitions of lithological units whose sequence and direction of dip are reversed on opposite sides of the fold axis.

Determine the relative ages and direction of dip of the rock units involved. If the dips of unfaulted adjacent rock units, or series of rock units, are in opposing directions, the rocks are folded. This becomes particularly clear if the outcrop of a rock unit forms a looped pattern.

Example Plate 1 shows a fold involving Devonian and Carboniferous rocks. The sequence of the Carboniferous rock units north of the Devonian outcrop is the reverse of that seen to the south, and the dips are in opposite directions.

5.B TO CONSTRUCT AN ACCURATE VERTICAL CROSS-SECTION OF A FOLD

The shape of a fold as it would be seen on a vertical plane cutting through the area is conveniently shown by means of a cross-section (also known as a structural section).

The technique of construction described below is conceptually simple and produces sections which are geometrically precise. Other methods (construction of lithological boundaries as concentric arcs, or as segments of parallel-sided units between boundary rays that mark changes of dip) are described by Roberts (1982).

The method is illustrated using an example of a simple fold; it is equally applicable to areas where the rock units dip in the same direction at constant or varying angles.

Constructions must not be extended across faults (Sect. 4.D) or unconformities (Ch. 6) which mark discontinuities in the structure. The construction of an extended cross-section through an area is described in Section 9.C.

5.B.1 Select a suitable line for the cross-section, using the following criteria:

(a) the line should be approximately perpendicular to the strike of lithological boundaries. If necessary, the line of section can be composed of segments at different angles to take account of differing strike directions. If a line of section oblique to the strike has to be used, see Section 5.B.13.
(b) there should be adequate dip information.
(c) the dips should be representative of the structure on a wider scale (i.e. avoiding locally anomalous dips which may represent small-scale local folding or faulting).
(d) the line can include topographically significant features (e.g. dip- and scarp-slopes that account for varying widths of outcrop of units).

Draw the chosen line of section on the map (see Plate 1). (To avoid marking the map, use a sheet of tracing paper or film overlay.)

5.B.2 Using a sheet of graph paper, construct the topographical profile (see Appendix 3). Use the same horizontal and vertical scales, unless there is good reason to use an expanded vertical scale (Sect. 5.B.12, below).

5.B.3 Mark on the profile the positions, directions, and amounts of dip measurements. Information should be projected along the strike provided that dips are not projected across faults, minor fold axes (Sect. 5.C), or unconformities. Make use of data from *both sides* of the line of section.

5.B.4 At each measured dip point, construct a faint line ('dip-line') 20–30 mm long at an angle equal to the dip angle, and extending above and below the topographical profile. Construct another faint line ('guide-line') 60–80 mm long perpendicular to the dip-line and also extending above and below the profile (e.g. for a dip of 26° to the south, draw a dip-line inclined at 26° to the south and a guide-line inclined at 64° to the north). If guide-lines intersect, terminate or erase both of them beyond the point of intersection.

5.B.5 Mark on the topographical profile the positions of the outcrops of lithological boundaries.

5.B.6 Begin the construction of lines to represent the lithological boundaries on the cross-section. The boundaries must be parallel to the nearest dip-line. To a first approximation, they maintain this dip up to midway between adjacent guide-lines; at this point they become parallel to the next dip-line (but see 5.B.7 below). Continue this process for all the lithological boundaries and extend the construction to a depth of 10–20 mm (depending on the degree of confidence in the interpretation of the structure) below the topographical profile.

5.B.7 At this stage, suspend construction in order to consider the thicknesses of lithological units. Data on the thickness of an individual unit are available:

(a) From the cross-section constructed so far.
(b) By determination of thicknesses from measurements on the map (Sect. 2.C).
(c) From published data on stratigraphical thicknesses in the margin of the map or in descriptions of the lithological units.

These three sources of information must necessarily yield compatible results. If there are gross differences between (a) and (b) or (c), the following procedures are possible (the choice of which to follow depends on the evidence on the map).

(i) Adjust the topographical profile so as to obtain the best fit of both the topographical data and the thickness data for the lithological unit.
(ii) Alter the position at which a change of dip occurs from the mid-point between two guide-lines to some position closer to one guide-line or the other so that a better fit to the thickness data is obtained.
(iii) Determine whether there are greater variations in thickness of the lithological unit than are indicated by (b) or (c). In this case, continuation of the construction will need to take account of such variations and of the lithology of the unit (Ch. 3) so as to produce a geologically reasonable interpretation.
(iv) Determine whether there could be subsidiary folding or faulting which locally makes the lithological unit appear to be thicker or thinner. In this case, continuation of the section construction should proceed on the basis of the thickness of the unit as indicated by (b) or (c), rather than the structurally-influenced thickness estimated by (a).

5.B.8 Using known, calculated, or constructed thicknesses of lithological units, extend the construction of the section to as far below ground level as possible. In doing this, make use of the following:

(i) The constant-thickness principle (except as indicated in Sect. 5.B.7 above).
(ii) The guide-lines; these indicate the sectors of the structure whose dips are the same as those at the surface. But note that if there is a lithological unit of irregular thickness, the guide-lines for the part of the section below it no longer apply; this

Plate 2. Extract from South Australia Geological Atlas Series Sheet SH54-9 (Copley). Scale 1 : 250 000. Caption on page x.

Dips and thicknesses shown at increased vertical scale must be calculated as follows:

dip: $\tan \alpha' = V \cdot \tan \alpha$
thickness: $d' = V \cdot d \cdot \cos \alpha' / \cos \alpha$

where:
α = true angle of dip
α' = exaggerated angle of dip
d = true thickness
d' = exaggerated thickness
V = exaggeration factor

For values of dip up to about 10°, the exaggerated thickness is approximately given by $d' = V \cdot d$ (error <5%). On a section constructed with increased vertical scale the thickness of a unit appears to decrease as the dip increases; serious distortions of the structure and stratigraphy result if the dip varies between low and medium to high values along the length of the section.

5.B.13 Where the dip direction is not the same as the line of section, (the line of section is not perpendicular to the strike direction), the inclination of the layers along the line of section is given by the apparent dip:

$$\tan \alpha^* = \tan \alpha \cdot \cos \varepsilon$$

where
α^* = apparent dip
α = true dip
ε = angle between true dip direction and apparent dip direction.

5.B.14 As a general rule, plot faults, intrusions, and unconformable rock units first on the vertical section. The angle of a fault plane as shown on the line of section may be determined from the form of its outcrop (Sect. 4.C.1). Discordant intrusions are plotted with vertical sides unless information on the map indicates a different shape. Unconformable rock units (Ch. 6) are plotted using the same procedure as for the rock units above them. Superficial deposits (Ch. 7) are plotted on the section if their thickness is sufficient to show. Alternatively, their outcrop can be indicated by drawing the relevant symbol just above the line of section.

5.C TO DESCRIBE FOLD GEOMETRY

5.C.1 TO DETERMINE THE FOLD AXIAL TRACE

The fold axial trace may be marked by a line with a distinctive symbol on a geological map along which the fold axial plane/surface intersects the ground surface. Inspect the legend of the map to identify the symbol used.

To determine the fold axial trace, the following method is accurate enough for most purposes. Locate the hinge zone of the fold by noting all the strike and dip symbols marked: the hinge zone runs between the points where opposing dips are closest together. Sketch in the fold axial trace as a line within, and parallel with, the hinge zone. Indicate the type of fold present with an appropriate symbol (cf. Fig. 5.7).

Fig. 5.6

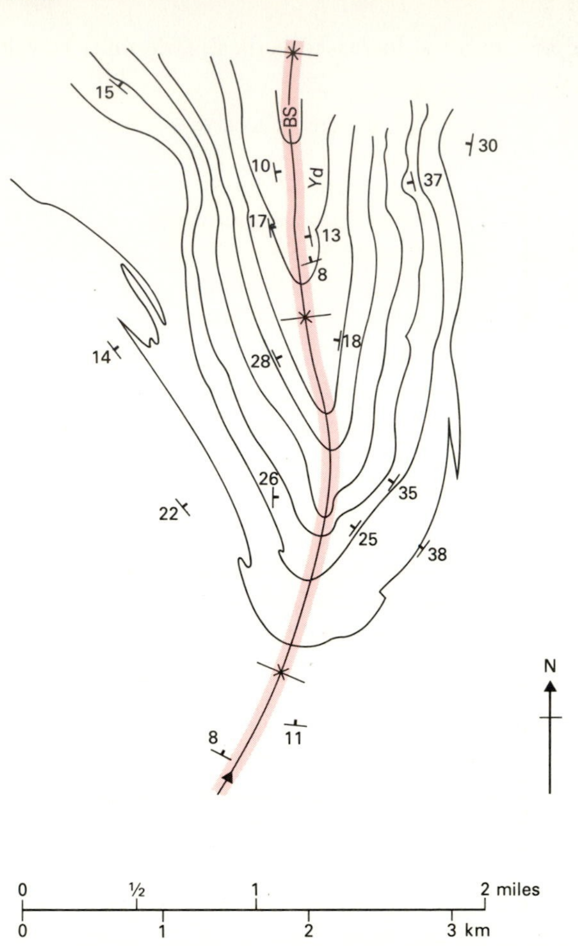

Fig. 5.6 Part of the west–central area of B.G.S. Sheet 111 of Buxton showing the location of the hinge zone and axial trace of a fold in Carboniferous rocks. The geology has been simplified by omitting faults and certain stratigraphical boundaries.

BS	Big Smut Coal	Westphalian	Carboniferous
Yd	Yard coal		

Other lithological units mainly Namurian

The hinge zone is shown as a pink band. The fold axial trace is shown as a black line with the symbol of a synclinal fold, plunging northwards (Sect 5.C.3 and 5.D).

5.C.2. TO CLASSIFY A FOLD ACCORDING TO THE DIP OF THE FOLD AXIAL PLANE/SURFACE

Note the dip angles recorded across the fold hinge zone along a line on the map perpendicular to the fold axial trace. Plot a diagrammatic cross-section of the fold and draw on it a line bisecting the inter-limb angle (Fig. 5.2). This line is the projection of the fold axial plane/surface. Measure the dip angle of this line and use the table below to classify the fold. If suitably oriented accurate vertical sections are available these may be used. If the axial surface projection line is curved determine an average dip by drawing a chord between its two ends.

Dip of axial plane/surface	Fold type	
90°–80°		upright
80°–60°	steeply	inclined
60°–30°	moderately	inclined
30°–10°	gently	inclined
10°–0°		recumbent

Examples The fold in Fig. 5.5 is a steeply inclined fold with the axial surface dipping at 65°. The folds in Fig. 5.10 are all either steeply inclined or upright.

5.C.3 TO RECOGNISE PLUNGING FOLDS, AND DETERMINE THE DIRECTION AND ANGLE OF PLUNGE

If the axis of a fold is not horizontal, the fold is said to plunge. (The word pitch has sometimes been used synonymously with plunge. However, it is now the convention to use pitch for the angle between an inclined line and the strike of the plane which contains the line.) In reality all folds plunge because eventually they must die out in the direction of the fold axis, although a map may contain only that part of a fold where the fold axis is essentially horizontal.

The direction of plunge of a fold may be indicated by a symbol on the published map. In the absence of such direct information the presence of plunging folds may be deduced from outcrop patterns and dip and strike information.

To recognise plunging folds from the outcrop pattern when the topography is flat or moderate, select the outcrop of a convenient stratigraphical boundary recognisable on both sides of the fold axis. If the boundary forms a nose- or V-shape as it crosses the fold axis, the fold is plunging. In mountainous topography a comparable pattern may be shown but other information is necessary to confirm plunging folds.

Fig. 5.7

To recognise plunging folds from dip and strike information, follow the outcrop pattern, noting dip and strike symbols in each rock unit, particularly in the closure of the fold where the outcrops of each limb converge. If the strike direction here is perpendicular, or nearly so, to that of the hinge zone, or fold axial trace, the fold is plunging and in a direction perpendicular to the strike direction.

Fig. 5.8

The direction and angle of plunge may be determined:

(a) By inspection of the strike and dip symbols in the closure of the fold.

Example In Fig. 5.8, the fold plunges west-southwest at angles varying from 40°–55°

(b) By calculation, using the relationship between the known thickness of a lithological unit and its width of outcrop (measured in the closure of the fold) as described in Section 2.C.

Example In Fig. 5.6 the width of outcrop of the Coal Measures, measured along the fold axial trace between Yd (Yard Coal) and BS (Big Smut Coal), is 1000 m, and the thickness measured from the key provided on the map is 95 m. The average plunge is, then, 5.5° north.

Once the direction and angle of plunge of a fold are known the map itself can be used directly to provide a picture of the profile of a fold, that is, the shape of the fold in a plane perpendicular to the fold axis.

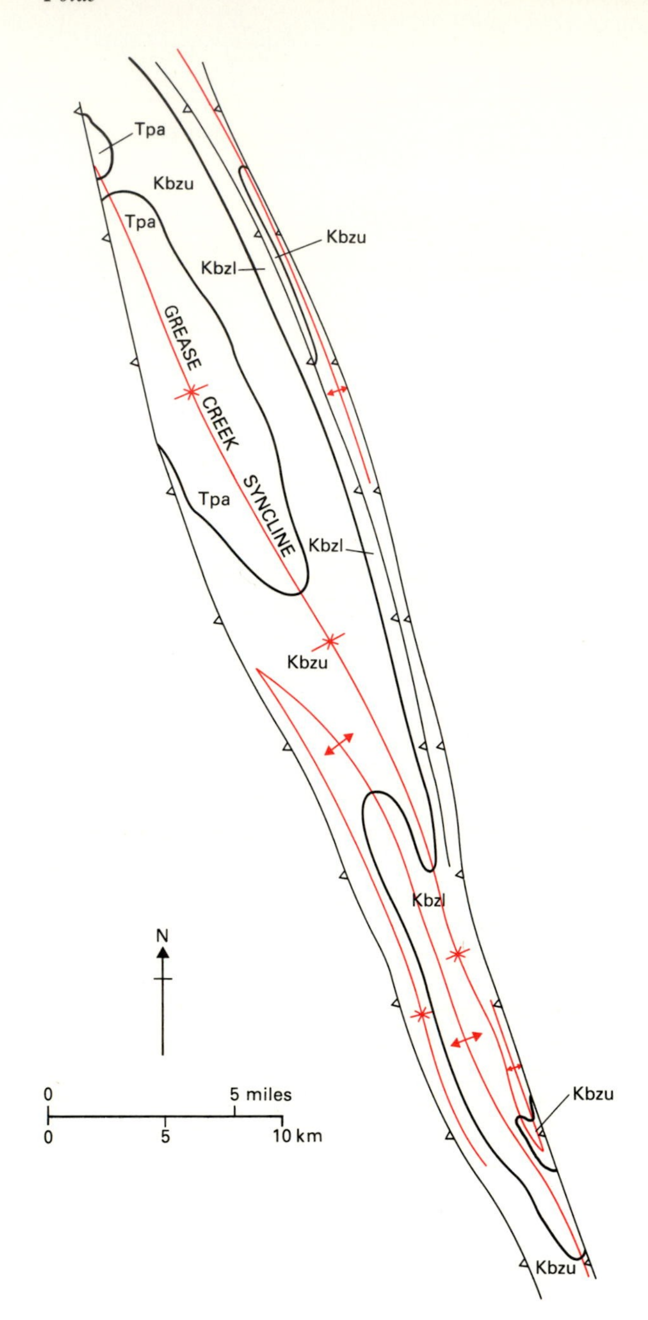

Fig. 5.7 Part of the east–central area of Geological Survey of Canada Map 1457A of Calgary showing the outcrop pattern of plunging folds.

The geology has been slightly simplified by omitting minor folds and faults and minor lithological units.

Tpa	Paskapoo Formation	Tertiary
Kbzu	Brazeau Formation, upper part	Cretaceous
Kbzl	Brazeau Formation, lower part	

Red lines are fold axial traces. Thrust faults are shown as black lines with saw-teeth on upthrust side.

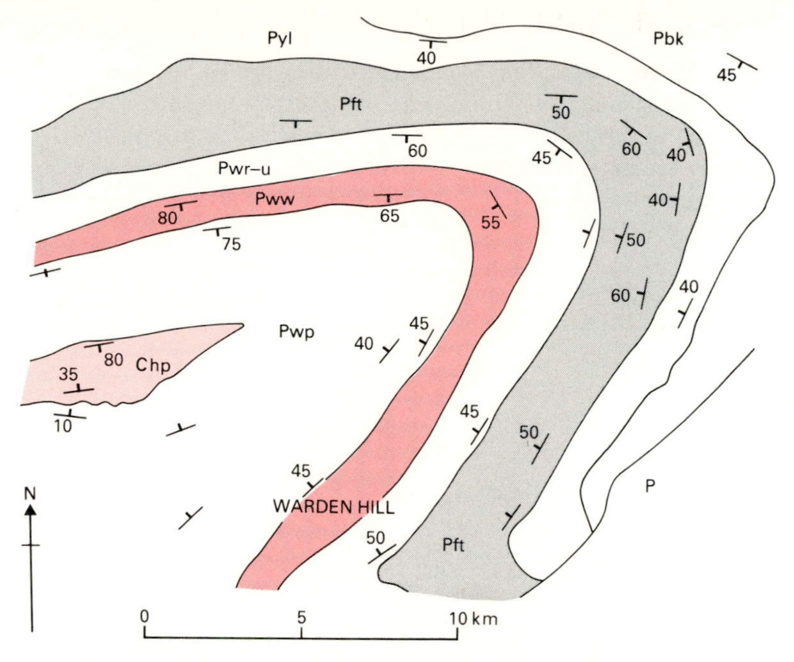

Fig. 5.8 Part of the east–central area of South Australian Geological Atlas series Sheet SH 54–9 (Copley) showing a plunging fold in Precambrian and Cambrian sediments, minor faults omitted. The area lies just to the north of that shown in Plate 2.

Chp	Parachilna Formation	Cambrian
Pwp	Pound Quartzite	Precambrian
Pww	Wonoka Formation	
Pwr-u	Brachina Formation	
Pft	Tapley Hill Formation	
Pyl	Tillite	
Pbk	Skillogalee Dolomite	
P	Other Precambrian units	

Experiment with Fig. 5.8. Turn the map so that your line of sight is in the direction of the plunge of the fold axis. Adjust the level of the map, or your own eye level so that you are looking along the fold axis at the same angle as the plunge (the 'down-plunge view'). The synclinal profile of the fold will now be seen at right angles to your line of sight.

Remember (i) this technique is suitable only for regions where the topography is no more than moderate; (ii) the plunge of a fold may vary in angle and direction: if the direction reverses then you must look down-plunge from the opposite direction. (This technique may also be used to ascertain the throw direction of dip-slip faults by viewing the structure in a down-dip direction.)

5.C.4 TO CLASSIFY A FOLD IN TERMS OF FOLD AXIAL SURFACE AND PLUNGE

Determine the dip of the fold axial surface as described in Section 5.C.2 and the angle of plunge as described in Section 5.C.3. Use the angles so determined and Fig. 5.4 to classify the fold.

Example The fold illustrated in Fig. 5.8 is a plunging inclined fold.

5.C.5 TO DETERMINE THE TIGHTNESS OF A FOLD

The tightness of a fold is a relative measure of the compressive forces that have produced it and is determined as the interlimb angle (Fig. 5.2).

Use the vertical cross-section provided with the map if this is suitable, or construct a cross-section as described in Section 5.B. For most purposes a sketch section using the dip angles will suffice. Measure the interlimb angle and classify the fold using the following values:

Interlimb angle	Class of fold
0°	isoclinal
0°–30°	tight
30°–70°	close
70°–120°	open
120°–180°	gentle

Example The fold in Fig. 5.5 is an open fold.

5.C.6 TO IDENTIFY ANTIFORMS AND SYNFORMS

Folds can be described according to the direction in which the limbs dip, relative to the hinge zone, or closure, of the fold.

Locate the fold hinge zone as described in Section 5.C.1, or examine the vertical cross-sections provided with the map; or examine the map with the 'down-plunge' view as described in Section 5.C.3.

If the limbs dip away from the hinge zone; fold closes upward: **antiform.**

Fig. 5.9 If the limbs dip toward the hinge zone; fold closes downwards: **synform.**

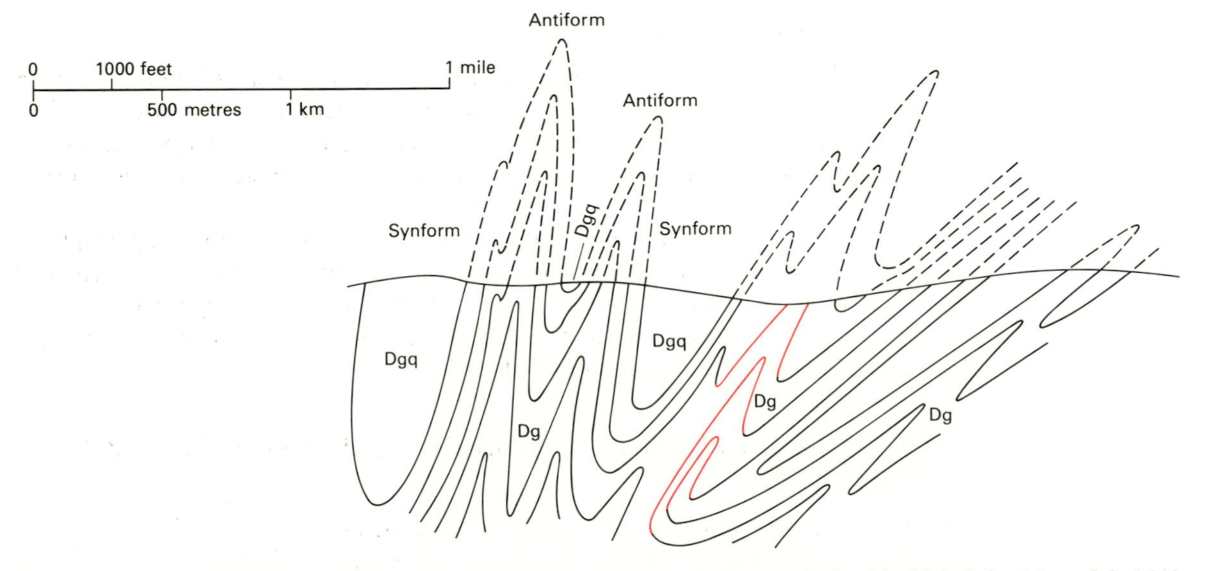

Fig. 5.9 Part of the vertical cross-section A–B provided with U.S.G.S. Map GQ-1561 (Goshen Quadrangle, Massachusetts), showing antiforms and synforms in Devonian rocks. Symbols as in Figs 5.12 and 5.17. Refolded folds are marked in red (cf. Fig. 5.17).

5.D TO IDENTIFY ANTICLINES AND SYNCLINES

The traces of anticlines and synclines may be indicated on the map by special symbols; examine the legend of the map to identify the conventions used.

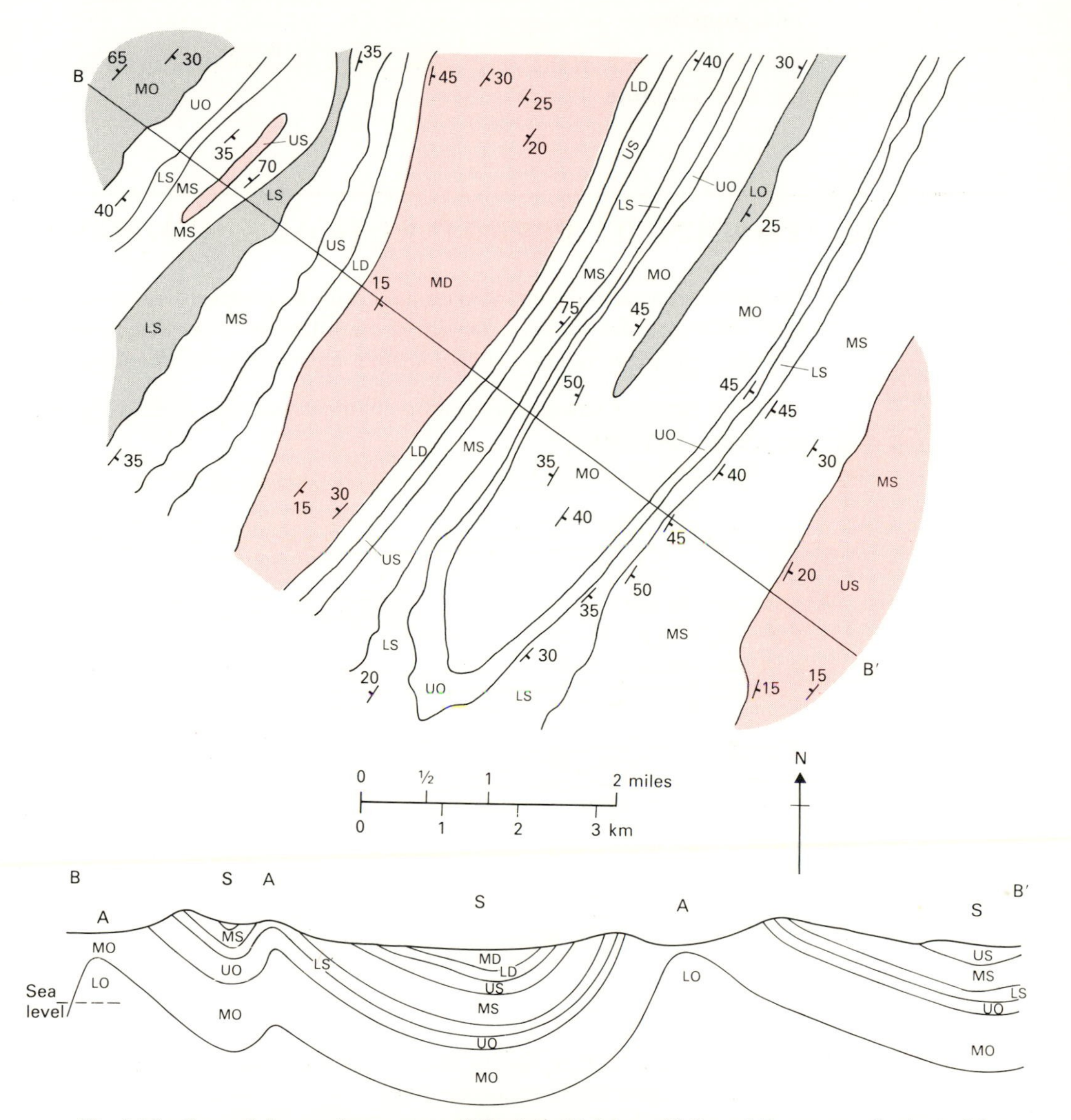

Fig. 5.10 Part of the northwest area of Virginia Division of Mineral Resources Geologic Map of Williamsville Quadrangle, showing the outcrop pattern of synclines (core rock unit coloured pink) and anticlines (core rock unit coloured grey).

The geology has been simplified by combining lithological subdivisions into major stratigraphical groupings as indicated below. Below the map is a vertical cross-section, constructed along the line B–B′ with synclines marked S, anticlines marked A.

MD	Middle Devonian	LS	Lower Silurian
LD	Lower Devonian	UO	Upper Ordovician
US	Upper Silurian	MO	Middle Ordovician
MS	Middle Silurian	LO	Lower Ordovician

Use the information in the margin of the map to determine the age of the rock unit in the core of the fold, as seen in outcrop on the map, or in the section provided with the map.

If the rock unit in the core of the fold is older than those in the outer parts of the fold: **anticline**.

Fig. 5.10 If the rock unit in the core of the fold is younger than those in the outer parts of the fold: **syncline**.

5.E TO IDENTIFY PERICLINES, DOMES, AND BASINS

Examine cases on the map of rock units with loop-shaped or closed outcrop patterns. Use strike and dip symbols, or any of the other methods described in Section 5.C.3, to verify the existence of a plunging fold, and determine the direction(s) of plunge.

Fig. 5.11 If the outcrop pattern is broadly elliptical and the fold axis plunges in opposite directions: **pericline**.

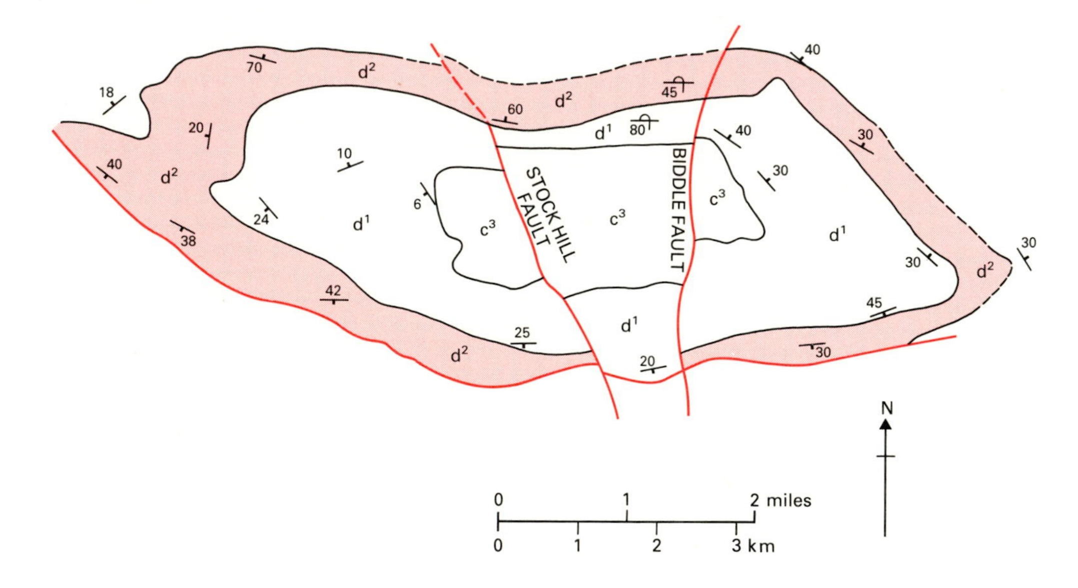

Fig. 5.11 Part of B.G.S. Special Sheet of Bristol District, showing the outcrop pattern and structure of a pericline in Palaeozoic rocks.

The geology has been simplified by showing only three stratigraphical boundaries and omitting Mesozoic and younger lithological units, and certain faults. The southern boundary of the pericline is formed by a thrust fault.

d^2	Vallis Limestone – Clifton Down Limestone	Carboniferous
d^1	Lower Limestone Shale and Black Rock Limestone	Carboniferous
c^3	Portishead Beds	Devonian

Fig. 5.12 If the outcrop pattern is circular or nearly so, dip directions are radially outwards, and the core is composed of older rocks: **dome**.

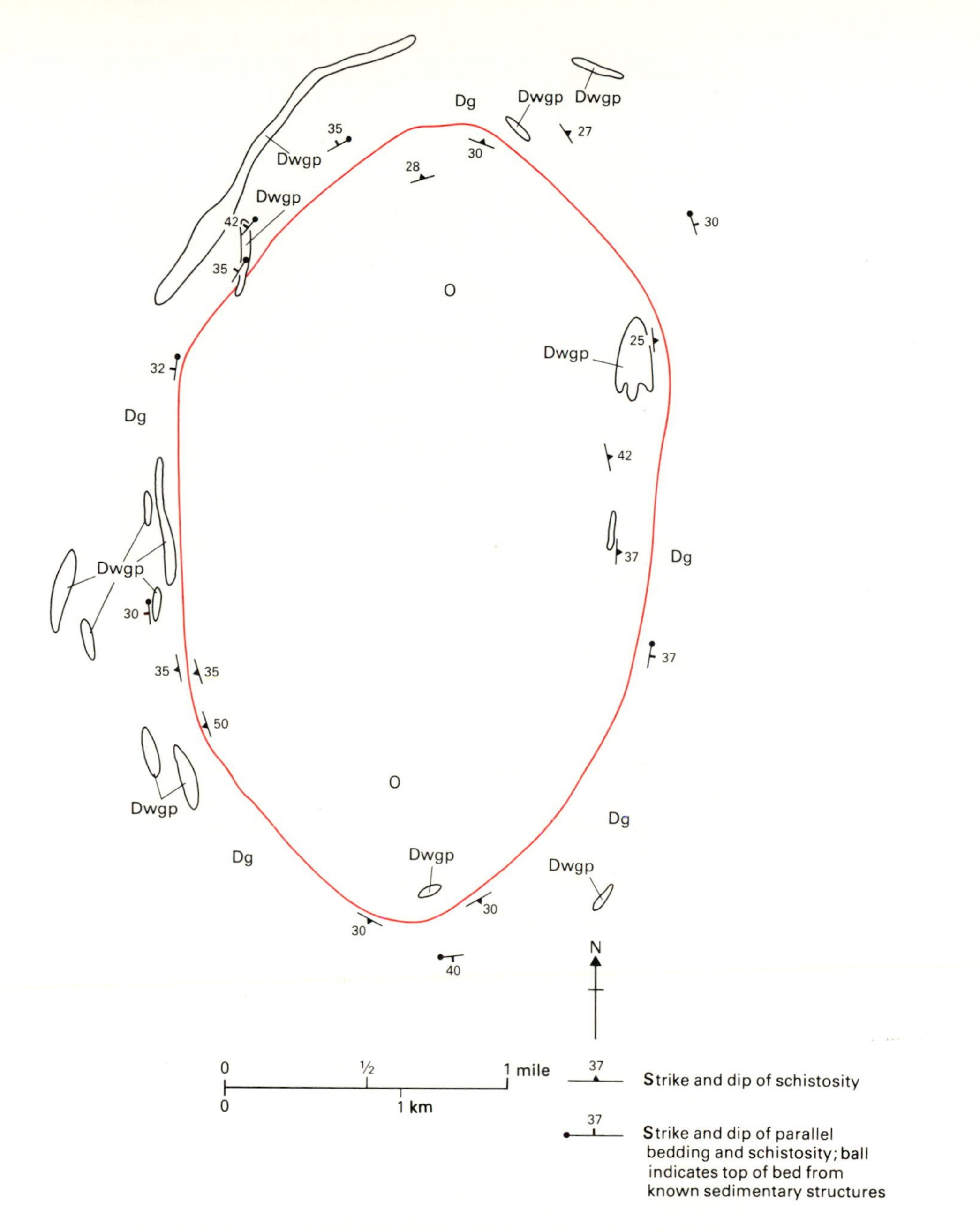

Fig. 5.12 Part of the central–south area of U.S.G.S. Map GQ-1561 (Goshen Quadrangle, Massachusetts) showing the outcrop pattern and structure of the Goshen Dome.

The geology has been simplified by grouping together the separate Ordovician rock units forming the core of the dome and showing only a selection of strike and dip symbols. The boundary between the Ordovician core of the dome and surrounding Devonian rocks is a décollement, that is, a structural plane of movement separating two sets of lithological units each of which has its own independent style of deformation.

Dwgp	Williamsburg Granodiorite	Devonian
Dg	Goshen Formation	
O	Cobble Mountain Formation	Ordovician

If the outcrop pattern is circular, or nearly so, the dip directions are radially inwards, and younger rocks form the core: **basin**.

Fig. 5.13

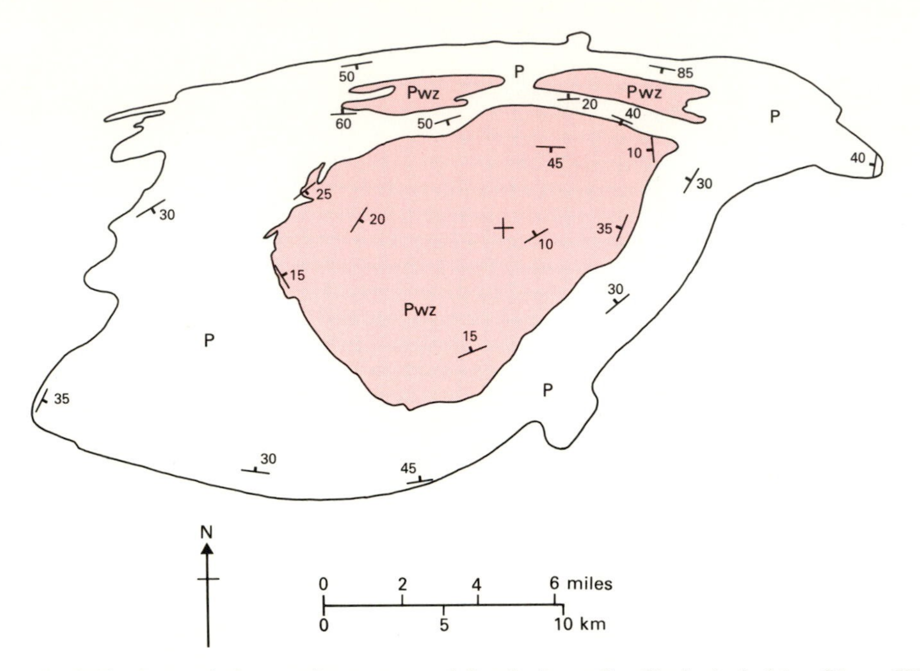

Fig. 5.13 Part of the northeast area of South Australia Geological Atlas Sheet SH 54–9 (Copley) showing the outcrop and structure of a basin in Preambrian rocks.

The geology has been simplified by reducing the number of stratigraphical boundaries shown, omitting faults and superficial deposits, and showing only a selection of strike and dip symbols. Minor folding occurs on the north side of the basin.

Pwz Billy Spring Beds
P Older Precambrian rocks

5.F TO DETERMINE THE TIME INTERVAL OF FOLD FORMATION

The time interval in which a phase of folding occurred may be expressed in stratigraphical terms or, using the data in Appendix 2, in millions of years. Large scale folding is associated with uplift and erosion of the exposed rocks so that when deposition recommences in the area the new sediments are deposited on a surface of unconformity (Sect. 6.B) eroded across the older and younger rock units.

Using the stratigraphical information in the margin of the map, determine the age of the youngest rock unit affected by the folding and the age of the oldest rock unit not affected by the folding. The difference between these two ages is the *maximum* time interval for the folding episode.

Fig. 5.14

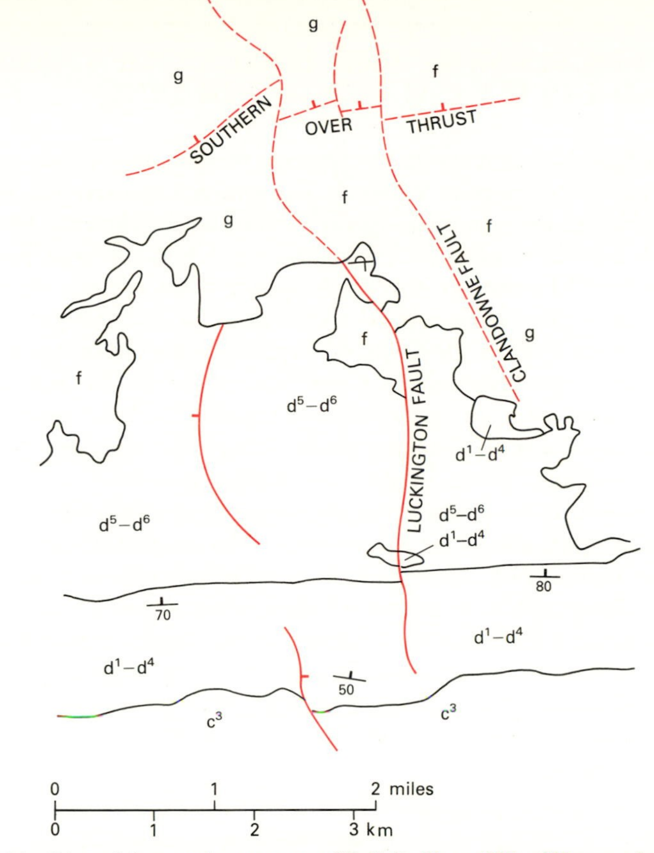

Fig. 5.14 Part of the southwest area of B.G.S. Sheet 281 of Frome showing folded and faulted Palaeozoic rocks overlain unconformably by flat-lying Mesozoic rocks.

The geology has been simplified by reducing the number of stratigraphical boundaries and omitting certain smaller outcrops and faults.

g	Lias	Jurassic
f	Dolomitic Conglomerate and Keuper Marl	Triassic
d^5–d^6	Coal Measures	Carboniferous
d^1–d^4	Carboniferous Limestone Series and Millstone Grit Series	Carboniferous
c^3	Portishead Beds	Devonian

Red dashed lines indicate the sub-crop of faults cutting folded Palaeozoic rocks below the Mesozoic cover.

The youngest rock unit affected by the folding is the Carboniferous Upper Coal Series (d^6) which is hidden below the Mesozoic cover, and the oldest rock unit not affected by the folding is the Triassic (Keuper) Dolomitic Conglomerate (f). The oldest *events* not affected by the folding are faults in the Carboniferous which do not affect the overlying Triassic rocks.

Using the data in Appendix 2, taking the top of the Coal Measures in the Frome area to be the top of the Westphalian stage of the Carboniferous, and the base of the Keuper to be the base of the Anisian stage of the Trias, the maximum time interval for the folding is calculated as follows:

$$300 \text{ m.y. (top of Coal Measures)} - 240 \text{ m.y. (base of Keuper)} = 60 \text{ m.y.}$$

Since there must also have been a period of erosion before deposition of the Keuper, the faults are older than the Keuper, and the estimated time interval for the folding is too large, but no dates are available for the age of the faults.

5.G TO DETERMINE THE AMOUNT OF LAYER SHORTENING AND UPLIFT CAUSED BY FOLDING

Readers are recommended to carry out the following simple experiment: hold one edge of a piece of paper on a smooth, flat surface and move the opposite edge towards the fixed edge; the paper will usually form a single antiformal fold with two half-synforms each side of it. (The fold shown in Fig. 5.1 was formed by an essentially analogous process.) The amount of layer shortening is the original width of the paper minus the present width, or the original width minus the **wavelength** of the fold (Fig. 5.2). The amount of uplift is the height of the crest of the antiformal fold above the flat surface, or twice the **amplitude** of the fold (Fig. 5.2). In real folds both measures are necessarily approximate: for example, part of the layer shortening may be produced by inter-layer slipping (bedding-plane sliding) and there may have been additional uplift or subsidence before, or after, the folding episode.

However, the calculations give meaningful values for the amounts of compression and of uplift due to the folding itself. The wavelength and crest-to-trough height are most conveniently determined from a vertical cross-section of the fold profile.

Folding results in erosion of the uplifted exposed rocks and important changes in depositional environment. Evidence of the former is found in unconformities (Ch. 6) and of the latter in changes in lithology of the folded stratigraphical units and those deposited after the periods of folding and erosion. The necessary stratigraphical and lithological information is provided on geological maps and accompanying descriptions of the geology.

(a) Layer shortening

Construct a vertical cross-section through the fold structures, or use the cross-section provided with the map, if suitable. Select a lithological boundary that can be followed continuously through the folds. Select two points on this boundary which should be as far apart as possible. Measure the following distances:

1. the present horizontal geographical distance, $\ell 1$
2. the distance along the trace of the lithological boundary, $\ell 2$. This is most conveniently done by measuring short, nearly straight, segments of the curved profile and summing them to obtain the total length.

The difference, $\ell 2 - \ell 1$, is the amount of layer shortening. The result can also be expressed as a percentage of the original distance, $(\ell 2 - \ell 1) \times 100/\ell 2$.

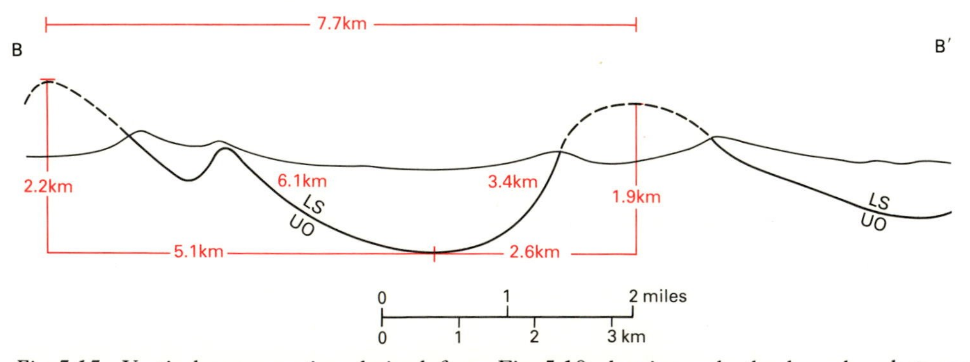

Fig. 5.15 Vertical cross-section derived from Fig. 5.10 showing only the boundary between Lower Silurian (LS) and Upper Ordovician (UO), extended above ground to give the complete profile.

(Caption of Fig. 5.15 cont.)

The horizontal distance between the crests of the two major anticlines is 7.7 km; the distance measured around the fold is 6.1 + 3.4 = 9.5 km; the amount of layer shortening is therefore 1.8 km in 9.5 km, or 19 per cent.

Note that, because the folds are asymmetrical, slightly different amounts of layer shortening are determined from different parts of the fold system, e.g. the west side of the main syncline (including the minor folding) gives a layer shortening of 1 km in 6.1 km, or 16 per cent, while the east side of the syncline gives 0.8 km in 3.4 km, or 24 per cent. These different determinations gives an indication of the amount of movement associated with each segment of the fold system.

Again, because the folds are asymmetrical, the vertical distance between the crests of the two major anticlines and the trough of the major syncline is different in each case, namely 2.2 and 1.9 km. Each of these is the local amount of uplift caused by the folding.

(b) Uplift

Use the same method of calculation, but measure the difference of altitude, or vertical distance, between the crest and the trough of an anticline – syncline pair of folds. *Fig. 5.15*

5.H TO DETERMINE THE RATE OF MOVEMENT ASSOCIATED WITH FOLDING

Use the calculation of the amount of movement (either layer shortening or uplift) associated with folding (Sect. 5.G) in conjunction with the determination of the time interval of the formation of the fold (Sect. 5.F) to calculate the rate of movement. *Fig. 5.16*

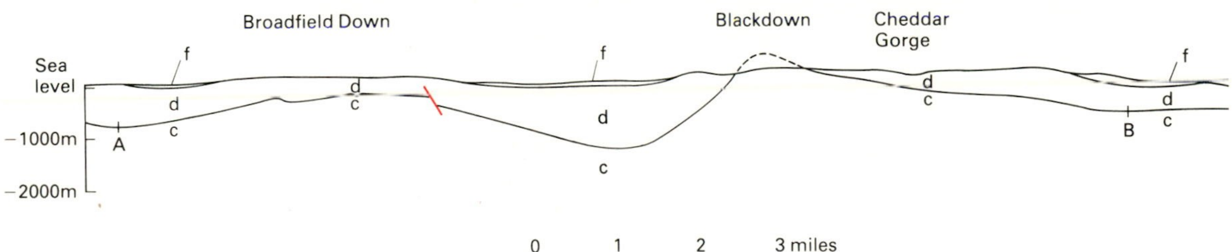

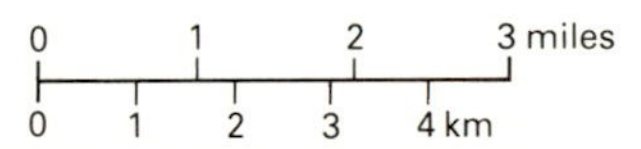

Fig. 5.16 Vertical cross-section, redrawn from Section 3 of the B.G.S. Special Sheet of the Bristol area so as to show equal vertical and horizontal scales, and showing only the boundaries between the rocks of the three stratigraphical Systems:

f Trias
d Carboniferous
c Devonian

The geographical distance between A and B is 19.5 km. The distance between A and B measured along the Carboniferous – Devonian boundary is 20.5 km. The time interval during which the folding occurred was not more than 60 m.y. (Fig. 5.14) and was probably much less. The *minimum* rate of horizontal movement associated with folding was, therefore (20.5 – 19.5 km)/60 × 10^6y = 0.017 mm/y

5.I SOME EXAMPLES OF FOLD PATTERNS

Plate-tectonics theory stresses the importance of horizontally-directed forces which impose strong compressional stress régimes on the thick piles of sediments and volcanics which accumulate at destructive plate margins, eventually elevating them into long, relatively narrow fold mountain belts. However, some of the folding can be attributed to vertically-directed forces which accompany or succeed the horizontal movements. Many of the factors, both regional and local, which influence the nature of the deformation cannot be determined from a standard geological map. Nevertheless, patterns of folding can be recognised which relate broadly to their plate-tectonic environments.

5.I.1 FOLDING AND THRUST FAULTING ON THE FORELAND MARGINS OF OROGENIC BELTS

The foreland margin of an orogenic belt is a zone of transition from a region of intense folding to one of unfolded, flat-lying, or tilted sedimentary layers. The sediment pile is relatively thin and distant from the centre of the orogenic belt and compressional forces are therefore relatively mild, consequently the rocks are not metamorphosed and the folds are open structures with occasional overfolds in which both fold limbs dip away from the foreland. Numerous thrusts cut the folds transporting the upper limbs towards the foreland. This shows that dominantly ductile stress conditions responsible for the folding were succeeded by dominantly brittle conditions leading to faulting, which can sometimes be related to the presence of incompetent, that is, more easily deformed, rock layers. The folded thrust sequence passes down abruptly through a surface of discontinuity into undeformed layers. The surface of discontinuity marks the level along which the upper sequence has moved relative to the underlying rocks and is known as a décollement.

Example Figure 4.15 shows the transition between a fold and thrust belt and flat-lying layers on the foreland. The region is on the east side of the Rocky Mountains. The Brazeau Thrust marks the décollement zone.

5.I.2 REFOLDED FOLDS WITHIN OROGENIC BELTS

The rocks forming the deeply eroded inner parts of ancient orogenic belts reveal a lengthy and complex history of several phases of deformation, metamorphism, and igneous activity which are ultimately the results of variations in the rate and direction of plate-tectonic movements, but modified considerably by local factors.

Fig. 5.17

Fig. 5.17 Part of the northern area of U.S.G.S. Map GQ-1561 (Goshen Quadrangle, Massachusetts) showing refolded folds in metamorphosed sediments.

The geology has been simplified by omitting superficial deposits, minor outcrops of some solid formations, and some fold axial traces.

Dwgp	Williamsburg Granodiorite		Devonian
Dgq	quartzites, schists	Goshen Formation	Devonian
Dg	schists	Goshen Formation	Devonian
Dgl	granulite, schists, marble	Goshen Formation	Devonian
Dw	schists	Waits River Formation	Devonian

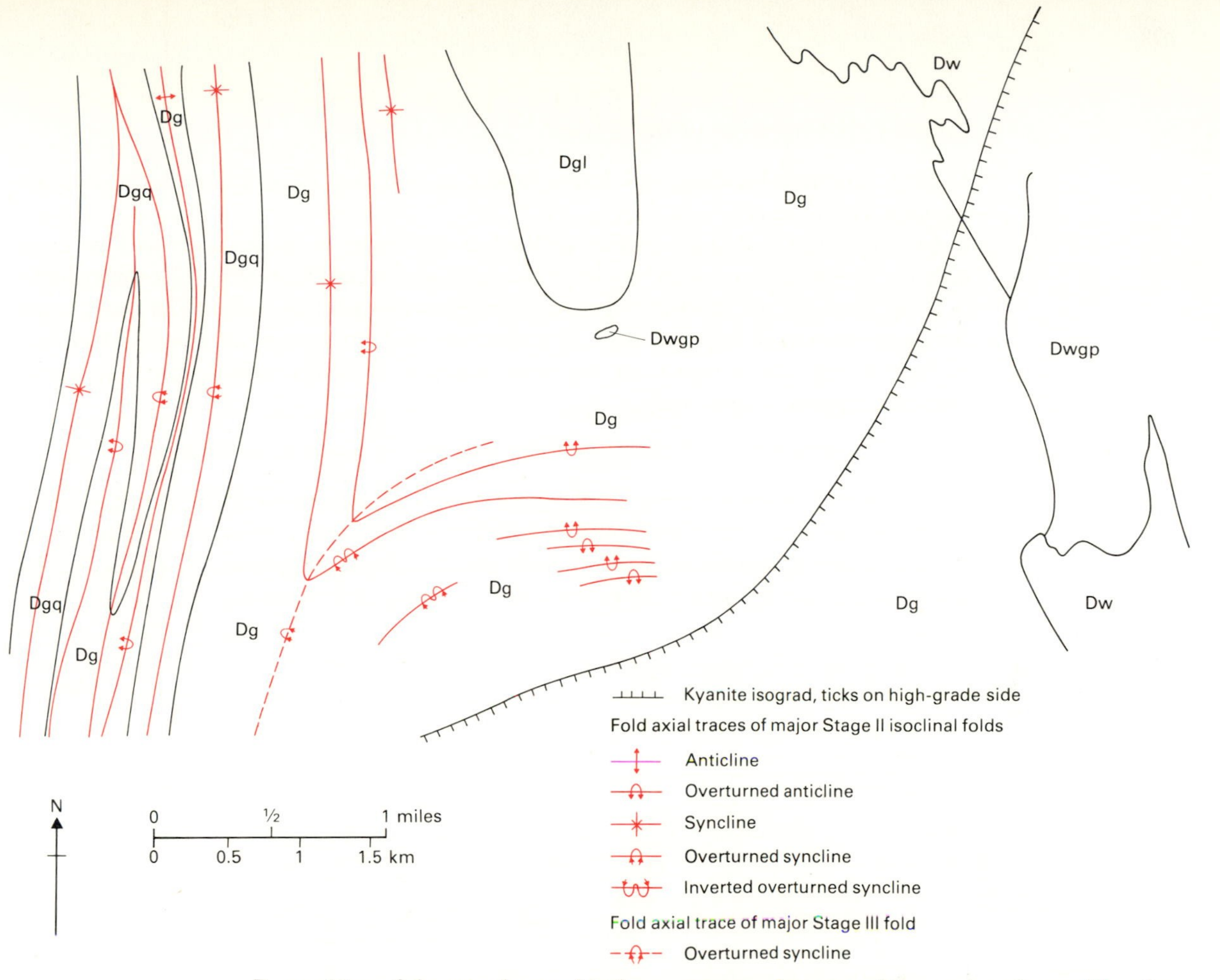

Recognition of the complex, multi-phase structure shown on this map is only possible after analysis of the results of detailed field work including the detection and measurement of smaller-scale structures such as schistosity, cleavage and minor folds, and of sedimentological features which give the way-up of beds. These observations provide the data for the initial identification of fold axial traces from which the effect of one phase of deformation on the structures of an earlier phase may be identified.

The map shows a number of features typical of eroded orogenic zones. Folding occurred on three separate occasions, referred to as Stages I, II and III, within the region, but only Stage II and Stage III folds are recorded in the map area. The earlier (Stage II) isoclinal folds with north–south trending axial traces in the western part of the map were re-folded by the later (Stage III) synclinal fold whose axial trace runs from south to north and then curves to the northeast near the centre of the map. The axial traces of the Stage II folds consequently swing from north–south to west–east. Two of the Stage II overturned synclinal folds have been inverted so that they are now antiformal synclines (folds which close upwards but have younger rocks in the core: cf. folds marked in red in Fig. 5.9).

The mapped line of the kyanite isograd marks the first appearance of this regional metamorphic mineral in the pelitic schists and gives an indication of the pressure – temperature conditions of metamorphism reached in the region (see Sect. 3.I and Fig. 3.3 for further explanation).

Igneous activity is represented by the Williamsburg Granodiorite, a rock typical of orogenic settings (cf. Table 3.2). Information provided with the map suggests a significant correlation between the first appearance of kyanite and the Williamsburg Granodiorite, meaning that both have possibly resulted from the same rise in temperature, with the granodiorite magma being formed by partial melting of pre-existing rocks.

5.I.3 FOLDS AND RELATED STRUCTURES CAUSED BY GRAVITATIONAL FORCES

During mountain-building rock layers folded and uplifted to considerable heights can become gravitationally unstable.

On the larger scale rock masses uplifted in the centre of mountain belts flow gravitationally outwards, forming very large overfolds and recumbent folds. The upper limbs of these folds may be transported many kilometres along thrust faults. The transported sheets are known as nappes, some of which may include deep-seated rocks metamorphosed during the orogeny, or previous orogenic episodes, while others comprise unmetamorphosed cover rocks.

Example Fig. 4.15 shows folds including overfolds transported along thrust planes.

At a higher level, sections of the upper parts of antiformal folds which become steeply inclined or overfolded towards adjacent synforms may collapse or slide down-dip across softer, underlying strata to form gravity collapse structures (cf. Fig. 5.1). The underlying strata frequently show evidence of strong deformation. *Fig. 5.18*

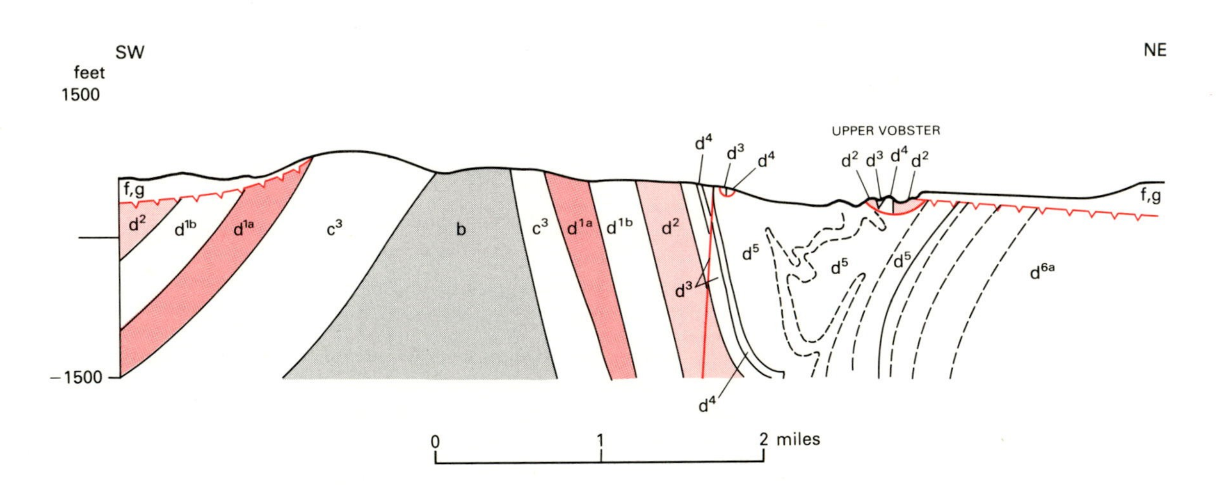

Fig. 5.18 Part of the vertical cross-section 1 provided with B.G.S. Map 281 (Frome) showing a suggested gravity collapse structure involving Carboniferous rocks.

f,g	Triassic, Jurassic	
d^{6a}	Pennant Series	Carboniferous
d^5	Lower Coal Measures	
d^4	Millstone Grit Series	
d^{1a}–d^3	Carboniferous Limestone	
c^3	Portishead Beds	Devonian
b	Silurian	
---	Coal seams in d^5 and d^{6a}	

The anticlinal fold becomes overturned farther to the NE as shown by the coal seams which are inverted. Contortion of the coal seams in d^5 shows strong deformation of softer strata during the folding. The outcrop of Carboniferous Limestone (d^2, d^3) and Millstone Grit (d^4) at Upper Vobster lies on top of the younger Lower Coal Measures (d^5). The structure formed by the collapse of part of the steeply inclined northern limb of the fold which slid along a curved fault plane onto the inverted Lower Coal Measures.

5.1.4 LARGE SCALE, SHALLOW LEVEL FOLDS AND ASSOCIATED STRUCTURES IN FORELAND REGIONS

Large areas in certain foreland regions are characterised by closely associated domes and basins, or doubly-plunging anticlines and synclines. The rocks show only low-grade metamorphism, or no metamorphism. Compression in the lower parts of the folds leads to the development of reverse faults, whereas normal faults in the upper parts indicate extensional strain. The frequency of faulting and the lack of metamorphism confirm that the folding, essentially a form of regional warping, occurred at shallow crustal levels. Many dome and basin areas contain salt dome or other diapiric structures, which may form part of the mechanism responsible for the deformation, although the fundamental cause is probably compressional forces exerted in two phases with the direction in the second phase more or less at right angles to that in the first phase.

Example Plate 2 shows a region of typical dome and basin structure in South Australia.

5.1.5 FOLDING CAUSED BY THE INTRUSION OF IGNEOUS ROCKS

The forceful emplacement of igneous intrusions often produces localised folding in the country rock, especially if the intrusion is almost solid at the time of emplacement. Doming caused by the intrusion can be recognised by dips in the country rock being directed radially away from the intrusion. Synformal and antiformal folds caused by the intrusion can be recognised by the fact that their axial traces are more or less parallel to the periphery of the intrusion.

Example Figure 3.8 shows up-doming of Palaeozoic sediments and meta-sediments caused by the intrusion of a granite.

6 UNCONFORMITIES

A significant break in the geological record is called an unconformity. Major unconformities occur when sedimentary strata lie upon an eroded surface of high-grade metamorphic rocks, or intrusive igneous rocks (**non-conformity**, Sect. 6.A.1), and when younger rock units rest with angular discordance upon an eroded surface of previously tilted or folded strata (**angular unconformity**, Sect. 6.A.2).

These kinds of unconformity preserve evidence of profound changes in geological environment, and are major features of the geological development of an area. The formation of a non-conformity indicates events on the scale of mountain-building processes, while an angular unconformity is the product of major earth movements such as occur on the periphery of an orogenic belt.

If there is uplift without tilting, and a period of erosion (or of non-deposition) occurs, and is followed by renewed deposition, a break in succession is formed with no angular discordance between the rock units. This is called **disconformity** (Sect 6.A.3). Breaks of this magnitude occur in areas where movements of the crust are entirely vertical (with no horizontal compression or tilting movements), and are common in basins of deposition where there is intermittent or irregular subsidence with periods of standstill or of temporary uplift.

The stratigraphical columns provided with maps frequently indicate the occurrence of unconformities, but it is useful to be able to identify them independently, and general criteria for their recognition are described in Section 6.A.

In an unconformity, the younger rocks are said to be unconformable on the older, whose eroded surface constitutes a **surface of unconformity.** This may be highly irregular, with strong relief resembling a **buried landscape** carved in the older rocks (Sect. 6.B.1). This is typically the result of non-marine conditions of erosion and deposition. On the other hand, an extensive regular surface of unconformity often indicates a **plane of marine erosion** (Sect. 6.B.2).

When a land surface of low relief formed across tilted or folded strata becomes a surface of unconformity the basal sediments of the younger series are deposited unconformably on the upturned and eroded edges of various members of the older rock series. This relationship between the two rock series is called **overstep** by British geologists (Sect. 6.C); American geologists commonly use the term **overlap** for this situation. At any one locality the angular discordance between the overstepped and overstepping rock units may be constant but when traced regionally may reveal variations which indicate a pattern of folding or faulting in the older series which is not obvious in a single outcrop. If the thickness or the present dip of the older overstepped series is known the pre-unconformity dip of this older series (sometimes called the angle of overstep) may be calculated (Sect. 6.C.2), using the relationship of dip angle, outcrop width, and bed thickness described in Section 2.C. The sub-crop of particular boundaries in the overstepped series can be determined by inspection in simple cases, and by the use of structure contours if the surface of unconformity itself is tilted (Sect. 6.C.3 and 9.D).

Rapid marine transgression across a land surface of very low relief characteristically leads to the deposition of basal sediments which are homogeneous and synchronous over a large area. On the other hand, gradual encroachment and/or a more irregular erosion surface results in successive strata extending beyond the limits of preceding rock units and wedging out progressively against the old land surface. This relationship of the strata which overlie a surface of unconformity is known as **overlap** (in American usage this is **onlap**) (Sect. 6.D).

If the geographical limits of the overlapping rock units can be measured on the map, chronometric data (Appendix 2) can be applied to estimate the **rate of overlap**, which approximates to the rate of transgression of the ancient sea (Sect. 6.D.2).

Using the criteria described in Section 6.A it is possible to determine the **relative age of an unconformity** (Sect. 6.E), and if radiometric or chronometric data are also available, to estimate the **time interval** represented by the unconformity (Sect. 6.F).

With the relative age of unconformities established, it becomes possible to summarise the sequence of events in a geologically complex region on a map of **chronostructural units**, that is, rock units affected by a distinct set of earth movements and bounded by unconformities (Sect. 6.G). By working out the nature and relative age of other

Fig. 6.1 Angular unconformity and surface of unconformity. Horizontal marine strata of Jurassic age lie on an eroded surface of tilted Lower Carboniferous marine strata. The surface of unconformity is seen in the foreground as a fairly even platform. The angular discordance between the Carboniferous strata and the pre-Jurassic erosion surface (in the middle right of the picture) is about 40°. See Section 6.E for further discussion. (Locality: Vallis Vale, Somerset, England.)

events recorded in the rocks within each chronostructural unit, **cycles of geological development** of the region, which reflect major plate tectonic movements, can be defined (Sect. 6.H).

6.A TO IDENTIFY UNCONFORMITIES ON A GEOLOGICAL MAP

Unconformity between two rock units can be identified by observing that geological structures, such as faults and folds, and other features (for example, igneous intrusions), which record events in the history of the older rock unit, are truncated at the surface of unconformity and do not appear in the overlying unconformable rock unit. The critical structures and other features are usually recorded only at outcrop but on some maps their **sub-crop**, that is, their position in the older rock unit below the surface of unconformity, is marked with a special symbol within the outcrop of the overlying rock unit (see also Sect. 6.C.3).

Fig. 6.2

To recognise particular types of unconformity, use the information in the margin of the map to identify sedimentary, igneous and metamorphic rock units, structural information such as strike and dip for bedding, foliation, etc., and faulting, and the ages of the formations. Remember that a fault is necessarily of more recent date than the younger unit affected by the fault, so evidence of unconformity must be sought only from boundaries which are not in faulted contact.

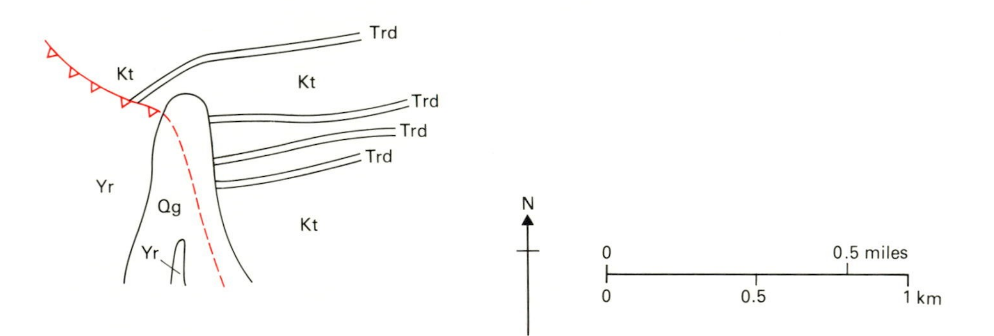

Fig. 6.2 Part of the southeast corner of U.S.G.S. Map GQ-1559 (Teapot Mountain Quadrangle) showing evidence of unconformity:

Qg	Older Alluvium	Quaternary
Trd	Rhyodacite dykes	Tertiary
Kt	Tortilla Quartz Diorite	Cretaceous
Yr	Ruin Granite	Precambrian

Thrust fault outcrop and sub-crop shown in red.

6.A.1 NON-CONFORMITY

Sedimentary formations rest directly upon plutonic igneous or metamorphic rock.

Example In Fig. 6.2 a sedimentary rock (Qg, Older Alluvium) rests directly upon plutonic igneous rocks (Yr, Ruin Granite, and Kt, Tortilla Quartz Diorite).

6.A.2 ANGULAR UNCONFORMITY

Difference of strike may be sufficient indication of angular discordance between the two rock units.

Examine the strike and dip symbols on each side of the lithological boundary. Consistent strong differences in dip angle and direction of dip in adjacent localities suggest angular unconformity between the two lithological units.

If topographical contours are sufficiently clear, draw structure contours for each lithological unit (Sect. 2.H) to enable an accurate determination of differences in strike and dip to be made.

If topographical information is limited, differences in dip angle can be deduced from differences in the response of outcrop pattern to topography (Sect. 2.A). *Fig. 6.3*

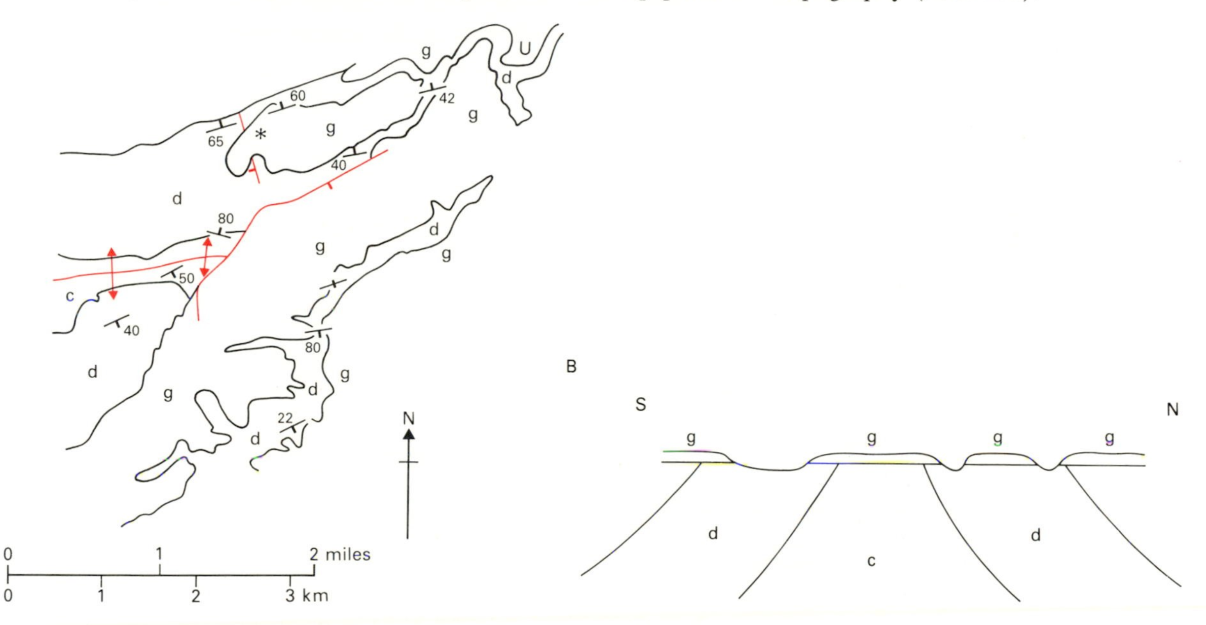

Fig. 6.3 Part of the southwest area of B.G.S. Map 281 (Frome) showing an east–west trending asymmetrical anticline of Palaeozoic rocks overlain unconformably by nearly horizontal Jurassic.

The map, Fig. 6.3(A), has been simplified by showing the Carboniferous as one unit, omitting outcrops of Triassic rocks, and by grouping the subdivisions of the Jurassic together so as to show only one unconformity. The more complex relationships of the rocks in this area will be demonstrated in Fig. 6.8. Certain faults have also been omitted.

The Palaeozoic rocks have high dips whereas the adjacent Mesozoic rocks are nearly horizontal, with lithological boundaries almost parallel with topographical contours.

The locality for the unconformity shown in Fig. 6.1 is marked U in the northeast part of this diagram.

g Jurassic
d Carboniferous
c Devonian
* Locality described in Section 6.C.3

Fig. 6.3(B) is a schematic generalised cross-section across the area of the map from north to south; symbols as in Fig. 6.3(A). Comparison of the map and section shows that the angular relationship between the strike direction of older structures (such as the stratigraphical boundaries and fold axial trace) and the outcrop of the unconformable series which truncates them is approximately the same when seen in outcrop as when seen in section.

6.A.3 DISCONFORMITY

Because disconformity occurs between parallel lithological units it cannot be recognised on the map by difference in dip and strike, or in outcrop pattern.

Study the information provided in the margin of the map. The occurrence of disconformity between two lithological units may be indicated in the stratigraphical column. Establish the stratigraphical succession in several areas (Sect. 2.B). If additional units are present in some areas, but not in others, there may be disconformity in the latter areas.

Example In the northeast part of Plate 2 (Constitution Hill–Campbell Bald Hill area) Cu forms the lowest unit of the Cambrian in parts of the synclinal structure but is absent elsewhere. This suggests, but does not prove, that the absence of Cu is due to disconformity. Disconformity is, however, confirmed by the legend on the map.

6.B TO RECOGNISE TYPES OF SURFACE OF UNCONFORMITY

The shape of the surface of unconformity is evidence of the environmental conditions in which it was formed. It can be recognised from the relationship between topography and outcrop pattern as described in Section 2.A.

6.B.1 TO RECOGNISE A SURFACE OF UNCONFORMITY WITH STRONG RELIEF ('BURIED LANDSCAPE')

Examine the outcrop of the unconformable lithological unit on the map and determine its strike and dip directions. Trace the outcrop of the base of this unit along the strike direction, noting any variations in the topographical height of the boundary. *Fig. 6.4*

A further **example** of a strongly irregular surface of unconformity may be studied on Plate 1. In this case conglomerates and sandstones were deposited subaerially on a deeply dissected old land surface of folded sedimentary rocks to form a buried landscape unconformity.

6.B.2 TO RECOGNISE AN EXTENSIVE, REGULAR SURFACE OF UNCONFORMITY: PLANE OF MARINE EROSION

Examine the strike and dip symbols on the lithological unit immediately above the surface of unconformity. If the dip is uniform in direction and in amount the surface of unconformity is probably more or less planar.

Construct structure contours on the contact between the two lithological units in question. If the structure contours are parallel and equally spaced the surface of unconformity is planar.

If several units of the younger rocks are mapped, parallelism of the outcrops of their boundaries with the surface of unconformity suggests that the latter was originally a planar surface. *Fig. 6.5*

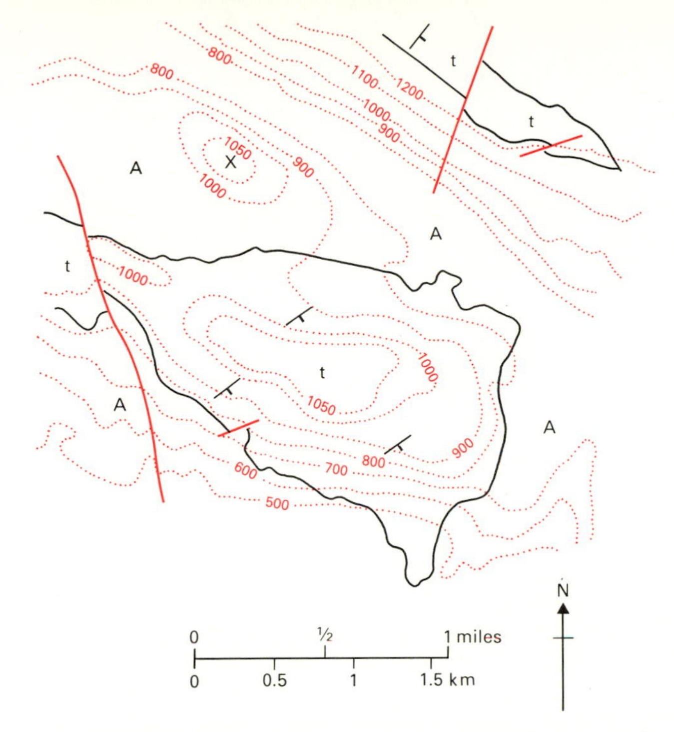

Fig. 6.4 Part of the west-central area of B.G.S. Special Sheet of Assynt, Scotland, showing the buried landscape unconformity between Torridonian (t) and Lewisian (A) rocks.

The geology has been simplified by omitting minor faults, and the topography simplified by selecting only certain critical topographical contours.

The symbols on the map show that the strike direction of the Torridonian is NE – SW. At the eastern end of the hill in the centre of the map the base of the Torridonian rises along strike from less than 500 feet (152 m) in the south to over 800 feet (244 m) farther north; the base of the Torridonian forming the outcrop in the northeast is at a height of more than 1200 feet (366 m). At locality X on the map the underlying Lewisian forms a hill which is higher than the base of the Torridonian along strike to the southwest.

The fact that non-metamorphosed sediments rest unconformably (nonconformably) upon a deeply eroded surface of high-grade metamorphic rocks proves that the metamorphism is older than the unconformity.

6.C TO RECOGNISE UNCONFORMITY WITH OVERSTEP (AMERICAN: OVERLAP) AND TO DETERMINE THE PRE-UNCONFORMITY DIP AND SUB-CROP OF THE OLDER UNITS

Overstep represents a change in the geological environment of a region in which a new phase of deposition succeeds a period of erosion after uplift and tilting or folding of older strata. The angle of overstep in effect is the dip that was imposed on the older rock unit before erosion to produce the surface of unconformity. If a second phase of tilting or folding occurred after the deposition of the overstepping rock unit, the original structure of the overstepped series can be deduced by calculation or by construction.

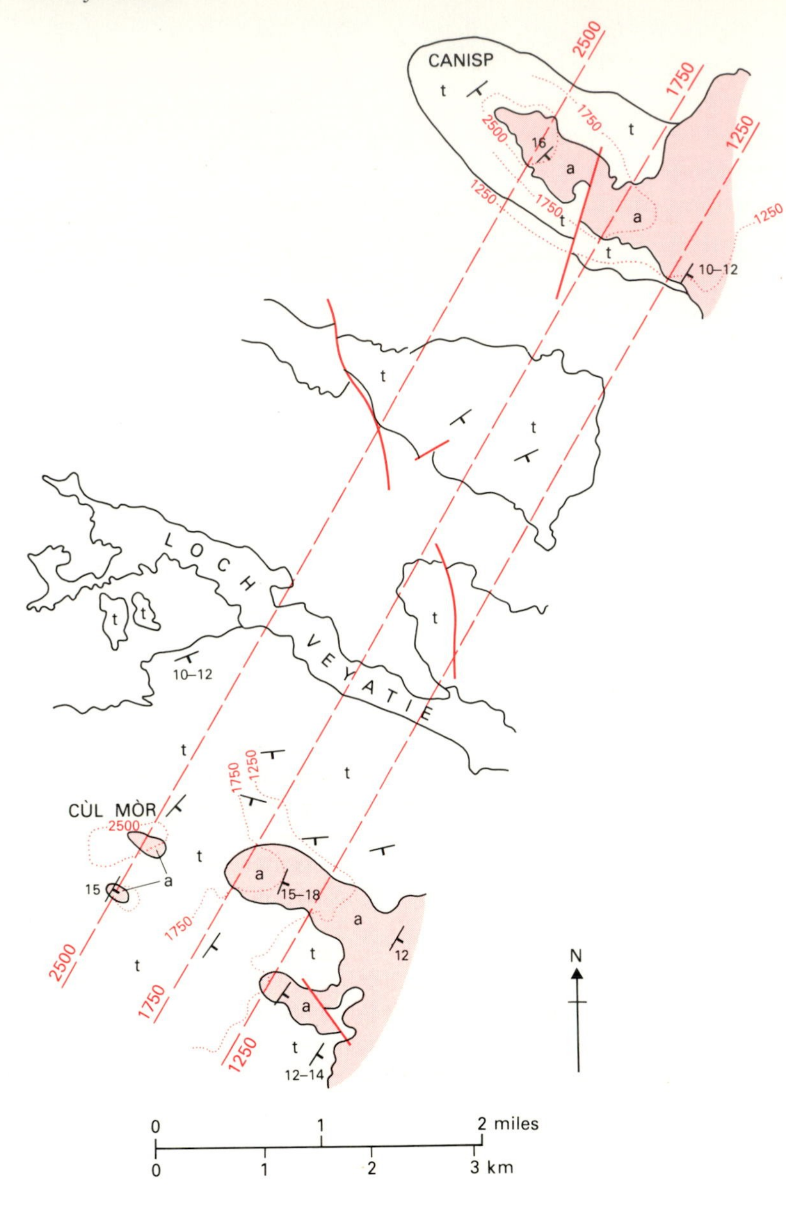

Fig. 6.5 Part of the southwest area of B.G.S. Special Sheet of Assynt, Scotland. The geology has been simplified to show only the principal outcrops of the Torridonian (t) and Basal Cambrian Quartzite (a). Structure contours have been constructed on the base of the Basal Quartzite for heights of 2500 feet (762 m), 1750 feet (534 m), and 1250 feet (381 m): only relevant lengths of the topographical contours for these heights are shown.

The dip of the Basal Quartzite is uniformly to the SE, contrasting with the variable directions of dip of the underlying Torridonian.

Structure contours drawn on the contact between the Basal Quartzite and the Torridonian are virtually parallel with each other over a distance of about 6 miles (9.6 km) along strike. The spacing of the structure contours indicates a decrease in the angle of dip from an average of 13° between the 2500- and 1750-foot contours to 11° between the 1750- and 1250-foot contours (in agreement with the stated dip values on the map).

The surface of unconformity is therefore slightly curved, probably resulting from post-Cambrian folding. The area of Fig. 6.4 lies in the northern part of this figure.

6.C.1 TO RECOGNISE OVERSTEP

If the outcrop of the base of the lowest lithological unit in the unconformable series cuts across the boundaries and structures of successive lithological units in the series below the surface of unconformity, the basal lithological unit of the upper series oversteps the lithological units below the unconformity. *Fig. 6.6*

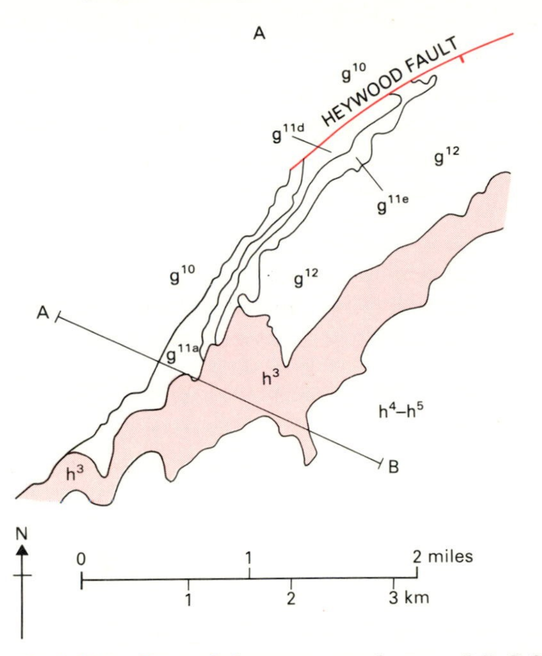

Fig. 6.6(A) Part of the east-central area of B.G.S. Map 281 (Frome) showing overstep of successive Jurassic lithological units by the basal lithological unit of the Cretaceous. The geology has been simplified by omitting superficial deposits.

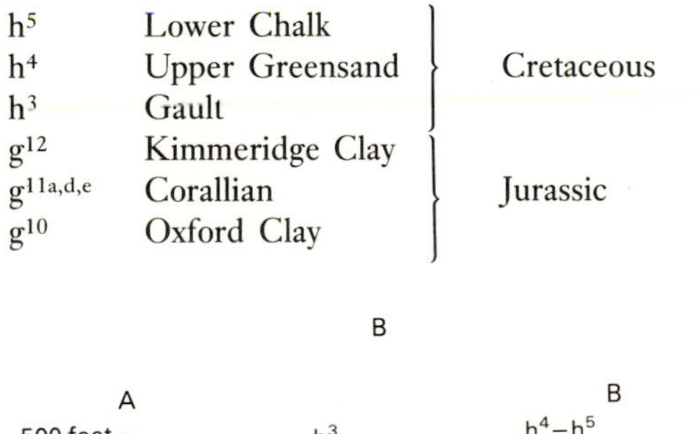

h^5	Lower Chalk	Cretaceous
h^4	Upper Greensand	
h^3	Gault	
g^{12}	Kimmeridge Clay	Jurassic
$g^{11a,d,e}$	Corallian	
g^{10}	Oxford Clay	

B

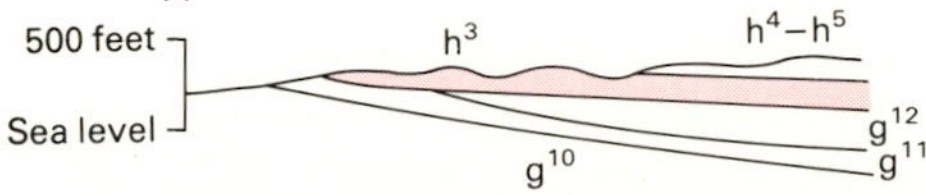

Fig. 6.6(B) Vertical cross-section along the line A–B in Fig. 6.6(A), derived from Section 2 of B.G.S. Sheet 281 (Frome) with vertical scale three times the horizontal scale. The geology has been simplified by grouping the subdivisions of g^{11} into one lithological unit.

Comparing the map and section shows how the representation of an unconformity is comparable in both – i.e. the low angle intersection of the Gault and Jurassic is similar on the map and on the section.

From northeast to southwest the base of the Gault (h^3) oversteps successively older lithological units of the Jurassic.

6.C.2 TO DETERMINE THE PRE-UNCONFORMITY DIP OF THE OLDER UNITS

1. If the surface of unconformity is horizontal:
   The present dip of the older rocks is the same as when the surface of unconformity first developed. If the dip is not recorded in the map it can be deduced from the outcrop width and stratigraphical thickness of the older rocks as described in Section 2.C.

Example In Fig. 6.6(A) the Gault (h^3) oversteps the Limestone facies of the Coral Rag (g^{11d}), which has a true thickness, d (stated in the stratigraphical column on the published map) of up to 35 feet (11 m) and a width of outcrop, w (measured on the map), of 75 m. The original angle of dip was then $\sin^{-1}(11/75) = 8°$ (maximum) to the east-southeast.

2. If the surface of unconformity is tilted:
   To determine the original angle of dip of lithological units below a tilted unconformity, use Fig. 6.7 as a guide and proceed as follows:

   1. Determine the angle of dip, α, of the unconformity (Sect. 6.B.2) and the azimuth (direction) of dip, ϕ_α.
   2. Determine the angle of dip, β, of the units below the unconformity and the azimuth of dip, ϕ_β.
   3. Determine the present difference of azimuth Δ_p ($= \phi_\alpha - \phi_\beta$) of the two dip directions.
   4. The original angle of dip, γ, of the units below the unconformity is then given by:

$$\cos\gamma = \cos\alpha \,.\, \cos\beta + \sin\alpha \,.\, \sin\beta \,.\, \cos\Delta_p.$$

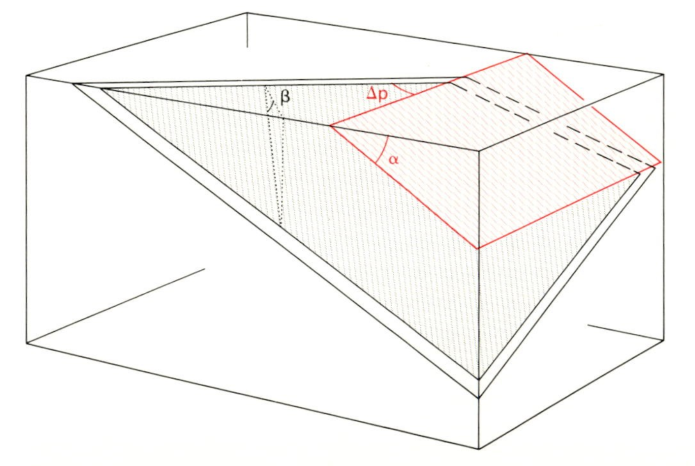

Fig. 6.7 Block diagram to illustrate the measurements required for determining the original angle of dip of lithological units below a tilted unconformity.

The diagram shows an unconformity (red lines) dipping at angle α to the right, and older units (black lines) dipping at angle β towards the front right lower corner. The difference in azimuth of the two dip directions, Δ_p, is equal to the difference of strike directions of the two surfaces. For the method of calculation of the original angle and direction of dip of the older units, see Section 6.C.2.

5. The difference of azimuth, Δ_o, between the original dip of the units below the unconformity, ϕ_γ, and the present dip of the unconformity, ϕ_α, is given by:

$$\cos \Delta_o = \frac{\cos \alpha \, . \, \cos \gamma - \cos \beta}{\sin \alpha \, . \, \sin \gamma}$$

The difference of azimuth, Δ_o, is measured in the same direction, positive or negative, as Δ_p.

Note that if the unconformity and the older units dip in the same, or opposite, directions ($\Delta_p = 0°$ or $180°$), the original angle and direction of dip can be determined by simple inspection (cf. Fig. 6.5).

Example On Fig. 6.9 the dip of the unconformity at locality A is 10° to 270°; the dip of the Lower Old Red Sandstone at locality B is 20° to 315°. The original dip of the Lower Old Red Sandstone before the tilting of the unconformity is then calculated as 15° to 345°.

6.C.3 TO DETERMINE THE SUB-CROP OF STRUCTURES BELOW AN UNCONFORMITY

Information on the map, such as sub-crop symbols (Sect. 6.A) or borehole data, may be sufficient to locate structures below an unconformity. Examine the map and any information provided with it for the necessary data. Alternatively, details of the extension of lithological units, and the structures affecting them, below an unconformable series may be determined by construction using strike and dip data recorded on exposed outcrops elsewhere on the map. Proceed as follows:

Determine whether the surface of unconformity is horizontal or tilted (Sect. 2.A, 6.B).

If the surface of unconformity is horizontal, the strike of an older lithological unit or structure, where it is exposed, is continuous with its sub-crop below the unconformity. The position of the sub-crop can then be determined by simple extension from the outcrop.

Example At the locality marked * in Fig. 6.3(A), a horizontal Jurassic lithological unit (g) lies unconformably on steeply tilted Carboniferous rocks which are cut by a pre-unconformity fault. The sub-crop of the fault may be drawn in by joining its two outcrops, north and south of the outcrop of g.

If the surface of unconformity is tilted, the sub-crop of an older lithological boundary must be determined from structure contour maps of the two surfaces, as follows:

1. Construct structure contours for the surface of unconformity.
2. Construct structure contours for the older lithological boundary.
3. Draw the sub-crop of the older boundary beneath the unconformity by joining the points of equal elevation of the two surfaces. For a description of structure contour construction, see Section 9.D.

For an **example** of sub-crop location see Fig. 9.8.

6.D TO RECOGNISE UNCONFORMITY WITH OVERLAP (AMERICAN: ONLAP) AND TO CALCULATE THE RATE OF OVERLAP

If each successive lithological unit in an unconformable series covers the outcrop of the previous lithological units, and extends farther across the old land surface, the unconformable series displays overlap. Overlap occurs typically during a marine transgression and the geographical limit of each lithological unit corresponds approximately to an ancient shoreline. Overlap also occurs between the successive lithological units accumulated during the infilling of continental basins of deposition.

Recognition and measurement of overlap are useful in palaeo-geographical reconstructions which are important for the determination of facies distribution, possible sources of sediment, and sometimes the location of economically important mineral and hydrocarbon resources.

6.D.1 TO RECOGNISE OVERLAP

Use the stratigraphical information provided in the margin of the map to determine the relative ages of the lithological units in the unconformable series. Examine the outcrop on the map of each lithological unit in the unconformable series and determine whether the outcrop of each successive lithological unit covers, and extends beyond, that of its preceding unit to lie upon the outcrop of the older series below the surface of unconformity.

Fig. 6.8

6.D.2 TO CALCULATE THE RATE OF OVERLAP

The rate of overlap is given by the original horizontal distance, ℓ, that a rock unit extended beyond the geographical limit of the preceding unit, per unit time, t. Hence the rate of overlap is ℓ/t. ℓ is readily measured on the map; t is given by chronometric data or can be approximately calculated from assumed depositional rate.

Example In Fig. 6.8(A), the rate of overlap of the Mesozoic rocks can be calculated as follows:

The Triassic and Jurassic rocks unconformably overlap against the old ridge of folded Palaeozoic rocks; but because of a very low regional eastward dip of the Mesozoic and dissected topography the outcrop patterns of the younger rocks are irregular. For simplicity assume that in Mesozoic times the anticlinal fold axis and the topographical ridge were coincident and that the feather edges of successive Mesozoic lithological units were parallel to the anticlinal fold axis (approximately east–west – see Fig. 6.3).

Draw lines parallel to the fold axis through B, the base of the Lias where it overlaps the Rhaetic, and through E, the most southerly outcrop of the top of the Lias overlapped by the Inferior Oolite on the north side of the old Palaeozoic ridge. The distance between these lines is 2000 m, corresponding to ℓ.

The time, t, taken to deposit the Lias (Appendix 2) is 205 − 180 Ma = 25 m.y.

The rate of overlap or the rate of advance of the Liassic sea over the old land surface is then given by ℓ / t = 2000 m / 25 m.y. = 0.08 mm/y (cf. Table 1.2).

A

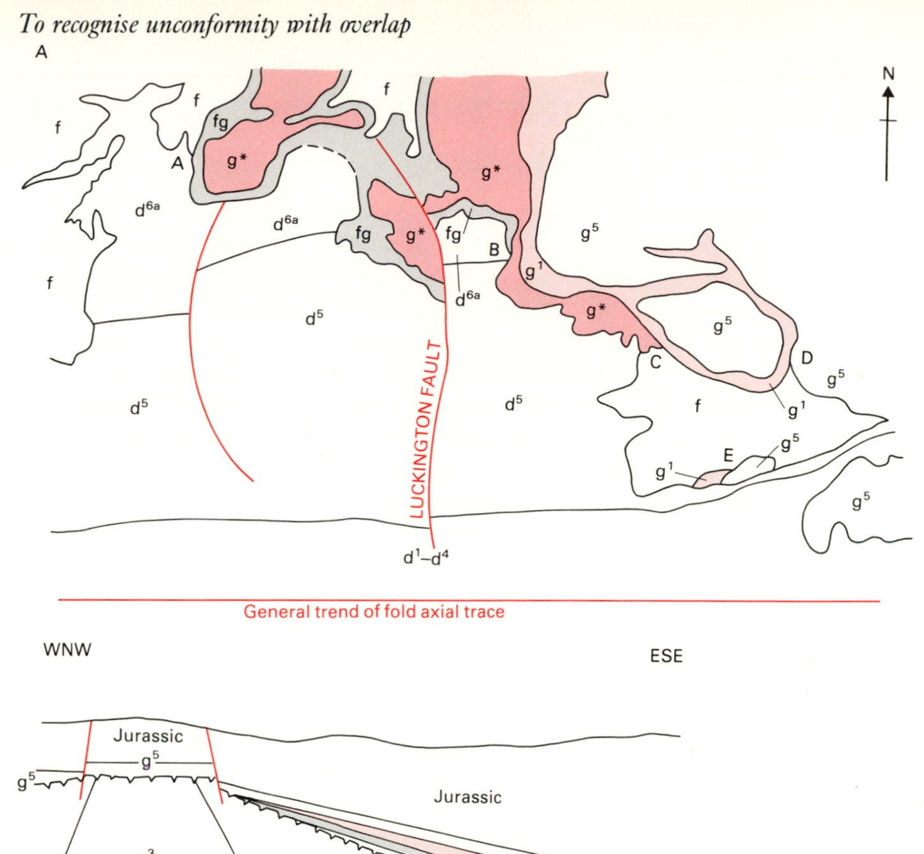

B

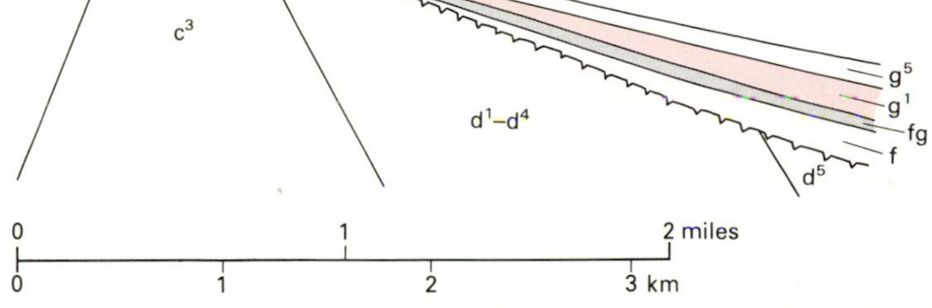

Vertical scale: three times the horizontal

Fig. 6.8(A) Part of the southwest area of B.G.S. Map 281 (Frome) showing overlap between Mesozoic lithological units unconformable upon Palaeozoic rocks.

The scale has been enlarged, and the geology simplified by omitting certain faults and minor outcrops, and by simplifying the effects of topography.

g^5	Inferior Oolite	Jurassic
g^1	Lower Lias	
g^*	White and Blue Lias	
fg	Rhaetic	Trias
f	Keuper: Marl and Dolomitic Conglomerate	
d^{6a}	Pennant Series	Carboniferous
d^5	Lower Coal Series	
d^1-d^4	Carboniferous Limestone – Millstone Grit	

Successive Mesozoic lithological units overlap as follows:

(1) fg overlaps f to rest on d^{6a} at A
(2) g^* overlaps fg to rest on d^{6a} at B
(3) g^1 overlaps g^* to rest on f at C
(4) g^5 overlaps g^1 to rest on f at D and E

Fig. 6.8(B) Part of Section 1 of B.G.S. Sheet 281 (Frome), showing overlap in the Mesozoic rocks on the south side of the anticline of Palaeozoic rocks.

6.E TO DETERMINE THE RELATIVE AGE OF AN UNCONFORMITY

In geological literature the relative age of an unconformity is commonly given an abbreviated phrase such as 'the Jurassic-Carboniferous unconformity'. However, many geological events must have affected the area in question during the time interval implied; information on the map can be used to identify these events, put them in chronological order and, hence, set closer limits on the relative age of the unconformity.

Select an area on the map where unconformity between two rock series can be proved by the criteria described in Section 6.A. Determine the latest identifiable geological event in the history of the older rock series: this may, for example, be faulting, or emplacement of intrusions. The latest datable event in the older rock series was necessarily followed by uplift and erosion, though the dates of these cannot usually be determined. Using information from the margin of the map determine the age of the first event in the history of the overlying unconformable rock series. The relative age of the unconformity is thus bracketed between the age of the last event in the older rock series and the age of the first event in the younger rock series. Finally, examine other parts of the map with the same procedure and determine whether the relative age of the unconformity can be more closely bracketed.

Example In Figs 6.1 and 6.3, outcrops of g (Jurassic) lie unconformably on d (Carboniferous Limestone), providing an example of a 'Jurassic-Carboniferous' unconformity. The last structural event in the history of the Carboniferous Limestone in this locality was faulting (followed by erosion); thus the relative age of the unconformity can be stated in a simple way as after the faulting of the Carboniferous Limestone and before the deposition of Jurassic sediments. Evidence elsewhere on the published map shows that during the time interval represented by this locality of the unconformity, Carboniferous rocks of Coal Measures age were folded, cut by thrust faults and later normal faults, and eroded to expose Silurian rocks in the core of the anticline (3 km west of the area of Fig. 6.3) – removing at least 4000 m of Carboniferous and Devonian sediments. The erosion period lasted throughout the Permian since no Permian rocks are identified on the map. Triassic continental sediments accumulated in places on the dissected land surface over which the sea slowly advanced in late Triassic and early Jurassic times, only reaching the locality of Fig. 6.3 at the beginning of the Middle Jurassic. The date of the unconformity at this locality can then be more precisely stated as younger than the folding and faulting of the Coal Measures and therefore at least as young as the end of the Late Carboniferous, while the land surface of which it forms part had been established by the time of the Triassic.

6.F TO ESTIMATE THE TIME INTERVAL REPRESENTED BY AN UNCONFORMITY

Ideally, chronometric dating of the latest event in one cycle of geological development, and of the earliest event in the succeeding cycle, would enable the length of time represented by the intervening surface of unconformity to be calculated precisely. However, this ideal situation is rarely encountered, and a combination of chronometric and stratigraphical data can be employed instead.

Example Following on from the **example** of Section 6.E, the time interval represented by the unconformity of Fig. 6.3 is not more than the interval between the end of the Carboniferous (290 Ma) and the end of the Triassic (205 Ma), or 85 million years. During this time at least 4000 m of Carboniferous and Devonian sediments were removed by erosion due to uplift resulting from fold movements (Section 6.E). The average rate of erosion was thus greater than 4000 m/85 m.y. = 0.047 mm/y.

6.G TO DISTINGUISH AND DEPICT CHRONOSTRUCTURAL UNITS ON THE BASIS OF UNCONFORMITIES

Major unconformities are significant, timed breaks in the geological development of a region and can be used to separate groups of rocks which contain evidence of often considerable differences in their conditions of formation and subsequent structural history. Unconformities thus provide a means of simplifying and summarising the sequence of events in a geologically complex region by subdividing its development into chronostructural units which are rock units affected by a distinct and datable set of earth movements, and bounded by unconformities.

Examine the information provided in the margin of the map and locate each unconformity recorded in the mapped area. Alternatively, identify the unconformities on the map itself using the methods described in Section 6.A. Place a sheet of tracing paper over the map and draw the outcrop of each surface of unconformity. A chronostructural unit is bounded by two unconformities and comprises all the rock units and structural features developed during the time elapsed between the younger limits of age (Sect. 6.H) of each unconformity, or by the younger limit of age of the most recent unconformity and the present-day land surface. On the tracing paper draw the outcrops of important structural features such as faults, fold axial traces, and igneous intrusions related to each chronostructural unit. *Fig. 6.9*

Structural movements during the development of a given chronostructural unit affect the previously formed structures of older units. The procedure for determining the original angle of dip of a lithological unit below a tilted unconformity is described in Section 6.C.2, with an example of the calculation using the dips at localities A and B on Fig. 6.9.

6.H CYCLES OF GEOLOGICAL DEVELOPMENT

The rocks within a chronostructural unit bear evidence of other events in their history, which, when chronologically ordered using the criteria listed in Section 3.K and chronometric data where available, reveal the sequence in the broader geological development of the region. Each surface of unconformity can then be seen as the ending of one series of events and the beginning of the next, hence the unconformities define cycles in the geological development in a region.

A cycle of geological development which ends with the formation of an angular unconformity necessarily comprises subsidence and deposition of sediments followed by uplift, tilting or folding, and erosion. Additional events which can be recognised

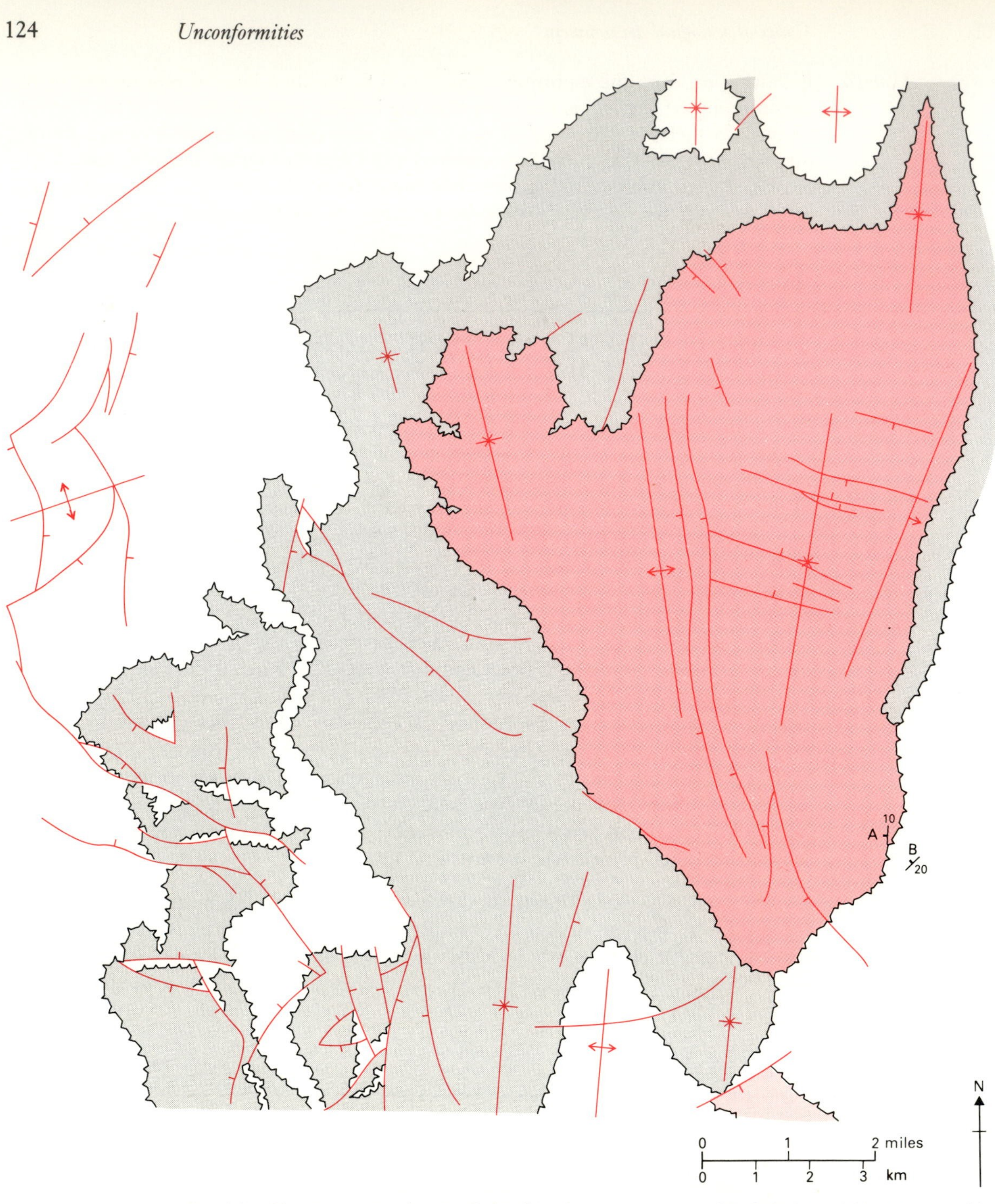

Fig. 6.9 Chronostructural units defined in the eastern part of B.G.S. Map 233 (Monmouth). Chronostructural units are defined by the unconformities on the map as follows:

4. Triassic (pink)
3. Upper Coal Measures (red)
2. Upper Old Red Sandstone – Carboniferous Limestone (grey)
1. Silurian – Lower Old Red Sandstone (white)

One aspect of differences in structure between the first three chronostructural units is brought out by locating fold axial traces (Sect. 5.C.1). These show a difference of direction between that in the oldest unit (WSW–ENE) and the succeeding two units (varying between NNE–SSW and NNW–SSE). There is tilting but no folding in the last unit.

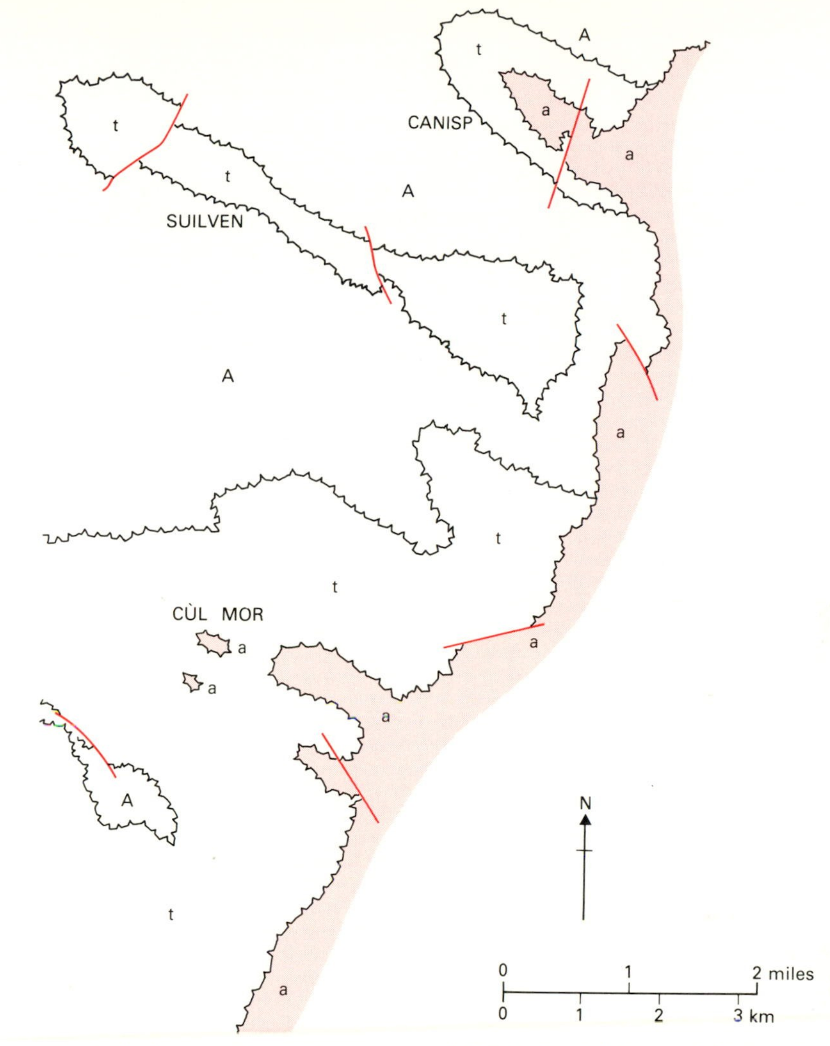

Fig. 6.10 Part of the southwest area of B.G.S. Special Sheet of Assynt, Scotland, simplified to show only the surfaces of unconformity between Lewisian, A, and Torridonian, t, and between Torridonian and Cambro-Ordovician, a.

The two unconformities separate three cycles of geological development. Using information provided on the map, chronometric data from Appendix 2, and further information from standard texts on stratigraphy (e.g. Anderton *et al.* 1979), the principal features of the cycles of geological development in this part of the Assynt district may be summarised as follows:

Name	*Type*	*Duration, m.y.*
Cambro-Ordovician	Shallow marine sedimentary (sandstones followed by limestones)	<75
Angular unconformity, planar surface		
Torridonian	Non-marine sedimentary (mostly coarse fluviatile clastics)	530
Non-conformity, buried landscape		
Lewisian	Metamorphic and igneous (gneisses, formed from igneous and sedimentary protoliths, and intruded by several series of dykes)	>1700

from the evidence on geological maps include the following: volcanic activity may accompany sedimentation; there may be a prolonged period of burial, regional metamorphism, and deformation; the period of uplift may be accompanied by cooling of metamorphic rocks and the emplacement of igneous intrusions, and followed by further uplift, cooling, and faulting. The final uplift and formation of an elevated land surface creates the conditions for erosion of the upper levels of the area. In the final stage not only is the surface of unconformity in the making, but erosion of the rocks of the first cycle is providing some of the materials for the earlier stages of the next cycle which may be of a different character, representing a new and different palaeogeographical environment from that of the first cycle. The unconformities separating (and linking) these cycles are on a scale which reflects global changes in plate tectonic movements.

Fig. 6.10

7 LANDFORMS AND SUPERFICIAL DEPOSITS

Present-day physiographical features and young rock deposits on the continental crust are the products of geological processes at the surface of the Earth. The physiography and lithology of the ocean floors are described in Chapter 8.

The characteristics of different landscape-forming environments are summarised in Table 7.1. Factors affecting the landscape are the pre-existing **lithology and structure** and the types of **erosional (or destructional)** and **depositional (or constructional)** features that are developed as a product of the **agents of erosion, transportation, and deposition.** All types of environment are subject to **change** over time, resulting from changes of sea level, vertical crustal movements, horizontal plate-tectonic movements, and climatic fluctuation – the latter is particularly evident at the present day in all climatic zones from the tropics to the poles as a result of the general warming of the Earth's surface since the last glacial period.

In order to understand the interaction of geology and landforms it is necessary first to **visualise the landscape of an area** (Sect. 7.A). In many instances the **topographical height** of a lithological unit or of a land surface is critical to interpretation of its significance (Sect. 7.B). The **relationship between lithological units and landforms** (Sect. 7.C) leads on to a full interpretation of the role of past and present geology in the formation of the present landscape.

Because of the wide range of landscape features that have been recognised, we use in this chapter an illustrative rather than comprehensive approach, and only deal with the larger-scale features that can be recognised on geological maps (Sect. 7.D). For more detailed accounts of geomorphology the reader is referred to Holmes (1978) and Press & Siever (1982); the relationship between landforms and tectonics is described in Ollier (1981); the environments and processes in the formation of present-day sedimentary and volcanic rocks are described respectively in Reading (1986) and Williams & McBirney (1979).

Observations of present-day processes and products are important for the interpretation of comparable rocks and structures in the geological past, and for understanding the environments of formation of sedimentary and igneous rocks (Ch. 2 and 3), faults and folds (Ch. 4 and 5), and surfaces of unconformity (Ch. 6). Such **actualistic** interpretations of ancient geological situations are essential to the working out of a fully geological interpretation of a map, as illustrated in Chapter 10.

Recently formed rocks are an important economic resource for certain commodities; these are noted where relevant in the examples in Figs 7.1 to 7.6.

Table 7.1 Characteristics of present-day landscape-forming environments.

	Volcanoes	*Recently glaciated and periglacial areas*
Predominant agents of erosion, transportation, and deposition	Magma, water, wind	Ice, water; wind in periglacial areas
Effect of pre-existing lithology and rock structure on landforms	Earlier constructional and destructional events of the same volcanic episode have strong influence. Pre-existing valleys may localise individual lava and pyroclastic flows	Very small influence in glaciated areas; moderate influence in periglacial areas
Large-scale erosional and destructional features	Explosion craters (maars); extensional rifts	Upland areas: glacial troughs, cirques, aretes, etc.; mountain tops may be free of ice and have gentler topography; glacial overflow channels of temporary lakes. Lowland areas: irregular low relief with numerous lakes
Large-scale depositional and constructional features	Strato volcanoes, cones, domes; lava flows and plateaus, ignimbrite sheets, ash falls	Unstratified boulder clay or till; till plains, drumlins, moraines dumped by ice. Stratified gravels, sands, varves, laminites, kames, eskers, kame terraces, kame deltas, glacial lake deposits. Poorly sorted solifluction deposits: head, rubble drift. In far-distant areas wind-blown deposits: loess
Comments	Volcanic chains are commonly aligned with plate boundaries and major faults Modification of pre-existing river systems is common	Modification of pre-existing river systems is common. Post-glacial isostatic rise of land-level affects later river and coast landforms
Predominant causes of change of environment	Variations in rate of production of magma; change of plate-tectonic situation	Climatic fluctuation
Examples	Fig. 7.1	Fig. 7.2

Deserts	*Rivers and lakes*	*Coasts*	*Shallow marine environments*
Wind; water after rainstorms	Water	Water; minor effects of wind	Sea water: waves, tides and currents; turbidity currents
Strong influence in areas of erosion: more resistant rocks form ridges, plateaus, mountains	Very strong influence: more resistant rocks form ridges, plateaus, mountains; thick limestones form distinctive 'karst' landscape	Strong influence: more resistant rocks form headlands	On local scale, generally small influence, but sea-mounts (in part volcanic) and structurally formed plateaus give shallower-water facies than adjacent areas. On global scale, active and passive continental margins are major features of Earth structure
Steep-sided plateaus and isolated hills (inselbergs) with piedmonts and pediplains; deflation hollows and depressions	Variety of drainage systems: sub-parallel, dendritic, radial, trellised, etc., according to structure and lithology of underlying rocks; incised drainage systems. Landslides	Cliffs and stacks. Wave-cut platforms. Submerged valley systems	Channels. Submarine canyons cut into continental slope
Alluvial fans (bajadas). Ergs with aeolian sands in dunes. Playas with temporary lakes, lacustrine deposits and evaporites. Duricrusts (calcrete, silcrete, gypcrete, etc.); laterite in tropical areas	Alluvial fans, fluvial deposits (including terraces). Landslip deposits. Lacustrine deposits	Beach deposits; sand dunes; salt marshes. Raised beaches. Deltaic deposits. Reef limestones (in latitudes $< \sim 30^{0}$); lagoonal deposits	Sand and mud sheets, sand patches, waves and ribbons. In deeper water turbidites and slumped deposits. Carbonate sands and muds; oolites. Reef limestones (in latitudes $< \sim 30^{0}$)
	Lakes are temporary features of landscape often resulting from earlier tectonic, volcanic, or glacial activity		
Climatic fluctuation	Local vertical crustal movements or eustatic changes of sea-level modify pre-existing river systems	Local vertical crustal movements or eustatic changes of sea-level cause submergence (drowned topography) or emergence (rejuvenation)	Local vertical crustal movements or eustatic changes of sea-level; plate tectonic movements
Fig. 7.3	Fig. 7.4	Fig. 7.5	Fig. 7.6

7.A TO ASSESS THE LANDFORMS OF AN AREA

Make use of topographical contours and (if available) photographs of the area, including oblique and vertical aerial photographs. If the topographical detail is hard to read on the geological map, the locations of lakes and towns, and the routes of rivers, railways, and roads are useful clues to the topography. Better still, use a topographical map of the same area.

Example In Plate 1 the highest ground, rising to over 1050 feet in the east of the area, is formed by the Devonian Portishead Beds; two ridges formed predominantly by the Carboniferous Black Rock Limestone and Burrington Oolite extend to the west margin of the area, with the intervening lower ground occupied by the Triassic Dolomitic Conglomerate. Deeply incised valleys, including Cheddar Gorge in the extreme southeast and Burrington Combe in the northeast, cut into the Carboniferous limestones.

7.B TO DETERMINE HEIGHTS AND SLOPES OF SURFACES OR OF LITHOLOGICAL UNITS

The heights and slopes of lava flows, river terraces and raised beaches, lava- and ice-dammed lake deposits, etc., are often critical for their recognition and for determination of their relationship to present-day or older land surfaces. Determine the height of a surface at any point by interpolation with topographical contours. The slope, σ, of a surface is given by $\sigma = \tan^{-1}(h/m)$, where h = vertical difference in height of two points on the surface and m = horizontal distance between the two points.

Examples In Fig. 7.1, the horizontal distance between the 1000-m and 800-m contours on the lava which flowed southeast from Puy de Lassolas/Puy de la Vache is 5500 m; the slope of the lava surface is then 2°.

Similarly in Fig. 7.6, the slope of the steepest part of the south side of the northernmost canyon is 20° and the slope along its axis is 4.5°.

In the northeast of Fig. 7.2, correlation of the three units of the Town Farm area sands and gravels (Qf1 to Qf3) with their respective spillways depends on the observation that the maximum topographical altitude of each set of deposits is the same as the height of the corresponding spillway.

7.C TO RELATE LITHOLOGICAL UNITS TO LANDFORMS

The boundaries of rock units may occur at breaks of slope (escarpments, etc.) or at changes in the type of topographical surface (rugged, smooth, ridged, etc.). Detailed study of the map is necessary to determine the extent of correlation of lithology and structure with landform. The landforms of recent sedimentary and volcanic rocks are often highly distinctive – for example, alluvial plains and deltas, conical and shield volcanoes, etc.; older rocks and structures produce features such as scarp- and dip-slopes, domes and basins, etc., and distinctive drainage patterns that correlate with the lithology and structure of the underlying rocks (see Holmes 1978, Ch. 3, 17–23; Press & Siever 1982, Part II).

Examples

In Fig. 7.1(B), study of the relative spacing of the topographical contours serves to recognise the trachyte of Grand Sarcouy as a dome, and the basalt volcano of Puy des Goules as a cone, slightly steeper at the top than at the base, and with a summit crater.

In Plate 4 and Fig. 9.1, the Ochil Fault marks the boundary between the hilly area of the Lower Old Red Sandstone volcanic rocks to the north and the plain occupied by the softer sediments of the Carboniferous to the south.

7.D SOME EXAMPLES OF PRESENT-DAY LANDFORMS

Figs. 7.1 to 7.6

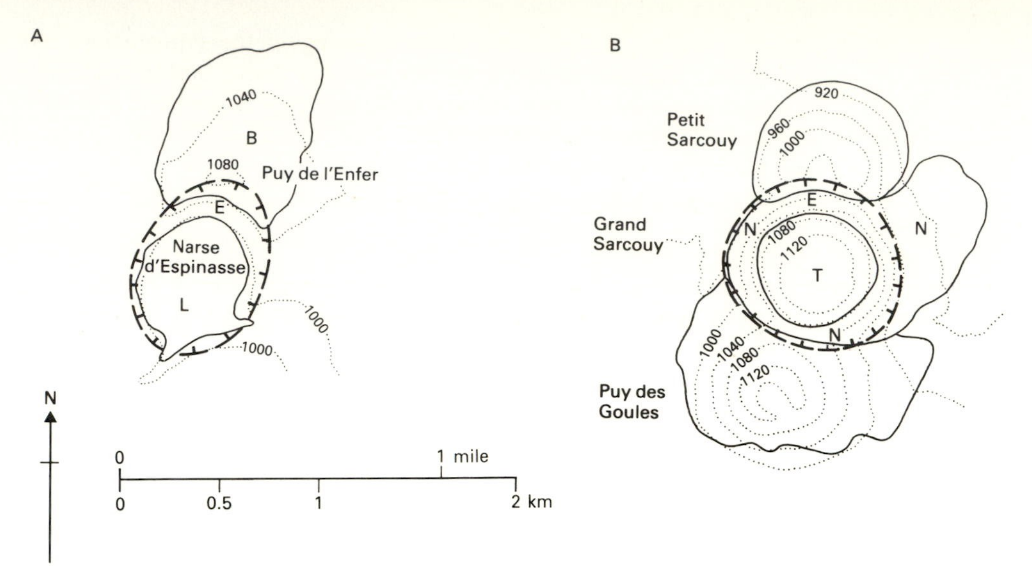

Fig. 7.1 Simplified extracts from Parc naturel regional des volcans d'Auvergne map of the Chaîne des Puys, showing features of the landscape of a volcanic region.

A1 to A3	Ash fall deposits
B	Basaltic maar deposits
E	Explosion crater; dashed line marks rim, with ticks on inner side
F	Basaltic lava flows
f	Basaltic lava flows covered by ash fall
L	Lake deposits
N	Nuée ardentes deposits
T	Trachyte lava

Volcanic cones indicated by shading – dashed outline indicates cones covered by ash fall.
Older rocks, including earlier volcanic products, not shown.
Selected topographical contours (heights in metres) are shown to illustrate the landforms.

(A) A basaltic explosion crater or maar (Narse d'Espinasse, now occupied by lake deposits) and a cone (Puy de l'Enfer) formed in part from a contemporaneous pyroclastic volcano (not separately mapped) and in part from the explosion products of the maar.

(B) Two basaltic pyroclastic volcanoes (Petit Sarcouy and Puy des Goules) with summit craters, intersected by a younger explosion crater which was subsequently occupied by a trachytic dome (Grand Sarcouy) and was also the source of nuée ardentes deposits.

(C) A group of basaltic pyroclastic cones with summit craters, some of them breached and acting as the source of basaltic lava flows. A lava flow occupies a former valley running southeast from Puy de Lassolas and Puy de la Vache (and continues a further 8 km to the east of the area shown); the flow has dammed side valleys to create two lakes (Lac d'Aydat and Lac de la Cassiè).

The relative dates of lava flows and ash falls (from Puy de Lassolas, Puy de la Vache, and from Puy Vasset) can be readily determined from the sequence of superimposition of the rock types. The basalt flow from Puy de Lassolas/Puy de la Vache has been dated, using ^{14}C in charred wood, at 7650 ± 350 years BP, and the ash from the same centre at 8100 ± 300 years BP.

Economic resources of the area include building stone, pozzolane (from volcanic ash), and diatomite (from lake deposits). The area has potential for geothermal energy derived from the heat of the postulated cooling magma chamber below the volcanic chain.

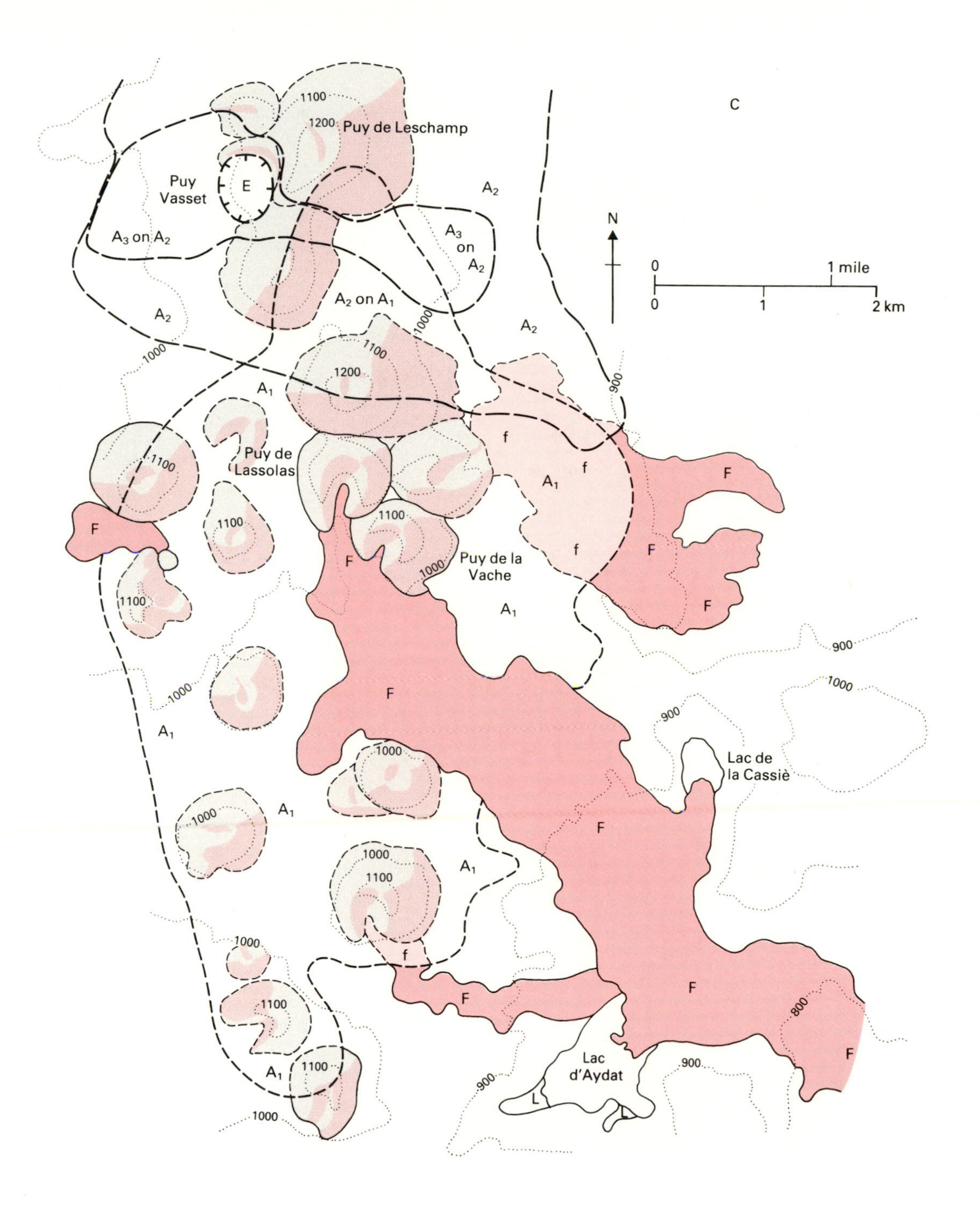
C
Puy de Leschamp
Puy Vasset
E
A_2
A_3 on A_2
A_3 on A_2
A_2 on A_1
A_2
A_2
N
0 1 mile
0 1 2 km
A_1
Puy de Lassolas
Puy de la Vache
f
f
f
A_1
F
F
F
F
F
A_1
A_1
A_1
A_1
A_1
F
F
F
F
F
f
Lac de la Cassiè
Lac d'Aydat
L
L
1000
1100
1200
900
800

Qt
Qrt
Qw
Qwr
Qb
Qal
Qs
Qf1
Qf2
Qf3
Gibbs Crossing
Pattaquatic Hill
Ware River
Whipples
710
750
675
N
0
½
1
2 km
1 mile
400
500
600
700
800
900
1000

Fig. 7.2 Part of the northeast corner of U.S.G.S. map GQ 1465 (Palmer Quadrangle, Massachusetts, surficial geology), slightly simplified, showing depositional features of a formerly glaciated area.

In stratigraphical succession:

Qal	Alluvium and swamp deposits	Holocene
Qrt	River terrace deposits	
Qb	Beaver Brook area fluvial gravels	Pleistocene
Qw	Glacial Lake Whipples fluvial sands and gravels and lake-bottom sands and clays. Heavy toothed line marks former position of ice front	
Qs	Summer Street–Swift River area fluvial gravels and sands	
Qwr	Ware River area fluvial gravels and sands. Chevron pattern represents esker ridges; chevrons point upstream	
Qf3, Qf2, Qf1	Town Farm area fluvial sands and gravels. Heavy toothed lines mark positions of ice fronts; arrows mark positions and heights of melt-water spillways.	
Qt	Till	

Till was deposited as ice advanced from north to south (the direction of elongation of the drumlin-shaped hills) during the last glaciation.

Stages of ice retreat are marked by:

1. Deposits of melt-water streams (Qf1 to Qf3) graded to the spillways shown.
2. Eskers and kame terraces (Qwr) deposited from streams flowing along the southeast side of a glacier in the Ware River valley.
3. Deposits (Qw), including kame deltas and kame terraces, of a lake ponded at a level of 440 feet by the kame terraces (Qs) of another valley glacier to the west.
4. Finally, kame plains (Qb) graded to about 350 feet.

   Terraces (Qrt) and alluvium (Qal) are deposits of the present-day Ware River.

The sands and gravels are important constructional mineral resources of the area.

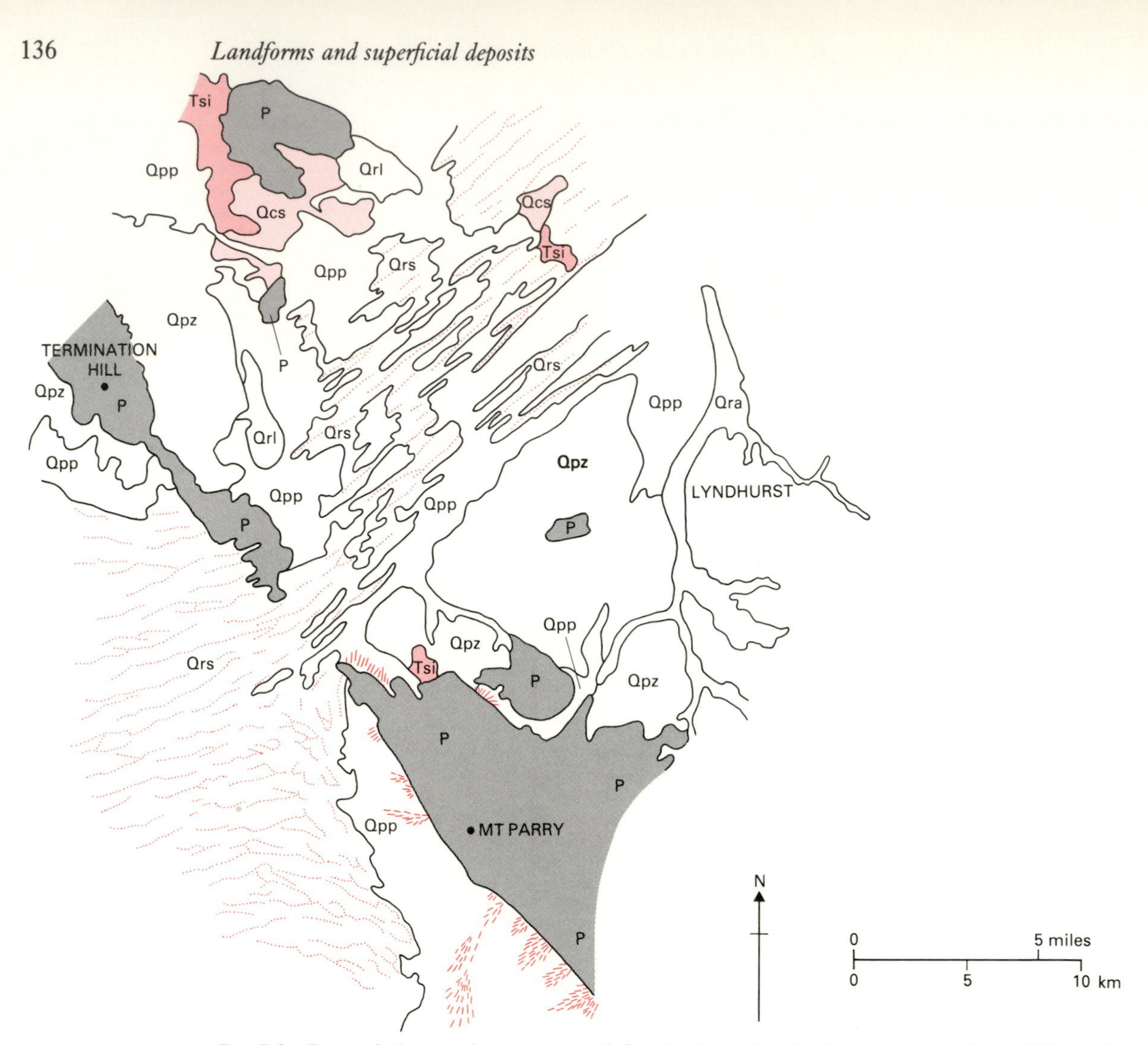

Fig. 7.3 Part of the northwest area of South Australia Geological Atlas sheet SH 54–9 (Copley), showing features of an arid landscape.

Qrl	Lake deposits } (sands, gravels, clays and gypsum)	Recent
Qra	Stream deposits } (sands, gravels, clays and gypsum)	Recent
Qrs	Fulham Sand equivalent (red-brown aeolian quartz sand, interdunal sand and clay; red dots indicate approximate lines of dune crests)	Recent
Qpp	Pooraka Formation (yellow to reddish-brown clay-sand; angular cobbles and pebbles near hill-ranges; red dashes mark alluvial fans)	Pleistocene
Qpz	Arrowie Formation (sub-angular lime-coated gravels in red-brown clay)	Pleistocene
Qcs	Gypcrete (massive, coarsely crystalline gypsum)	Pleistocene
Tsi	Silcrete (columnar and nodular grey quartzose duricrust)	Tertiary
P	Precambrian	

The area is part of the arid desert region of central Australia. Ridges and isolated hills of Precambrian (P) are surrounded by the deposits of alluvial fans (Qpz and Qpp). Silcrete and gypcrete (Tsi and Qcs) formed on an older land surface show that desert conditions have persisted in the area for at least several million years. The aeolian dune field (Qrs) of the Lake Torrens basin to the west becomes strongly directional to the northeast of the windgap in the mountain range. Rare rain storms produce ephemeral streams and lakes with detrital sediments and evaporites (Qrl and Qra).

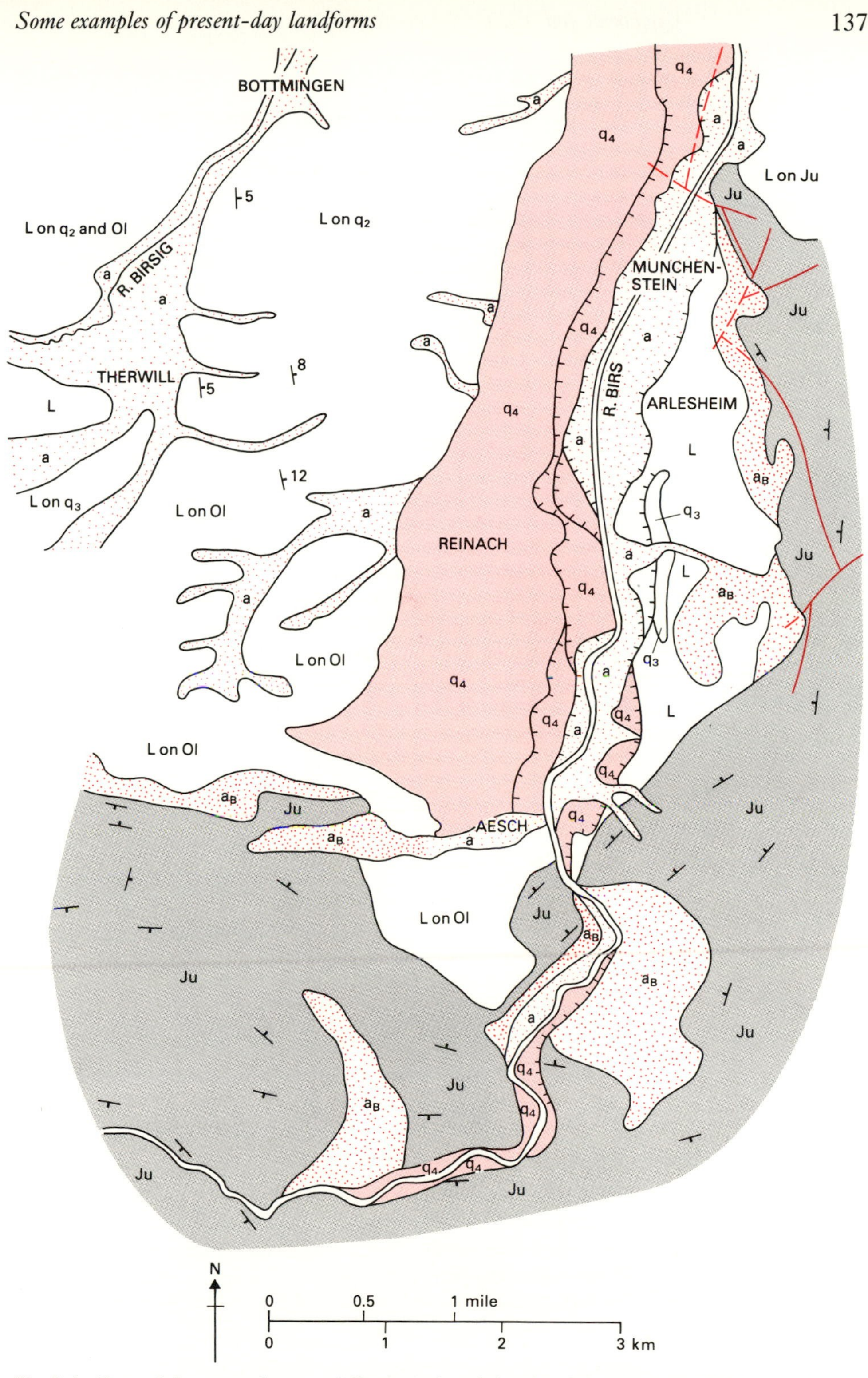

Fig. 7.4 Part of the central area of Geologischer Atlas der Schweiz sheet 1067 (Arlesheim), simplified and reduced in scale, showing features of a recently developed fluvial landscape. The area is immediately to the south of the great bend of the River Rhein at Basel, and lies within the area of Fig. 1.1, 15 km to the northwest of the centre of the photograph.

(*Caption continues on p. 138*)

(*Caption of Fig. 7.4 cont.*)

a	Alluvium (edges of terraces marked by lines with dashed ornament)	Holocene
a_B	Boulder debris, screes, landslip deposits	Holocene
L	Loess	Pleistocene-Holocene
q_4	Lower Terrace Gravels	Pleistocene
q_3	Higher Terrace Gravels	Pleistocene
q_2	Younger Plateau Gravels	Pleistocene
q_1	Older Plateau Gravels	Pleistocene
Ol	Oligocene (molasse)	Tertiary
Ju	Jurassic	Mesozoic

The topography is low-lying in the northwestern one-third of the map, and mountainous in the south and east where the Jurassic rocks outcrop. The River Birs is a strike stream flowing through folded Jurassic rocks of the Jura Mountains in the south of the area, but turns north through a gorge at Aesch to flow for 10 km over an alluvial plain and join the Rhein at Basel. Its course over this stretch is partly determined by faults forming the south end of the east side of the Rhein graben (see Fig. 4.17). To the west the River Birsig is a strike stream in gently eastward-dipping Tertiary molasse; alluvium marks the course of subsidiary streams flowing into the Birs and Birsig. This northwestern area is extensively covered with plateau gravels and loess, dating from the Pleistocene glaciation. The Lower Terrace Gravels are outwash deposits of the final (Würm) stage of the Alpine glaciation. Terraces in the valley of the River Birs mark earlier stages of cutting down of the present river. Extensive screes and boulder debris indicate the rapid erosion of the mountains at the present day.

Towns are constructed predominantly on the Lower Terrace Gravels and alluvium; the former is also an important groundwater reservoir and source of gravel for the construction industry.

Fig. 7.5 Part of the southeast corner of Geological Survey of Japan sheet 6–89 (Matsushima), slightly simplified, showing features of a coastal landscape.

a	Flood plain and coastal plain gravels, sands, and muds	Holocene
l	Levée deposits – sands and muds	Holocene
b	Beach deposits – sands	Holocene
d	Beach ridge deposits – sands	Holocene
TT	Tertiary rocks; detail not shown, but the axis of an active anticlinal fold is marked	

The area is in the northeast part of Honshu; it shows drowned topography resulting from eustatic rise of sea-level after the Pleistocene glaciation, with present-day and Recent river and coastal alluvium, including levée and beach ridge deposits. The wide expanse of coastal plain deposits in the east of the area is a consequence of 20 metres relative subsidence compared to the area of the active Oshio anticline in the centre of the map.

Naruse River
Oshio anticline
Ishinomaki Bay
Matsushima Bay
N
0 1 2 miles
0 1 2 3 km

9° 30′W
48°30′N
S
SW
S
S
S
200
400
600
800
1000
1500
150
0
5
10 miles
0
5
10
15 Km

Fig. 7.6 Part of the central area of B.G.S. sheet 48 °N–10 °W (Little Sole Bank–Sea Bed Sediments), showing an area approximately 300 km SW of Land's End, with the edge of the continental shelf and part of the continental slope. Bathymetric contours are depths below sea-level in metres.

S Shelly sand, gravelly sand, and muddy sand forming the top few centimetres of the sea-bed sediments, covering Pleistocene sandy sediments
SW Sea-bed sediments with sandwaves resulting from present-day water movements and tidal currents
Uncoloured area: sea-bed sediments not sampled
Line with barbs: landward limit of erosion on the upper continental slope associated with mass movement and headward erosion in the canyons

The submarine topography is the product of:

1. The formation of the continental margin in the mid-Cretaceous as a result of the opening of the North Atlantic by ocean-floor spreading.
2. Subsequent submarine erosion of the continental margin along fault lines to form the major canyons.
3. Erosional and depositional processes associated with lowered sea-levels during the Tertiary and Pleistocene as a result of:
   (a) eustatic changes of sea-level resulting from the Pleistocene glaciation.
   (b) epeirogenic movements of the north European continental crust.

The shelf break (the boundary between the near-horizontal sea floor of the continental shelf and the much steeper continental slope) in this area is at about 180–200 m below sea-level; it marks the oceanward limit of the effects of erosional episodes associated with Quaternary lowstands of sea-level. The difference between this level and the maximum eustatic lowering of sea-level (about 130 m) during the Pleistocene glaciation indicates post-Pleistocene subsidence of the area by some 50–70 m.

The low mound indicated by the 150-m bathymetric contour lies at the southwest end of a major tidal sand ridge 100 km long and up to 20 m high oriented N 30° E; this was formed in shallower water depths during the late Pleistocene lowstand of sea-level and extended landwards (northeastwards) during the marine transgression which accompanied the melting of the Pleistocene glaciers.

The geology of the deep ocean floors, on the oceanward side of the continental slope, is the subject of Chapter 8.

8 OCEAN-FLOOR GEOLOGY

The preceding chapters have been concerned with geological maps of the continental crust, which covers only 40 per cent of the Earth's surface, and one-quarter of which is submerged below sea-level as the continental shelf. The remaining 60 per cent of the Earth's surface is oceanic crust and is the topic of this chapter. The balance of space devoted in this book to the two types of crustal rocks reflects both the smaller amount of knowledge and the simpler structure of oceanic rocks compared with continental rocks, though ocean-floor geology is of supreme importance for the understanding of Earth processes on the largest scale.

Geological maps of the ocean floors differ from those of the continents in three major respects. Firstly, the rocks are relatively inaccessible, so that most of the information about them comes from geophysical observations, with a limited amount of direct sampling by submersibles, dredging, or drilling; both types of data are shown on the geological maps. Secondly, the rocks are of different lithology, structure, and history from those that form the continents, and different mapping techniques are used to represent them. And thirdly, ocean-floor geology is fundamental to the theory of plate tectonics, and the interpretation of the relatively sparse amount of information about these huge areas of the Earth is strongly influenced by the theory.

The following description of the development of oceanic rocks, which (for brevity and simplicity) combines geological observation and theoretical modelling, sets the scene for the understanding of geological maps of the ocean floors. Fuller accounts are given in modern textbooks (e.g. Press & Siever 1982, Ch. 19; Kennett 1982, Part I; Cox & Hart 1986).

The plates that form the surface of the Earth have divergent (constructive), transform (conservative), and convergent (destructive) margins, each type being characterised by distinctive patterns of topography, vulcanicity, and seismicity (Sect. 8.B).

New oceanic crust of basaltic composition is formed at **divergent plate margins,**

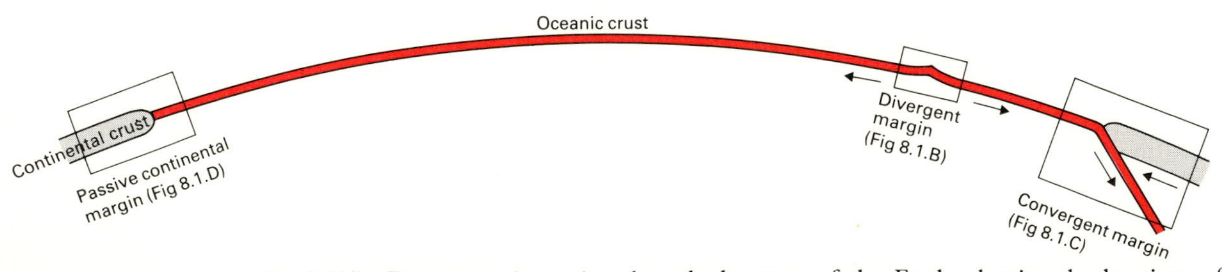

Fig. 8.1(A) Diagrammatic section through the crust of the Earth, showing the locations of Figs 8.1(B) to 8.1(D).

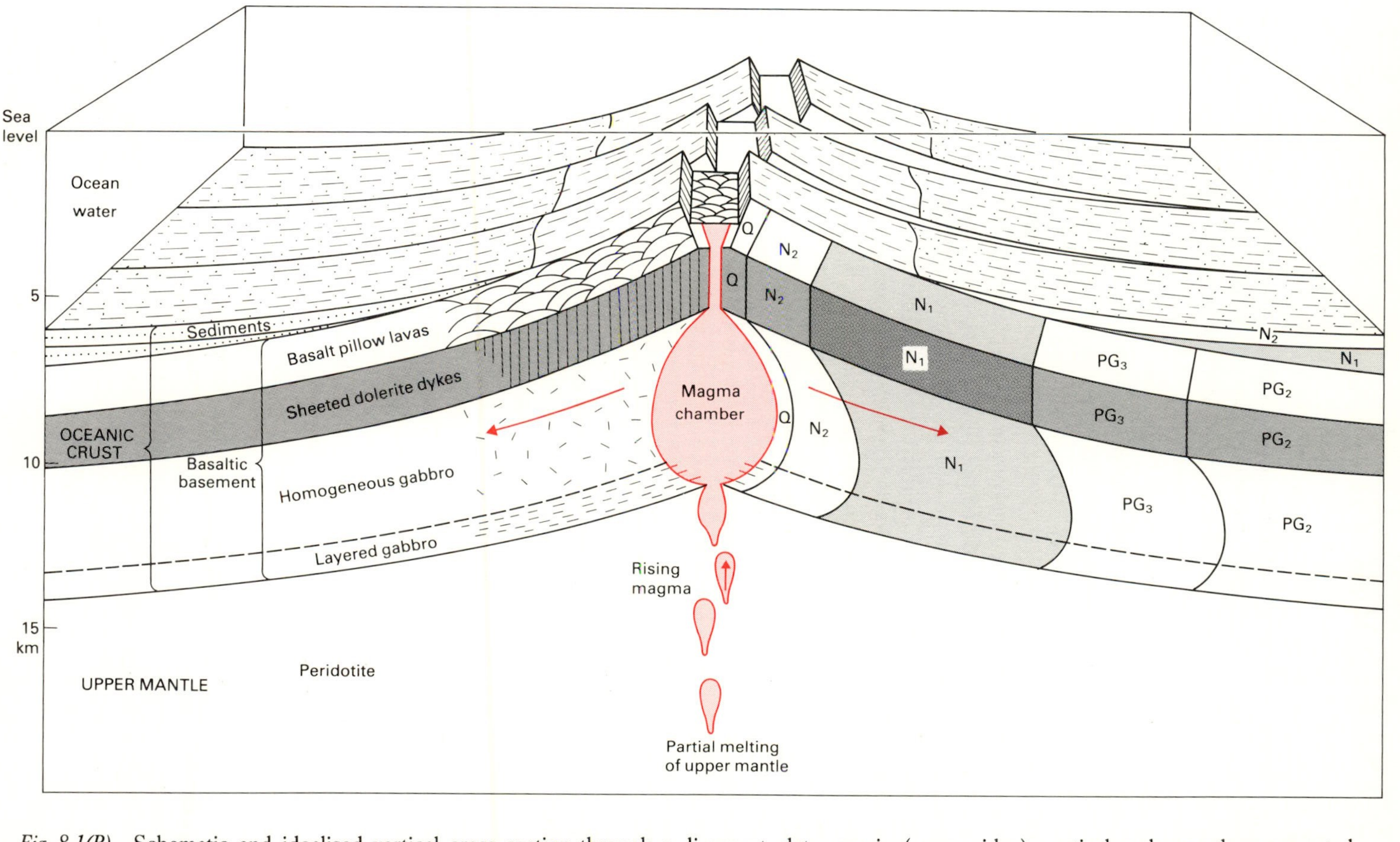

Fig. 8.1(B) Schematic and idealised vertical cross-section through a divergent plate margin (ocean ridge); vertical scale greatly exaggerated. The left side of the diagram shows the structural features of the oceanic crust. The right side represents the ages of crystallisation and magnetisation of the basaltic basement and of deposition of the sediments, labelled using the same conventions as in Figs 8.5 to 8.11: Q Pleistocene, N_2 Pliocene, N_1 Miocene, PG_3 Oligocene, PG_2 Eocene. The different cooling rates of the three layers are indicated by the shapes of the boundaries between the dated sections – all three layers are formed concurrently (and not in sequence from bottom to top as in other layered rocks); the age of the basement increases away from the ridge. The newly-formed oceanic crust travels some distance away from the ridge before it begins to be covered over with sediments (though some sediment may be ponded in an axial rift), so there is usually a time interval between each unit of basalt and the sediments immediately above.

which are generally located along submarine topographic ridges, such as the Mid-Atlantic Ridge and the East Pacific Rise (Fig. 8.1(B), Sect.8.B.1). Normally, the emplacement of the magma along ocean ridges is a geologically quiet process, with few visible volcanoes and only shallow low-intensity earthquakes (focal depths less than 70 km). (Exceptionally, as in Iceland and the Afar triangle (southern Red Sea), the supply of magma is so great that volcanic rocks build up to above sea-level and the divergent margin outcrops on land.) As further magma is emplaced, in the form of pillow lavas on the ocean floor, fed by dykes from an underlying source of liquid basalt, the newly-formed crust spreads laterally in each direction away from the ridge. The rate of spreading is typically between 1 and 10 cm per year. Slower spreading rates are associated with ridges having an axial rift zone, faster rates with ridges lacking a rift. As the basalt cools it is imprinted with the orientation of the Earth's magnetic field. Reversals of the field orientation at irregular intervals of about 100 000 to 1 million years are revealed geomagnetically as stripes of oceanic crust about 1 to 100 km wide with alternately normal and reversed magnetism. The symmetrical mirror-image relation of the patterns of stripes each side of the ridge shows that the rates of ocean-floor spreading are generally about equal in the two directions. The dates of the Earth's magnetic reversals have been determined from the radiometric ages of rocks of known polarity, and so the date of formation of identifiable stripes can be determined, and the rate of ocean-floor spreading can be calculated (Sect. 8.C). The dates can also be correlated with the stratigraphical time-scale (Appendix 2), so that the rocks of the ocean floor can be mapped in terms of the stratigraphical age of their formation.

During the cooling of the newly-formed oceanic crust the basic rocks become metamorphosed at low to medium temperatures (zeolite, greenschist, and amphibolite facies, Fig. 3.3) and hydrothermally altered by convective circulation of sea water with the formation of mineral deposits (locally ores), principally of zinc and copper sulphides.

Cooling and thermal contraction of the basaltic crust as it spreads away from the divergent plate margin leads to a decrease in elevation of the sea floor and an increase in bathymetric depth from about 2–3 km at the ridges to about 5–6 km in the ocean basins. The depth increases in proportion to the square root of the age of the ocean floor. The ocean floor is gradually covered over by an increasing thickness of marine sediments, which include both deep marine (pelagic) deposits and clastic sediments derived from land and transported into the deep oceans by turbidity flow and other mechanisms (cf. Table 2.1). Deep-sea drilling shows that the age of the oldest sediment above the basaltic basement increases away from the spreading ridges and that the oldest sediment at any point is younger than the age of the basaltic crust beneath (see Fig. 8.1(B). The geological map used as an example of oceanic geology in this book shows the stratigraphical age of the basaltic basement. In this respect the map can be likened to a geological map of bedrock geology with the superficial cover of recent sediments removed, with the difference that the 'superficial' rocks extend in age as far back as the Jurassic.

Numerous volcanic islands and submerged seamounts occur in the ocean basins. Some of them form linear chains in which the ages of the islands increase in age away from a present-day active volcano (e.g. the Hawaiian chain); these are assumed to represent **hot-spots** or sources of magmatism arising from deep in the mantle. Hot-spots have maintained nearly constant positions relative to each other over tens of millions of years, and thus they appear to have remained geographically fixed in position while the overlying plate travelled across them (Sect. 8.D).

Calculations based on the rate of movement of plates relative to hot-spots and on

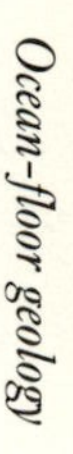

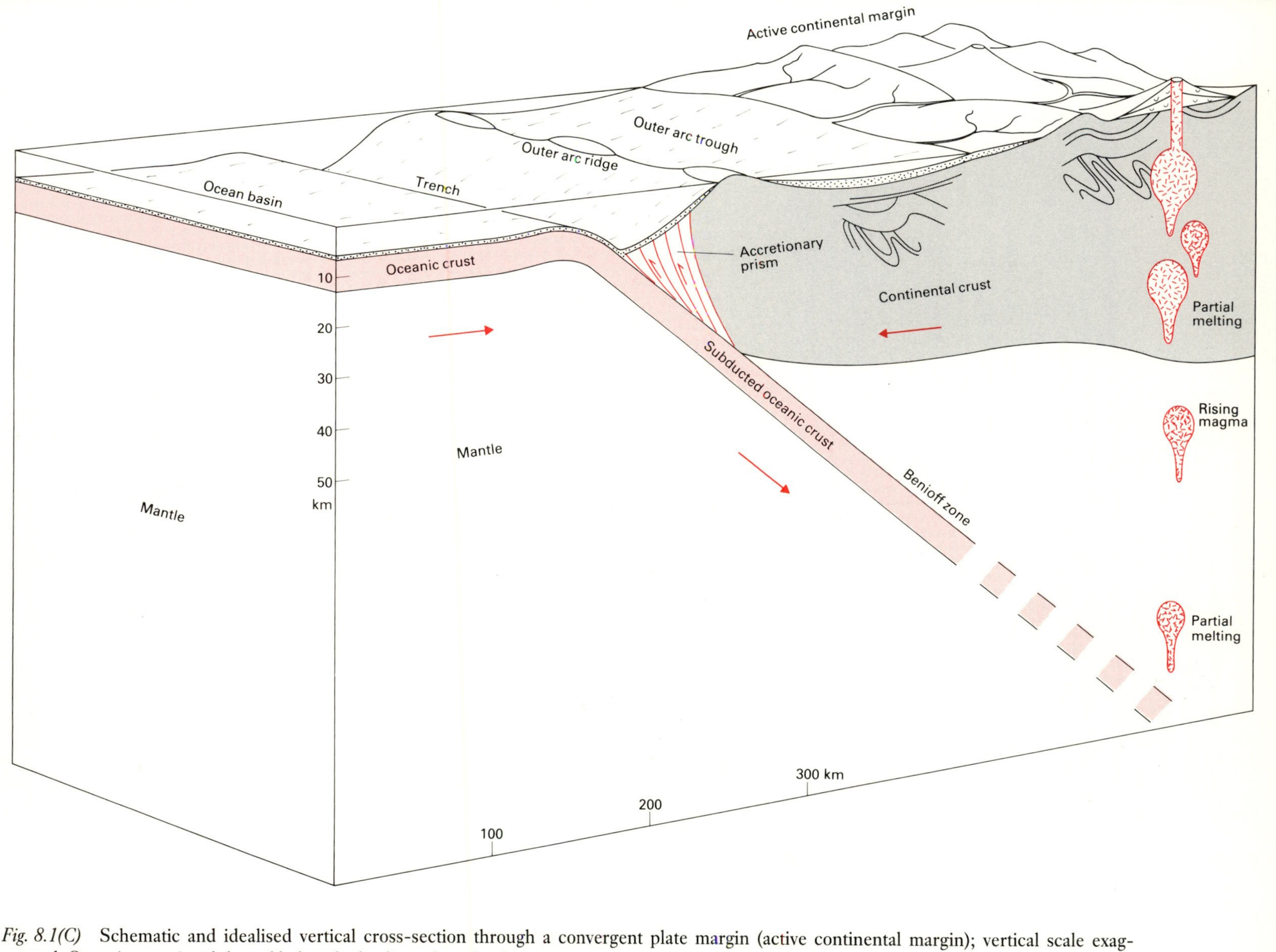

Fig. 8.1(C) Schematic and idealised vertical cross-section through a convergent plate margin (active continental margin); vertical scale exaggerated. Oceanic crust is subducted below the leading edge of a plate carrying continental crust. Deformation, magmatism, metamorphism, erosion, and sedimentation occur in the region of the active continental margin.

the rate of ocean-floor spreading lead to the determination of the rates of movement of plate boundaries relative to hot-spots and of plates relative to each other; illustrative calculations are given in Sect. 8.D to 8.F.

Transform plate margins are marked by fracture zones or transform faults, with usually strong topography of ridges and furrows, low seismicity, and virtually no vulcanicity; here the ridge axes, with their associated magnetic stripes and oceanic crust, are displaced from one position to another (Sect. 8.B.2, 8.G). Despite their superficial similarity to strike-slip faults, which are later than the structures they cut, transform faults pre-date the adjacent rocks and are an integral part of the sea-floor spreading process (see Fig. 8.2). The San Andreas fault complex (Fig. 4.19) is an example of a transform fault system occurring on land and cutting continental rocks.

Convergent plate margins are expressed topographically by an ocean trench up to 11 km deep where the oceanic plate bends to pass below another plate. If the other plate is also oceanic an island arc is formed on its leading edge, while if it carries continental crust an Andean-type mountain range is formed (Fig. 8.1(C); Sect. 8.B.3). An inclined surface of strong seismicity – the Benioff Zone – with earthquakes extending to 700 km in depth marks the plane of relative movement between the two plates, on the lower side of which the subducted oceanic plate descends into the mantle.

Many important rock-forming processes occur in the parts of the crust and mantle close to and above the Benioff zone, including plutonic and volcanic igneous activity of calc-alkaline type, the formation of an accretionary prism of sediments deposited in an ocean trench and subsequently scraped off the descending oceanic plate by the leading edge of the other plate (cf. Fig. 4.18), the formation of new sediments by erosion of the rising island arc or cordillera, and deformation and metamorphism of both sediments and igneous rocks. Porphyry copper and Kuroko-type ores are important mineral deposits in the region above the Benioff zone. Further processes may be operative if, as in the western USA, an ocean ridge is driven by plate movements below a continent; additional vulcanism, transform fault movements, and extension and thinning of the continental crust are related to the continued production of basaltic ocean-floor type rocks with crustal spreading from the subducted but still active divergent margin.

Extensional forces on the concave side of an island arc can lead to back-arc spreading with the formation of new oceanic crust parallel to and a few hundred kilometers from the convergent boundary, as at the Mariana Rift between the Philippine plate and the small Mariana plate.

Intersections of plate margins produce various kinds of triple junction where the absolute and relative motions of the three adjoining plates must necessarily be compatible. Examples on Fig. 8.3 are the junctions of the Chile Rift with the East Pacific Rise and with the Peru–Chile Trench.

The plate-tectonic process is essentially based on convection, with rise of hot mantle material below the ocean ridges, cooling in the stage of horizontal ocean-floor spreading, and sinking of cold, dense oceanic crust below the Benioff zone. The processes in the return flow deep within the mantle cannot yet be determined, and in any case are not revealed on conventional geological maps.

The rocks of an oceanic plate show little or no deformation between their creation at ocean ridges and their subduction at trenches, a few tens or hundreds of million years later; in this respect the structure and history of oceanic crust is shorter and simpler than that of continental crust. Because of its higher density, oceanic crust nearly always sinks below continental crust at convergent plate margins; rarely, however, a slice of oceanic crust is thrust (obducted) over continental rocks and is then seen as an **ophiolite complex**, revealing a cross-section through old oceanic crust for direct geological observation (Sect. 8.H, 8.I). Magmatic chromite and hydrothermal sulphide mineral deposits commonly occur in ophiolites. Mappable lithological units known as *mêlanges* may occur above or below the ophiolite; these consist of large and small blocks of sedimentary, igneous, and metamorphic rocks from both the continental and the oceanic environments in a fine-grained sedimentary or tectonically-sheared matrix. They may be the product of gravitational slumping of unconsolidated erosion fragments on an over-steepened continental slope or the result of shearing and mixing of rocks in the accretionary prism.

The remaining type of convergent plate margin involves a **collision** of two plates both carrying continental crust, and creates an orogenic belt of Alpine-Himalayan type, with rock types and structures such as those described in the first seven chapters of this book.

A further type of boundary is represented by **passive continental margins** between continental and oceanic crust where there is little or no relative movement between the two types of crust because they both belong to the same plate (Fig. 8.1(D), Sect. 8.B.4).

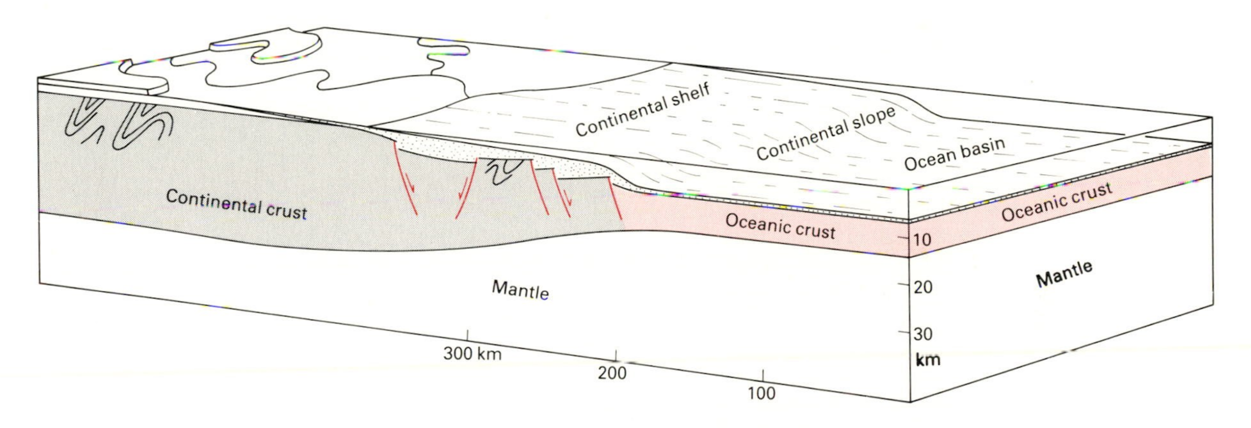

Fig. 8.1(D) Schematic and idealised vertical cross-section through a passive continental margin; vertical scale exaggerated. The boundary between continental and oceanic crust is approximately at the position of the continental slope.

This brief account of plate tectonics emphasises the dynamic aspects of plate movement and the processes of rock formation and deformation which are going on at the present day. The magnetic stripes of the ocean floor provide one of the most direct and important indications of the rates of relative movement of major units of the Earth's crust. The operation of similar processes can be recognised in older rocks, at least back to the beginning of the Phanerozoic, 570 million years ago. The use of geological maps for evaluating such processes will be illustrated in Chapter 10.

The relative positions of continents in the geological past are now well established, and maps showing such palaeogeographical reconstructions are available (Smith, Hurley & Briden 1981; Owen 1983), though they will not be discussed in this book.

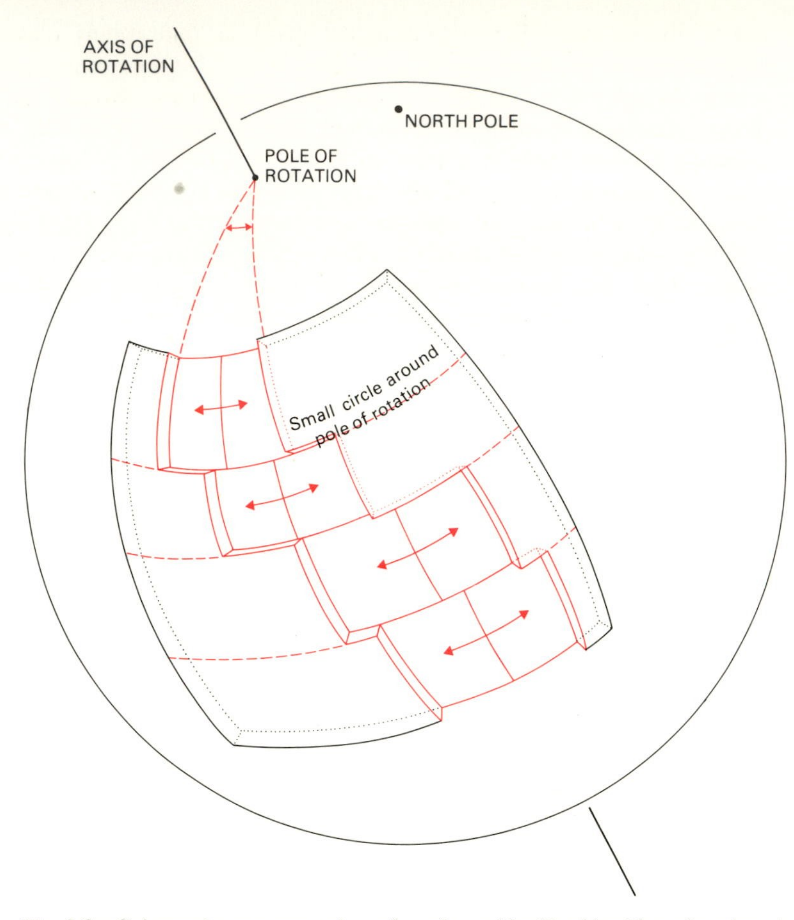

Fig. 8.2 Schematic representation of a sphere (the Earth) with a plate (continent) on its surface. The plate is split by a spreading mechanism (ocean ridge) whose position is displaced by transform faults (transform plate boundaries). The relative movement of the parts of the plate can be described as rotation about an axis which intersects the Earth's surface at the poles of rotation. The transform faults form small circles (see Appendix 4) about the pole of rotation, and their orientation can be used to define the position of the pole. The shapes of the rifted continental margins replicate the shape of the ocean ridge and transform faults.

The rate of rotation of a plate can be expressed as an angular velocity (e.g. in degrees per million years), which is the same for every point on the plate. The velocities of individual points on the plate (e.g. in centimetres per year) depend on their distances from the pole of rotation (see Sect. 8.A.3).

Fig. 8.3. Index map for Sheet 20 of the Geological World Atlas. Names of plates and plate boundaries, relative velocities (double arrows) and absolute velocities (single arrows) from '*Plate-Tectonic Map of the Circum-Pacific Region*' (American Association of Petroleum Geologists, 1981).

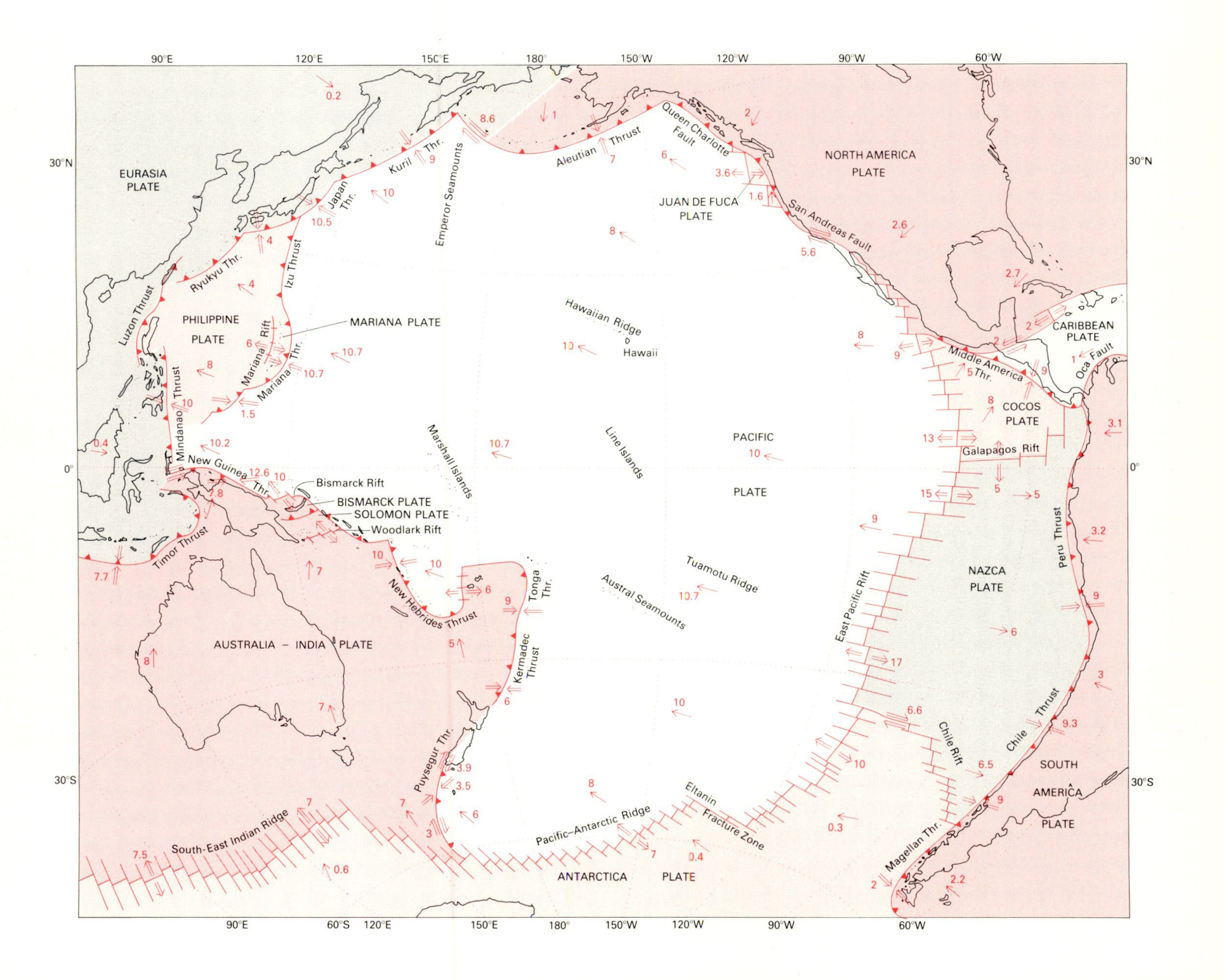

90°E
120°E
150°E
180°
150°W
120°W
90°W
60°W
30°N
0°
30°S
60°S
EURASIA PLATE
PHILIPPINE PLATE
MARIANA PLATE
NORTH AMERICA PLATE
CARIBBEAN PLATE
COCOS PLATE
NAZCA PLATE
SOUTH AMERICA PLATE
JUAN DE FUCA PLATE
PACIFIC PLATE
ANTARCTICA PLATE
AUSTRALIA – INDIA PLATE
BISMARCK PLATE
SOLOMON PLATE
Luzon Thrust
Ryukyu Thr.
Izu Thrust
Japan Thr.
Kuril Thr.
Emperor Seamounts
Aleutian Thrust
Queen Charlotte Fault
San Andreas Fault
Middle America Thr.
Oca Fault
Galapagos Rift
Peru Thrust
Chile Thrust
Chile Rift
Magellan Thr.
East Pacific Rift
Eltanin Fracture Zone
Pacific–Antarctic Ridge
South-East Indian Ridge
Puysegur Thr.
Kermadec Thrust
Tonga Thr.
New Hebrides Thrust
Woodlark Rift
Bismarck Rift
New Guinea Thr.
Mariana Thr.
Mariana Rift
Mindanao Thrust
Timor Thrust
Marshall Islands
Line Islands
Hawaiian Ridge
Hawaii
Tuamotu Ridge
Austral Seamounts

8.A TO DETERMINE DISTANCES ON THE SURFACE OF THE EARTH

Large-scale maps show areas of the Earth's surface small enough to be treated as essentially flat; such maps have constant scale and negligible distortion of angular relationships, so that distances and directions are simply determined with a ruler and protractor. However, on small-scale maps showing areas as large as continents and oceans the representation of the curved surface of the Earth on a flat sheet of paper introduces distortions. Distances and directions must then be determined by the methods of spherical trigonometry (Appendix 4).

8.A.1 TO CALCULATE DISTANCES AND DIRECTIONS ALONG GREAT CIRCLES

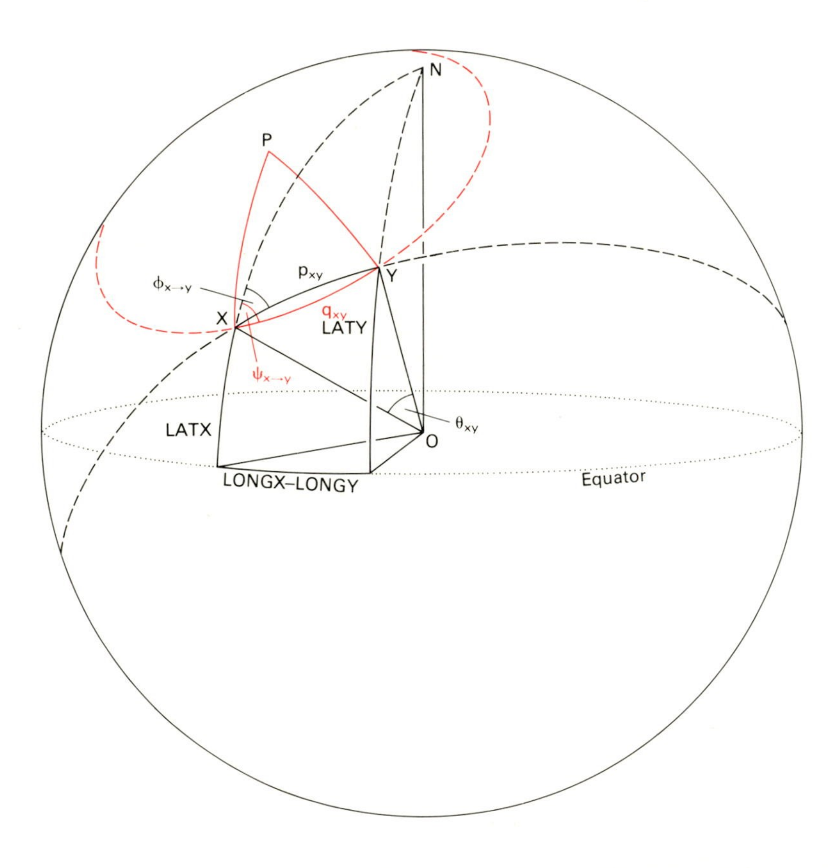

Fig. 8.4 A sphere, representing the Earth, with two points X and Y on its surface. O is at the centre of the Earth, and the angle $\angle XOY$ is θ_{XY} in equation 1, Section 8.A.1. p_{XY} is the geographical distance between X and Y measured along the great circle (black dashed line on the surface of the sphere). X and Y also lie on a small circle about the pole P (red dashed line), and q_{XY} is the geographical distance between X and Y measured along the small circle. See Sections 8.A.1 and 8.A.2 for further explanation.

If the latitude and longitude of two points, X and Y, are LATX, LONGX and LATY, LONGY, the angle, θ_{XY}, subtended by the two points at the centre of the Earth (Fig. 8.4) is given by:

$$\cos\theta_{XY} = \sin \text{LATX} \,.\, \sin \text{LATY} + \cos \text{LATX} \,.\, \cos \text{LATY} \,.\, \cos(\text{LONGX} - \text{LONGY}) \quad \ldots [1]$$

It is convenient to give a negative sign to longitudes west of Greenwich and to latitudes south of the equator; the equation can then be applied directly to points in any relative positions on the surface of the Earth.

The distance, p_{XY}, measured along the great circle, between the two points is then given by:

$$p_{XY} = 2\pi \,.\, 6371 \,.\, \frac{\theta_{XY}}{360°} \text{ kilometres} \quad \ldots [2]$$

θ_{XY} is in degrees; the mean radius of the Earth is 6371 kilometres.

The direction of point Y from X, expressed as the azimuth or angle $\phi_{X\rightarrow Y}$ between the north meridian through X and the arc of the great circle through X and Y, is given by:

$$\cos\phi_{X\rightarrow Y} = \frac{\sin \text{LATY} - \cos\theta_{XY} \,.\, \sin \text{LATX}}{\sin\theta_{XY} \,.\, \cos \text{LATX}} \quad \ldots [3]$$

The direction of Y from X is east of north if LONGX $<$ LONGY, and is west of north if LONGX $>$ LONGY.

8.A.2 TO CALCULATE DISTANCES AND DIRECTIONS ALONG SMALL CIRCLES

Calculations of ocean-floor spreading or movement of plates require the determination of distances and directions along small circles about the pole of spreading or rotation (Fig. 8.2). In Fig. 8.4, the distance q_{XY} from X to Y along the small circle about the pole P is calculated as follows:

1. In the (great circle) triangle PXY, calculate the angles of the sides θ_{XY}, θ_{PX}, and θ_{PY} (formula [1] above).
2. Calculate $\angle$XPY from:

$$\cos\angle XPY = \frac{\cos\theta_{XY} - \cos\theta_{PX} \,.\, \cos\theta_{PY}}{\sin\theta_{PX} \,.\, \sin\theta_{PY}} \quad \ldots [4]$$

3. Then $q_{XY} = 2\pi \,.\, 6371 \,.\, \frac{\angle XPY}{360°} \,.\, \sin\frac{(\theta_{PX} + \theta_{PY})}{2}$ kilometres $\quad \ldots [5]$

The azimuth $\psi_{X\rightarrow Y}$ of the small circle from X towards Y is given by:

$$\psi_{X\rightarrow Y} = \phi_{X\rightarrow P} \pm 90° \quad \ldots [6]$$

(whether 90° is to be added or subtracted can be determined by inspection).

The difference between p_{XY} (calculated from equation [2]) and q_{XY} (from equation [5]) is usually smaller than the error in determining the geographical co-ordinates of the points from a small-scale map, except:

(i) if the distance between X and Y is large ($>\sim$ 1500 km),

or (ii) if X and Y are close to P ($<\sim$ 30° or about 3000 km).

8.A.3 TO CALCULATE THE VELOCITY OF ROTATION OF A POINT ABOUT A SMALL CIRCLE

If the angular velocity of a plate about the pole of rotation, P, is ω degrees per million years, the velocity of a point, X, on the plate at angle θ_{PX} from the pole is $2\,\pi\,.\,6371\,.\,\frac{\omega}{360^\circ}\,.\,\sin\theta_{PX}$ kilometres per million years (or numerically the same value in millimetres per year).

8.B TO RECOGNISE THE TYPES OF PLATE BOUNDARY AND CONTINENT – OCEAN MARGIN

The distinguishing features of the different types of boundary as shown on ocean-floor geological maps are the topographical shapes, the patterns of seismicity and vulcanicity, and the presence or absence of symmetrical patterns each side of the boundary.

8.B.1 DIVERGENT PLATE BOUNDARIES

An ocean ridge is characterised by symmetrical (mirror-image) distribution of bathymetric contours deepening away from the ridge (which may have a central rift valley),

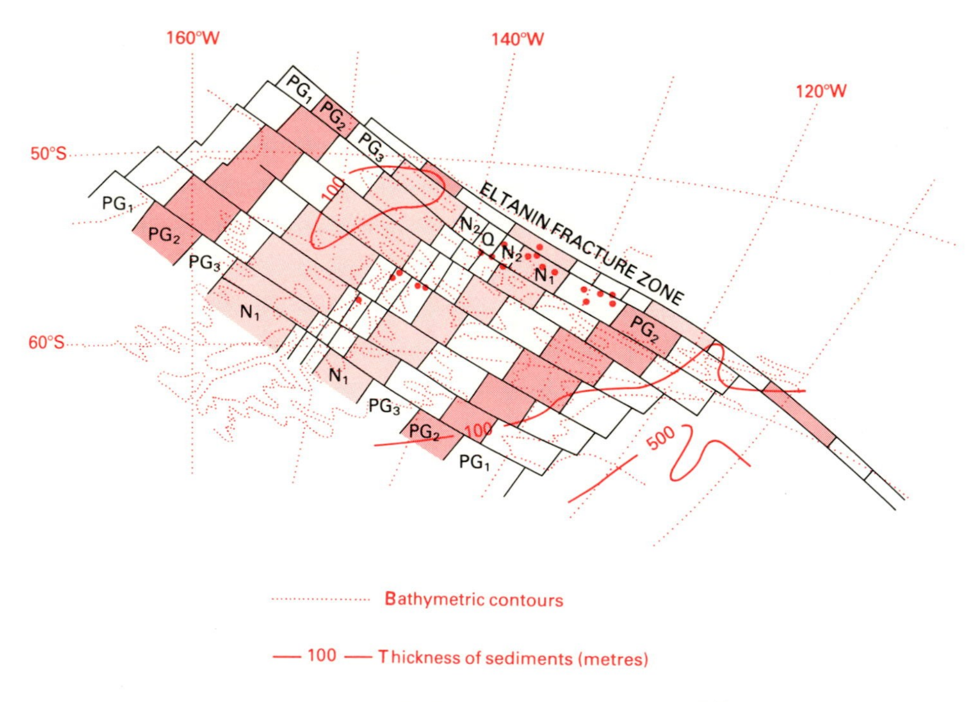

Fig. 8.5 Part of the south-central area of Sheet 20 of the Geological World Atlas, showing the Pacific-Antarctic Ridge mid-way between New Zealand and Cape Horn. Ages of basaltic basement on this diagram and on Figs 8.6 to 8.11: Q Pleistocene, N_2 Pliocene, N_1 Miocene, PG_3 Oligocene, PG_2 Eocene, PG_1 Paleocene, K_2 Upper Cretaceous, K_1 Lower Cretaceous, J Jurassic. Bathymetric contours simplified. The geological characteristics are summarised above.

thickening sediment cover, and increasing ages of magnetic anomalies and basaltic basement. The youngest rocks form the axis of the ridge. Ridges are off set by fracture zones (transform boundaries, Sect. 8.B.2). Shallow-focus earthquakes occur along the ridge axis and fracture zones. Volcanic islands are rare and are probably related to hot-spots (Sect. 8.D).

Fig. 8.5

Very young spreading axes within formerly intact continental plates are similar in character to ocean ridges; in addition, the adjacent coastlines, which are newly-formed passive continental margins (Sect. 8.B.4), are parallel to the divergent boundary and transform faults, except where they are covered over by rocks younger than the date of formation of the ocean.

Fig. 8.6

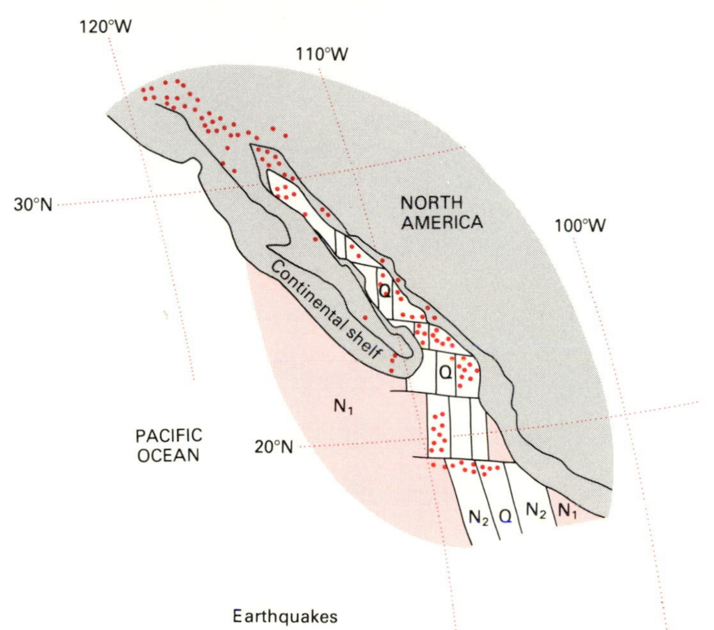

Fig. 8.6 Part of the northeast area of Sheet 20 of the Geological World Atlas, showing the west coast of California and Mexico. Ages of basaltic basement as in Fig. 8.5. The East Pacific Rise, off set by fracture zones, extends along the mid-line of the Gulf of California, which is actively spreading to form a new ocean.

8.B.2 TRANSFORM PLATE BOUNDARIES

Fracture zones or transform faults off set ocean ridges (Sect. 8.B.1). Ocean-floor topography is often strongly furrowed parallel with the fracture zone. Shallow-focus earthquakes occur along the fracture zone close to the ocean ridge. Volcanic activity is generally absent.

Examples Figs 8.5, 8.6.

8.B.3 CONVERGENT PLATE BOUNDARIES

(a) Ocean/ocean boundaries

Subduction of one oceanic plate below another produces an island arc on the overriding plate. The boundary is non-symmetrical, with different ages of ocean crust on each side. A deep ocean trench marks the plate boundary and the approximate outcrop of the Benioff zone. Increasing depth of earthquakes on one side only of the trench marks the Benioff zone and the surface of subduction. Volcanoes (with andesite as the predominant magma type) occur above the Benioff zone at a geographical distance usually not less than 125 km from the ocean trench.

Fig. 8.7(A) and (B)

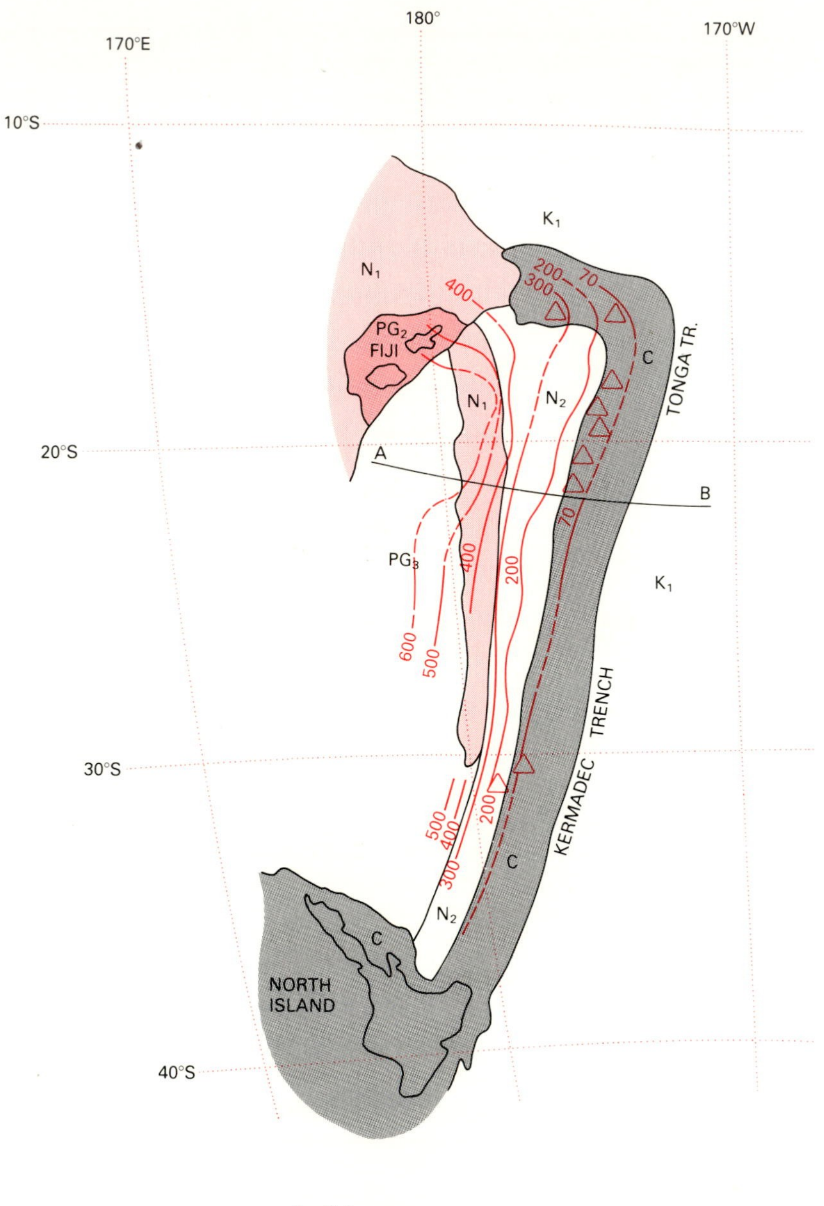

Fig. 8.7(A) Part of the central area of Sheet 20 of the Geological World Atlas, showing the southwest Pacific Ocean north of New Zealand. Ages of basaltic basement as in Fig. 8.5; C continental crust (New Zealand) and rocks of the island arc (Tonga and Kermadec Islands). Contour lines of focal depths of earthquakes (in kilometres) derived from *Plate-Tectonic Map of the Circum-Pacific Region* (American Association of Petroleum Geologists 1981). The increasing age of basaltic basement westwards from the island arc is the result of back-arc spreading.

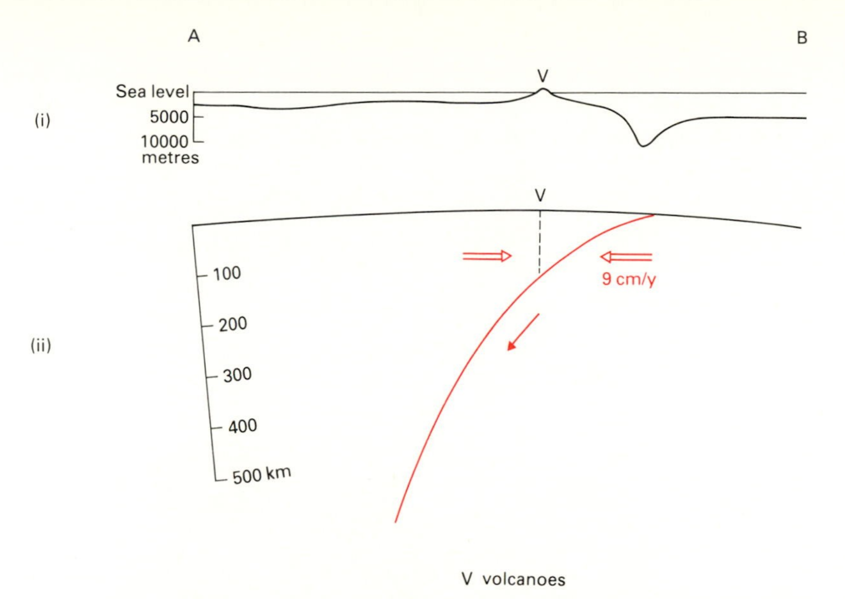

Fig. 8.7(B) Vertical cross-section along the line A–B on Fig. 8.7(A). (i) with exaggerated vertical scale to show the topographical profile, (ii) with equal horizontal and vertical scales to show the shape of the Benioff zone. Volcanoes occur where the Benioff zone is at a depth of about 120 km and indicate that the rocks in this region are at a high enough temperature for partial melting and the generation of magma to occur. The relative rate of plate movement is from Fig. 8.3.

(b) Continent/ocean boundaries = Active continental margins

The pattern is similar to that of ocean/ocean boundaries, but the side above the Benioff zone is overlain by continental crust, and carries an Andean-type mountain chain on its leading edge.

Fig. 8.8(A) and (B)

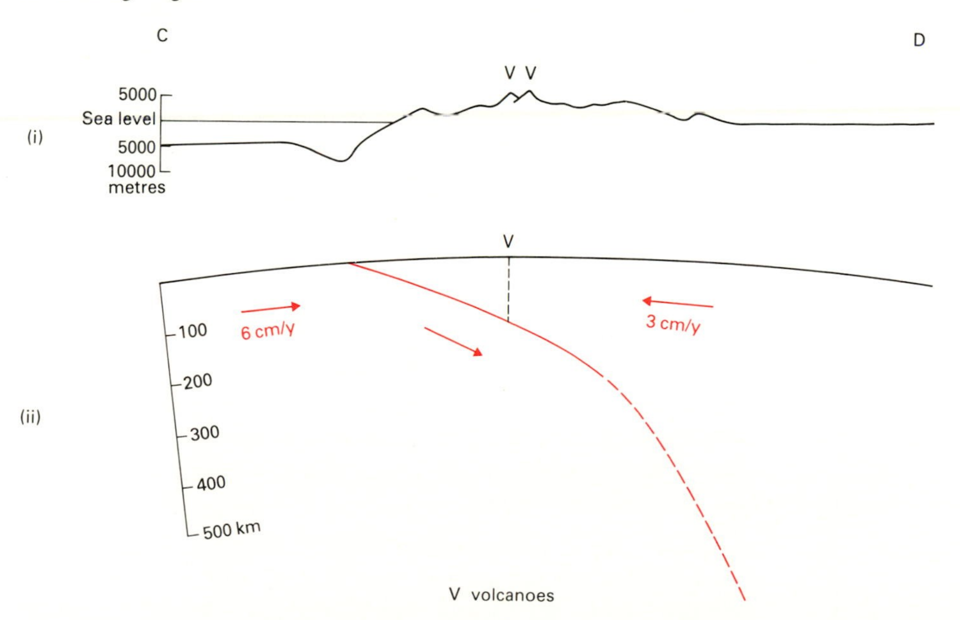

Fig. 8.8(B) Vertical cross-section along the line C–D in Fig. 8.8(A) (see next page). (i) with exaggerated vertical scale to show the topographical profile, (ii) with equal horizontal and vertical scales to show the shape of the Benioff zone. Volcanoes occur where the depth to the Benioff zone is about 110–150 km (compare with Fig. 8.7(B)). The relative rate of plate movement is from Fig. 8.3.

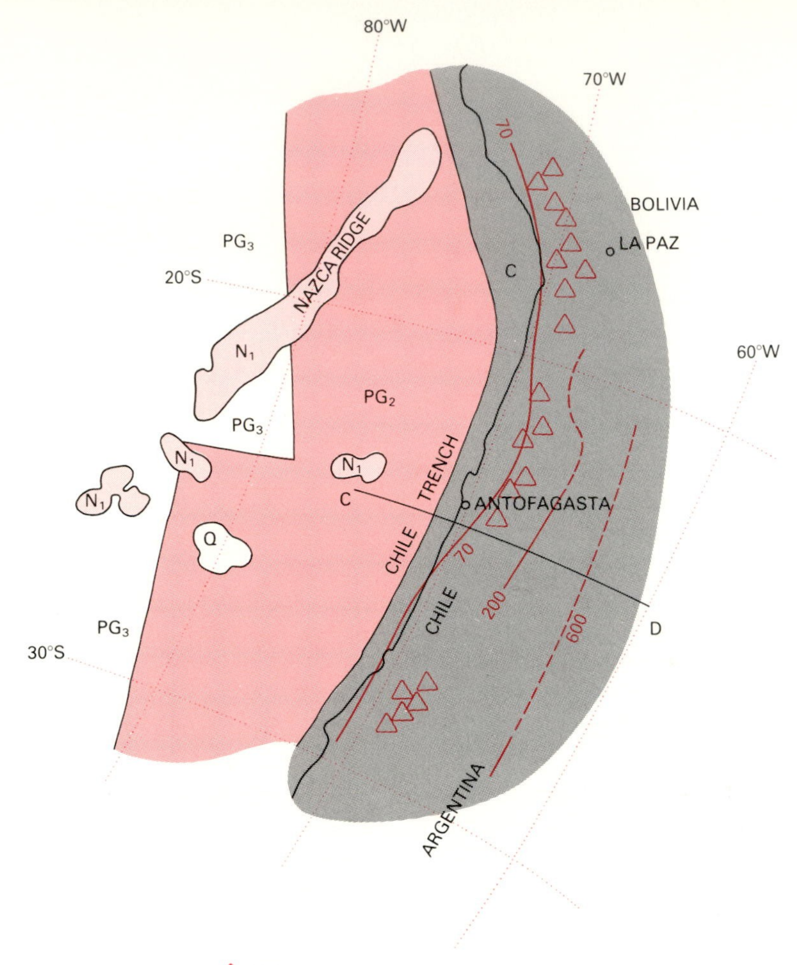

Fig. 8.8(A) Part of the southeast corner of Sheet 20 of the Geological World Atlas, showing the southern part of South America. Ages of basaltic basement and sea-mounts as in Fig. 8.5. Contour lines of focal depths of earthquakes (in kilometres) as in Fig. 8.7(A).

8.B.4 PASSIVE CONTINENTAL MARGINS

The boundary between the continental and oceanic crust is approximately marked by the continental slope. There is no seismicity or vulcanicity. The shape of the continent corresponds to that of an ocean ridge and another continent, possibly thousands of kilometres distant; the age of the oceanic crust adjacent to the continent indicates the date of continental rifting. *Fig. 8.9*

Further example Fig. 7.6

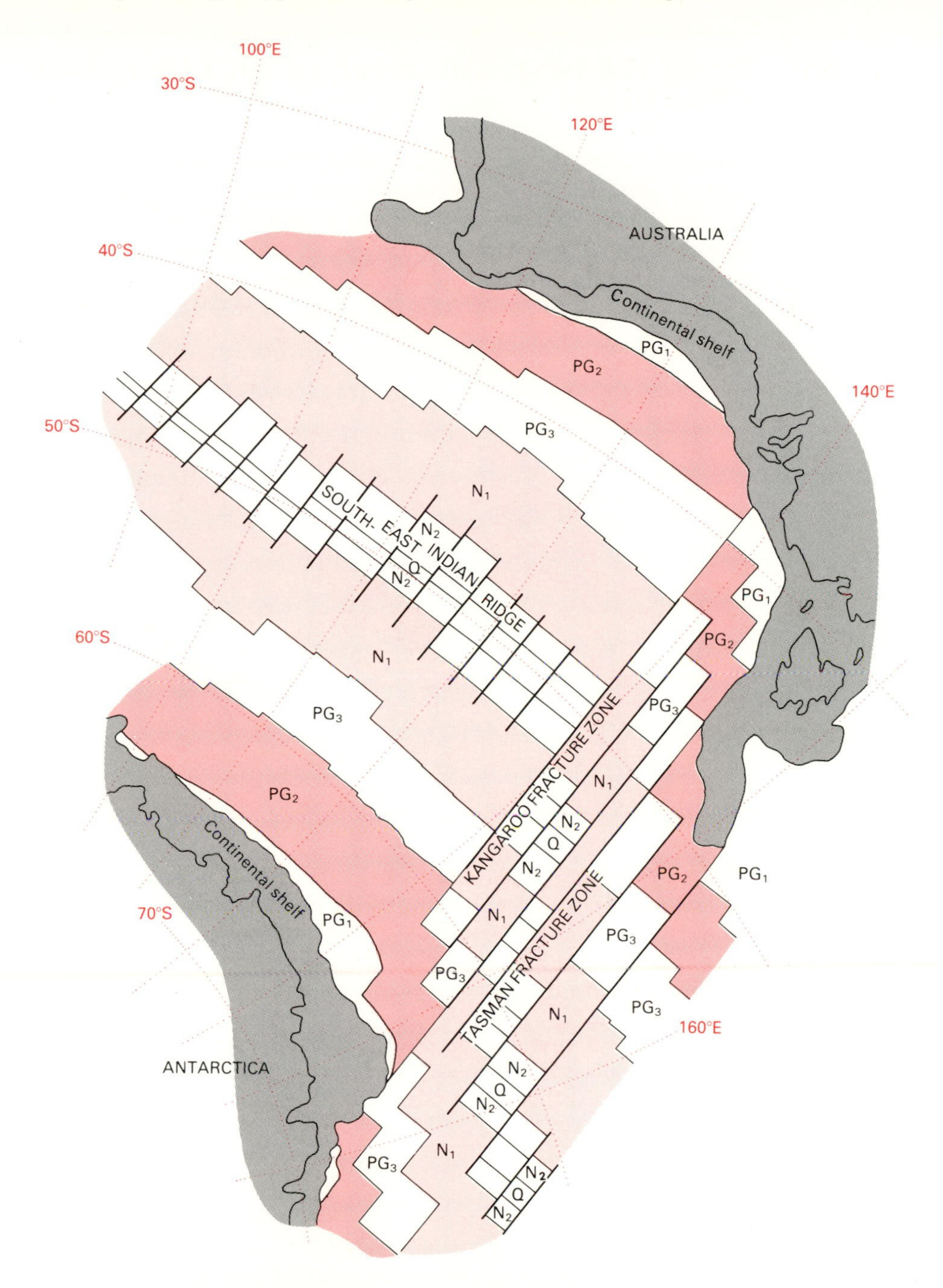

Fig. 8.9 Part of the southeast corner of Sheet 21 of the Geological World Atlas, showing parts of Australia, the Southeast Indian Ridge, and Antarctica. Ages of basaltic basement as in Fig. 8.5. The date of continental rifting between Australia and Antarctica was in the Paleocene. (The same area is shown in a different projection in the southwest corner of Sheet 20 of the Geological World Atlas.)

8.C TO DETERMINE THE RATE OF OCEAN-FLOOR SPREADING

Identify points on the ocean ridge and on datable magnetic anomalies or basaltic basement, on a line parallel to the spreading direction as indicated by fracture zones. Check that the age of the basement increases continuously away from the ridge. Determine the geographical co-ordinates of each point, and calculate the distances between them (Sect. 8.A). Divide the distance from the ridge to each point by the age of the point to obtain the rate of ocean-floor spreading.

Fig. 8.10

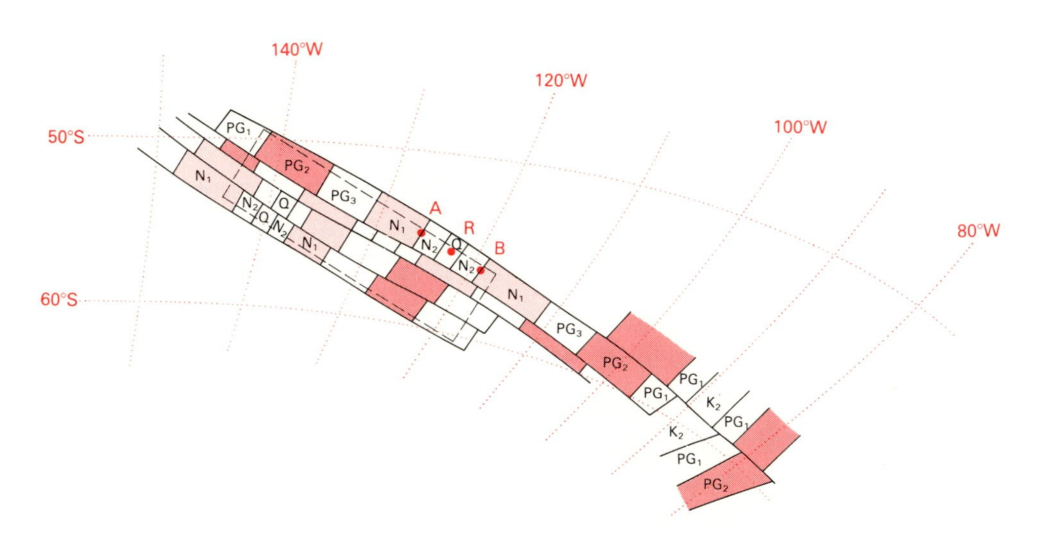

Fig. 8.10 Part of the south-central area of Sheet 20 of the Geological World Atlas, showing the Pacific–Antarctic Ridge and East Pacific Rise in the region of the Eltanin Fracture Zone (the northeast part of Fig. 8.5). Ages of basaltic basement as in Fig. 8.5. (Note the decrease in age of basement at the east end of the fracture zone, due to ocean-floor spreading from the Chile Rift 1500 km to the north.) The rectangle (dashed lines) outlines the area shown in the block diagram, Fig. 8.15.

Points A and B are on the Pliocene/Miocene boundary, age 5.3 Ma. (The chronometric timescale in Appendix 2 has been used in preference to the older scale printed in the margin of the published map.) R is on the ridge axis. The geographical co-ordinates are:

	Lat.	*Long.*
A	−54.8°	−125.8°
R	−55.5°	−122.5°
B	− 56.2°	−119.2°

giving great-circle distances:

RA 224 km, RB 220 km

and average spreading rates for the last 5.3 million years:

R → A 4.2 cm/y, R → B 4.2 cm/y

8.D TO DETERMINE THE RATE OF PLATE MOVEMENT RELATIVE TO A HOT-SPOT

Ocean-floor spreading (Sect. 8.C) is a relative movement of ocean floor away from a divergent plate boundary, but plates and their boundaries themselves move relative to the Earth as a whole. Linear volcanic chains, which mark the tracks of plates over hot-spots that are assumed to have remained fixed in position, provide a means of determining the rate and direction of plate movement relative to geographical co-ordinates. (**Note**: The calculations in Sect. 8.D, 8.E, and 8.F are to illustrate the principles of the determination of rates of plate movements. It is not expected that students will need to perform this type of calculation at the beginning of their study of ocean-floor geology.)

Determine the geographical co-ordinates of a datable point on the volcanic chain and of the present-day active volcano. Calculate the distance between them, and the direction of the chain from the hot-spot; divide the distance by the age of the point to obtain the rate of plate movement.

Fig. 8.11

Fig. 8.11 (next page) Part of the north-central area of Sheet 20 of the Geological World Atlas (slightly simplified), showing the northern Pacific Ocean with the linear volcanic chains of the Hawaiian Islands and the Emperor Seamounts. Ages of basaltic basement and of volcanic islands as in Fig. 8.5. A marks the point on the Pacific plate which was over the hot-spot (now marked by the active volcanoes on Hawaii, H) at the date of the Miocene/Oligocene boundary (23.7 Ma). R and J are the points on the East Pacific Rise and on the Miocene/Oligocene boundary of the basaltic basement from which the rate of ocean-floor spreading from R towards Hawaii can be calculated (see Sect. 8.E). The geographical co-ordinates of all the points and of the pole of rotation of the Pacific plate, P (see Fig. 8.13) are:

	Lat.	*Long.*
A	+29.6°	−178.6°
H	+19.3°	−155.6°
R	+22.7°	−107.9°
J	+23.0°	−121.8°
P	+61.7°	−82.8°

Calculation, using the procedures of Section 8.A, gives: θ_{AP} = 66.8°, θ_{HP} = 65.0°, ∠APH = 25.5°; the small-circle distance from H to A is 2580 km in the direction 300° from H. So the average rate of movement of the Pacific plate relative to the Hawaiian hot-spot for the last 23.7 million years is 10.9 cm/y, or an angular velocity about the pole of rotation of 1.07°/m.y. Calculations based on many more observations (Minster & Jordan, 1978–see reference in Appendix 5) give a best-fit angular velocity of the Pacific plate of 0.967°/m.y., and a rate of movement from H towards A of 9.8 cm/y (cf. Fig. 8.3).

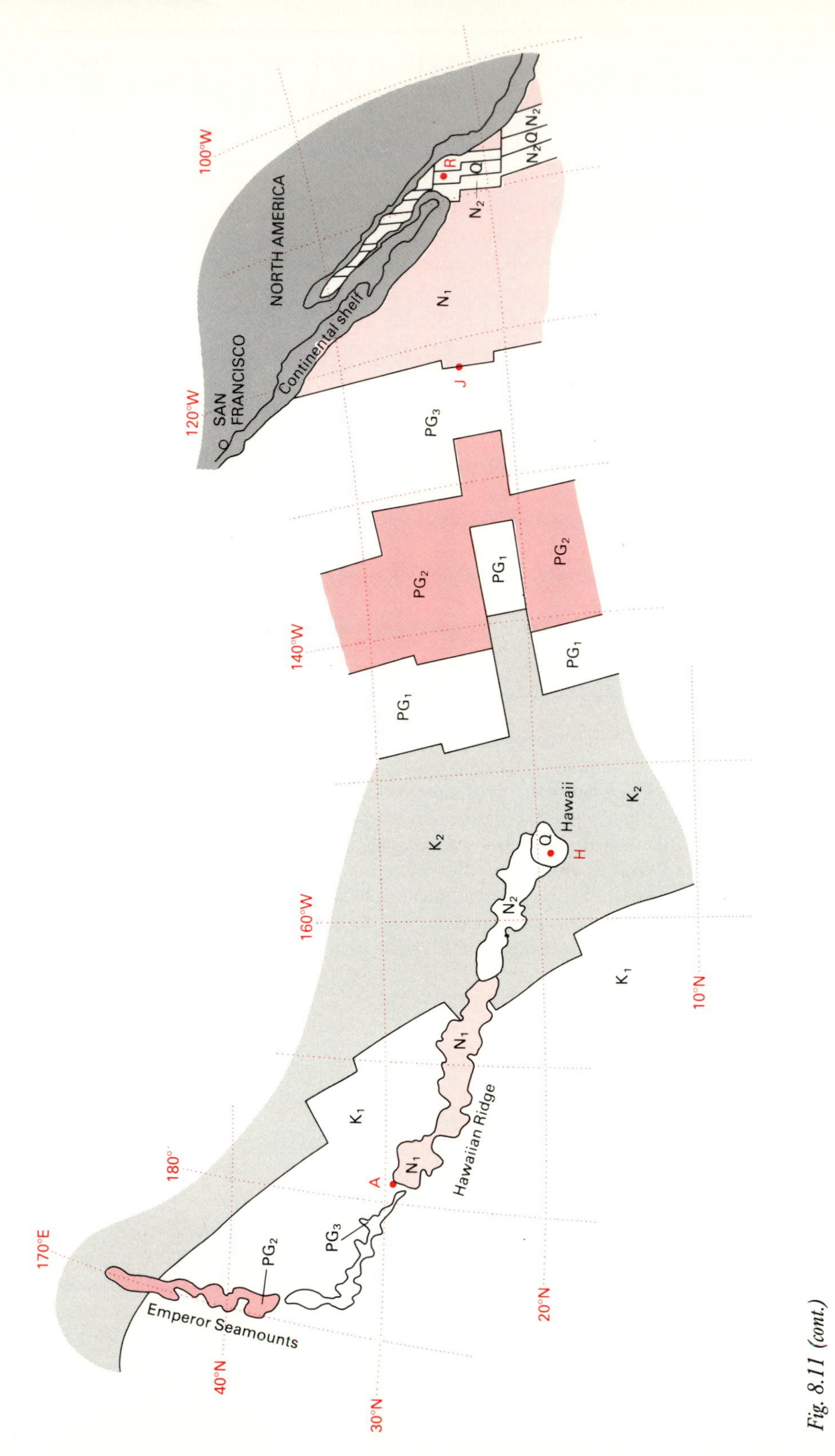
100°W
120°W
140°W
160°W
180°
170°E
10°N
20°N
30°N
40°N
NORTH AMERICA
SAN FRANCISCO
Continental shelf
R
Q
N$_2$
N$_2$Q N$_2$
N$_1$
J
PG$_3$
PG$_2$
PG$_1$
K$_2$
K$_1$
Hawaii
H
A
Hawaiian Ridge
Emperor Seamounts

Fig. 8.11 (cont.)

8.E TO DETERMINE THE RATE AND DIRECTION OF MOVEMENT OF A DIVERGENT PLATE BOUNDARY RELATIVE TO A HOT-SPOT

Calculate the rates and directions of plate movement and of ocean-floor spreading at the position of the plate boundary. The rate and direction of plate movement is the resolved component of two velocities – that of ocean-floor spreading and that of movement of the plate boundary (see Fig. 8.12).

Fig. 8.13

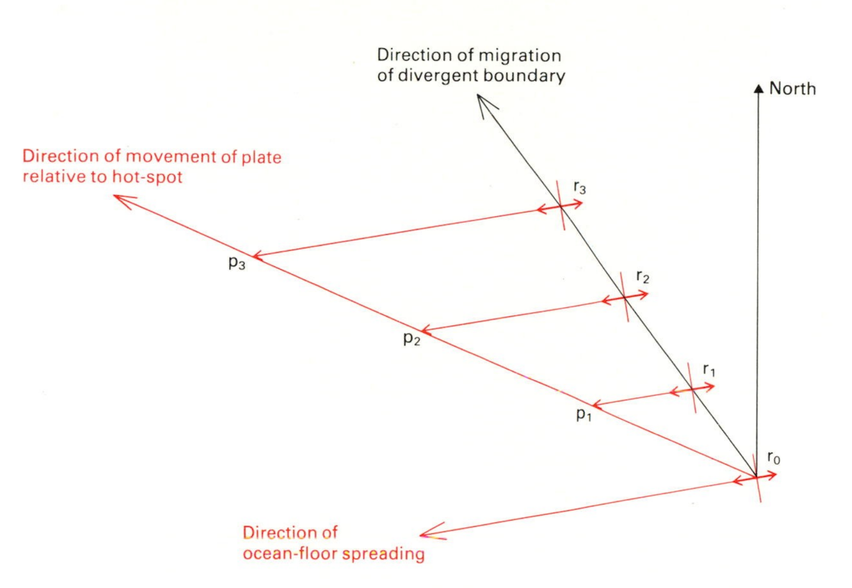

Fig. 8.12 Schematic diagram to represent migration of a divergent plate boundary, ocean-floor spreading, and the resultant plate movement.

A small segment of ocean floor is generated on a divergent plate boundary at position r_0 at time t_0. At time t_1 the point from which the segment was generated has migrated to r_1, while ocean-floor spreading perpendicular to the boundary has moved the segment to p_1. Relative to fixed geographical co-ordinates (e.g. a hot-spot) the segment has moved from r_0 to p_1. Continued ocean-floor spreading and migration of the divergent plate boundary carry the initial segment and its source respectively to p_2, p_3, . . . and to r_2, r_3, ... The rate of ocean-floor spreading is the distance the segment has travelled, p_1r_1, divided by the time interval, $(t_0 - t_1)$: $p_1r_1/(t_0 - t_1)$; the rate of plate movement relative to the hot-spot is $p_1r_0/(t_0 - t_1)$; the rate and azimuth of migration of the plate boundary can be determined by calculation or by geometrical construction.

Fig. 8.13(A) A globe showing part of the Pacific plate and adjoining areas. Points A, H, R, and J as in Fig. 8.11. P is the pole of rotation of the Pacific plate, co-ordinates +61.7°, −82.8°.

The movement of the Pacific plate from H to A involves a rotation along a small circle about the pole, P, at an angular velocity of 0.967°/m.y. or 22.9° in 23.7 m.y.; in the same period of time, with the same angular velocity, a point starting at R would have moved to S, a distance of 1725 km, at an average rate of 7.3 cm/y in the direction 287°. The average rate of ocean-floor spreading from R towards J is 6.0 cm/y in the direction 274°.

(The steps of the calculation give θ_{RP} = 42.6°, $\phi_{R \to P}$ = 17.3°, $\psi_{R \to S}$ = 287.3°, q_{RS} = 1725 km; θ_{RJ} = 12.8°, p_{RJ} = 1420 km, $\phi_{R \to J}$ = 274°.)

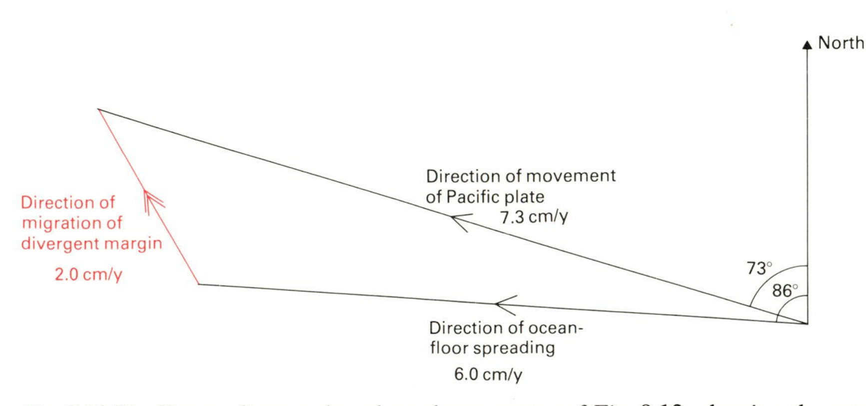

Fig. 8.13(B) Vector diagram based on the geometry of Fig. 8.12, showing the rates of Pacific plate movement and of ocean-floor spreading at point R. The rate of migration of the divergent margin relative to the Hawaiian hot-spot can be calculated as 2.0 cm/y in the direction 331°.

8.F TO DETERMINE THE RATE OF MOVEMENT OF TWO PLATES RELATIVE TO EACH OTHER

Using the data on poles of rotation and angular velocity in Appendix 5, determine the rate and direction of movement of each plate at a point on their mutual boundary. Construct a vector diagram (cf. Fig. 8.13(B)) to determine the rate and direction of relative movement of the two plates.

Example The poles of rotation, P and N, and angular velocities, ω, of the Pacific and North America plates respectively, and of a point, F, on their boundary on the San Andreas Fault 330 km southeast of San Francisco, are:

	Lat.	Long.	ω
P	−61.7°	+97.2°	0.967°/m.y.
F	+35.5°	−120.0°	
N	−58.3°	−40.7°	0.247°/m.y.

Calculation of rates and directions of movement of the two plates at F gives:

Pacific plate 6.2 cm/y in direction 300°
North America plate 2.5 cm/y in direction 235°

(The steps of the calculation give $\theta_{FP} = 145.0°$, $\phi_{F\to P} = 210.0°$; $\theta_{FN} = 114.5°$, $\phi_{F\to N} = 145.4°$.)

A vector diagram, Fig. 8.14, then gives the rate and direction of relative movement of the two plates as 5.6 cm/y in the direction 324°. This closely matches the displacement and trend of the San Andreas Fault at this point in California (see Fig. 4.19).

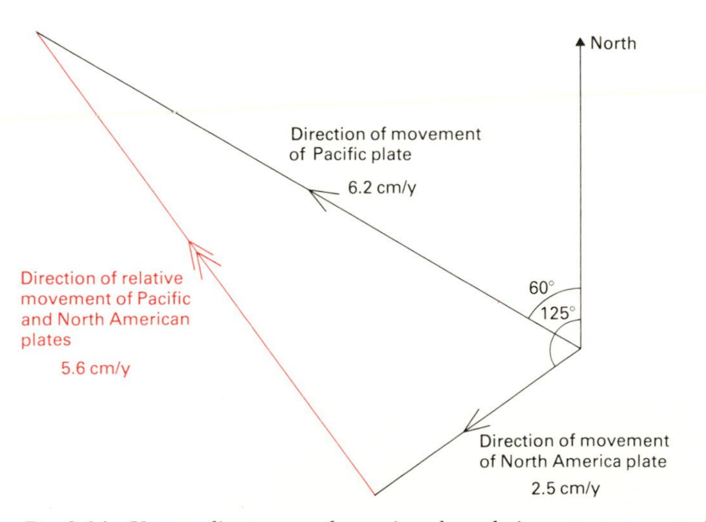

Fig. 8.14 Vector diagram to determine the relative movement on the San Andreas Fault.

8.G TO DETERMINE THE DIRECTIONS OF RELATIVE MOVEMENT AT TRANSFORM PLATE BOUNDARIES

There are two components of relative movement of the rocks of the ocean floor: (i) horizontal ocean-floor spreading and (ii) vertical decrease in elevation due to cooling and contraction of the basaltic basement.

Determine the direction of increasing age of the basaltic basement – this indicates both the direction of horizontal movement relative to a ridge axis and the direction in which there is progressive downward vertical movement. (The rate of relative horizontal movement can be calculated as in Sect. 8.C.)

Fig. 8.15

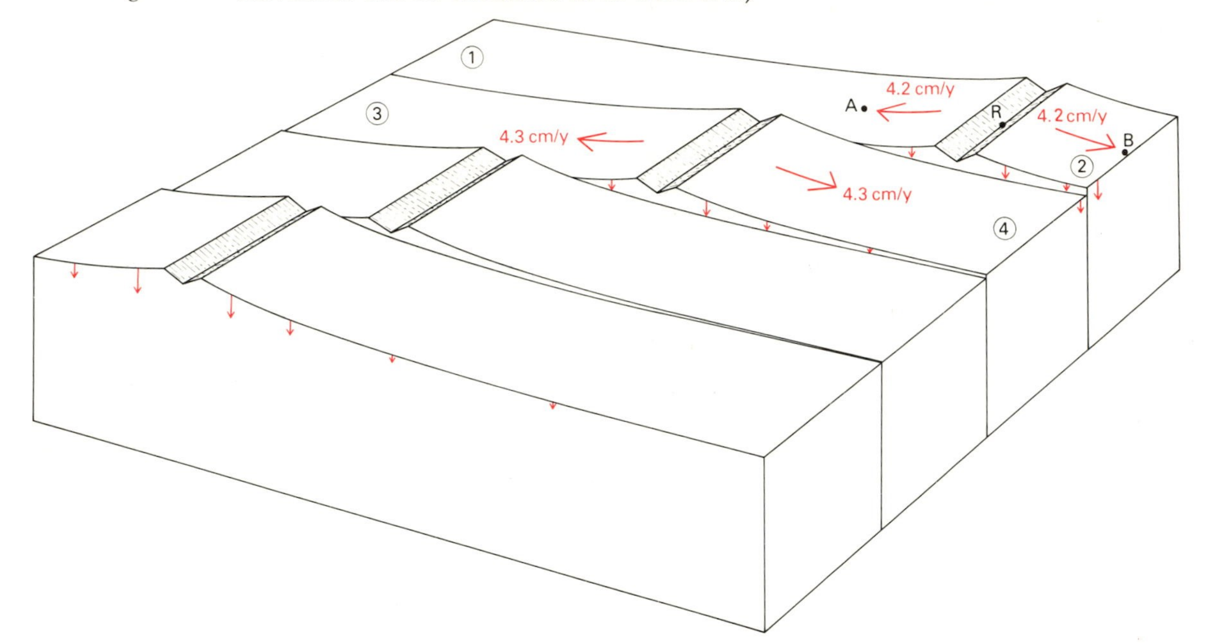

Fig. 8.15 Schematic block diagram of the area marked by the rectangle in Fig. 8.10, showing directions of relative movement of segments of the ocean floor (a) relative to the ridge axis (long arrows), with rates calculated as in Fig. 8.10, and (b) relative to each other (short arrows). The rates of vertical movements due to thermal contraction (not readily calculable from data on the map) are about 4 mm/y at the ridge crest, diminishing exponentially to about 0.1 mm/y in crust of age 10 Ma. The sector of the constructive plate boundary between the ridge axes of segments 1 and 4 is a transform fault with left-lateral movement of 4.2 + 4.3 = 8.5 cm/y and significant though smaller vertical movement; earthquakes occur predominantly in this region of strong horizontal motion. Outside the inter-ridge area there is little or no horizontal movement between adjacent segments; downward vertical movement due to cooling is greater in the segment which is closer to its associated ridge, and gives a maximum relative vertical movement of adjacent segments of ocean floor of about 1 mm/y, diminishing to negligible values as the distance from the ridge increases.

8.H TO RECOGNISE AN OPHIOLITE SEQUENCE

Look for part or all of the sequence of rocks that forms oceanic crust (Fig. 8.1(B)). In particular (since gabbro and pillow lavas can occur in other geological environ-

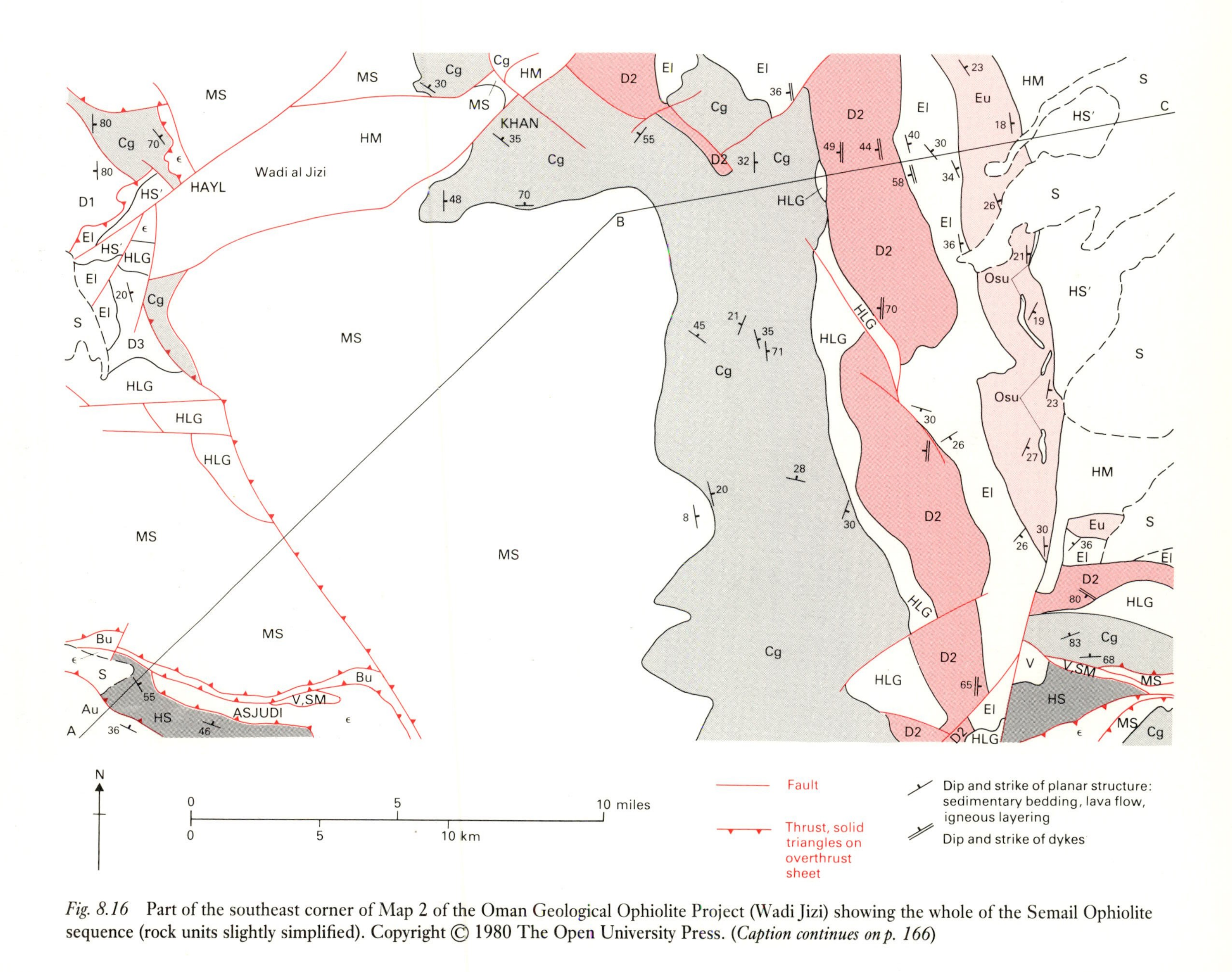

Fig. 8.16 Part of the southeast corner of Map 2 of the Oman Geological Ophiolite Project (Wadi Jizi) showing the whole of the Semail Ophiolite sequence (rock units slightly simplified). Copyright © 1980 The Open University Press. *(Caption continues on p. 166)*

(*Caption of Fig. 8.16 cont.*)

Code	Unit	Age
S	Superficial deposits	
HM	Hawasina mélange, containing blocks of serpentinite, mica- and graphite-schists, exotic reef limestones, and HS′: Hawasina Sediments	Late Cretaceous
Osu	Metalliferous and pelagic sediments	Semail Ophiolite (Late Cretaceous)
Eu	Upper lavas: pillow and flow basalts and acid volcanics; metamorphosed to zeolite facies	
El	Lower lavas: pillow and flow basalts; metamorphosed to greenschist facies	
D3	Sheeted intrusive complex: D3 pillow lava host rock	
D2	Sheeted intrusive complex: D2 no host rock	
D1	Sheeted intrusive complex: D1 plutonic host rock; metamorphosed to greenschist facies	
HLG	High-level gabbro: gabbro, diorite, tonalite, and trondhjemite; metamorphosed to greenschist and amphibolite facies	
Cg	Gabbro cumulates: layered gabbro, troctolite, peridotite, and dunite	
MS	Mantle sequence: serpentinised harzburgite, lherzolite, dunite, and chromitite	
Bu	Banded unit: sheared rocks of the Mantle sequence	
------	Semail Thrust ------	
ε	Basal serpentinite and metamorphic rocks	
SM	Sedimentary mélange: mudstones with exotic blocks	Late Cretaceous?
V	Haybi Volcanic Sequence; predominantly pillow basalts	Trias
HS	Hawasina Sediments: limestone and sandstone turbidites	mid-Trias to mid-Cretaceous
Au	Sumeini(?) Limestone: continental shelf and shelf-edge limestones	mid-Permian to Early Cretaceous

The southern part of the map and the line of section A – B – C (Fig. 8.17) show the ophiolite and underlying thrusts as a complete sequence from the lowest structural level in the west to the highest in the east. The same lithological units occur in a more complex faulted structure in the northwest of the map area.

The Semail Ophiolite represents oceanic crust formed at a spreading ridge and has a thin cover of pelagic sediments; the petrological equivalent of the Mohorovicic discontinuity occurs between Cg and MS. The metamorphic rocks below the ophiolite are basalts and sediments scraped off the ocean floor at the beginning of thrust movement, and deformed and metamorphosed during the thrusting. The Sumeini(?) Limestone, Hawasina Sediments, the metamorphic rocks, and the ophiolite form a sequence of segments from continental shelf, through submarine fan, to ocean-floor environments that were successively thrust southwestwards onto the Arabian Shield.

ments), the diagnostic features are peridotite or serpentinite, sheeted dykes, and pelagic sediments, associated with mélange rocks and occurring in a major thrust zone. (Because of the continuation of plate movement and active margin processes after their emplacement, ophiolites are nearly always structurally complex.)

Fig. 8.16

8.I TO CONSTRUCT A VERTICAL CROSS-SECTION THROUGH AN OPHIOLITE AND TO DETERMINE THE THICKNESSES OF UNITS

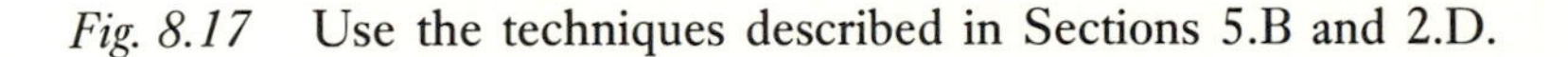

Fig. 8.17 Use the techniques described in Sections 5.B and 2.D.

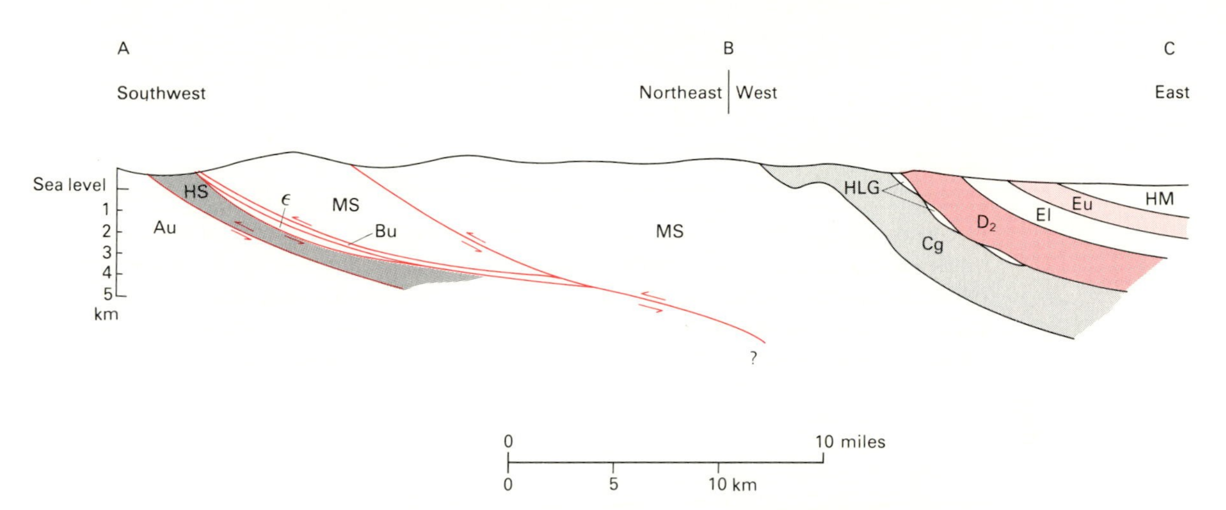

Fig. 8.17 Vertical cross-section through the Semail Ophiolite along the line A–B–C on Fig. 8.16. Units labelled as in Fig. 8.16. The lack of recorded primary features in the Mantle sequence makes its structure and thickness uncertain. Dykes in D2 happen to have the same strike direction as the boundaries of the unit so that the dip of the (formerly vertical) dykes can be used to construct the section. Thicknesses of the units, measured from the section, are:

Eu	Upper lavas	1000 m
El	Lower lavas	1100 m
D2	Sheeted dyke swarm	1900 m
HLG	High-level gabbro	0–175 m
Cg	Gabbro cumulates	~ 2500 m
MS	Mantle sequence	? 7000 m

9 SYNTHESIS OF THE GEOLOGY OF AN AREA: 1. GEOMETRICAL INTERPRETATION

Previous chapters have shown how the three-dimensional geometry of *individual* rock units and structures can be deduced from their outcrop patterns on the approximately two-dimensional surface of the Earth. In this chapter we describe procedures for understanding and representing the three-dimensional geometry of the whole area; in Chapter 10 we shall show how the essentially geometrical interpretation of this chapter

Fig. 9.1 Aerial photograph of part of the area of B.G.S. Sheets 39W and 39E, looking towards the east–northeast from a position southwest of Stirling. In the centre is Stirling Castle built on an outcrop of the Midland Valley Sill. The Ochil Fault runs along the base of the hills in the distance, separating the low ground of the folded Carboniferous sediments on the right from the rugged topography of the Lower Old Red Sandstone volcanic rocks on the left. Reproduced by permission of the Director, British Geological Survey: Crown copyright reserved.

is converted into a fully geological understanding in terms of geological time, rates, environments, processes, and products.

The procedures to be described provide alternative methods for representing the geological relationships of the area, and they must therefore be mutually self-consistent. The different types of construction are complementary, and it is often convenient to work on two methods simultaneously, using the results of one to provide the data needed for the other.

It is useful to make some **preliminary observations** on the map (Sect. 9.A) before beginning the formal interpretation; these can be conveniently summarised on a sketch map of the area. A **structural** or **tectonic map** (Sect. 9.B) is used to emphasise what is structurally significant about the area. A **vertical cross-section** is the most generally effective interpretative method for map work because it shows how the two-dimensional rock relationships on the present-day erosion surface continue into the third dimension of depth; consequently this method is described in detail (Sect. 9.C).

The three-dimensional geometry may be further displayed by a **structure contour map** showing the elevation, relative to sea-level, of lithological boundaries of interest (Sect. 9.D). Cross-sections, structure contour maps, and thickness contour maps are used for the assessment of **economic mineral resources** (Sect. 9.E).

All but one of the examples used to illustrate Chapters 9 and 10 are drawn from the British Geological Survey 1 : 50 000 Sheets 39W and 39E (Stirling and Alloa) showing part of the north side of the Midland Valley of Scotland (Fig. 9.1). The area includes a wide variety of rock types and structures, mapped in sufficient detail to allow in-depth interpretation both of the three-dimensional geometry (Ch. 9) and of the geological processes (Ch. 10). One additional example is based on the Forties oil-field of the North Sea hydrocarbon basin.

9.A PRELIMINARY OBSERVATIONS

The intention is to obtain an overview of the area which will serve as a framework for the synthesis that follows.

9.A.1 REGIONAL SETTING

Fig. 9.2

Study the location of the area on a small-scale geological map, noting in particular the extent of rock units and structures outside the area of interest.

9.A.2 PRELIMINARY STUDY OF THE MAP

Study the whole of the printed map, including all the marginal information. The names of lithological units, and especially descriptions of them, give an impression of the rock types; stratigraphical columns provide information on thicknesses of units and lithological variations (facies changes) in the area. A structural map identifies the major structures of the area and cross-sections show the geometry of folds, faults, unconformities, and igneous rocks. Many modern maps include a description of the geology of the area in the margin of the map.

The observations can be summarised on a simple sketch map of the area. A convenient scale is about 1/3 to 1/4 of the scale of the published map. A grid (reduced from the geographical grid printed on the map) can be used to aid precision of construction. Unconformities should be delineated because they separate chrono-

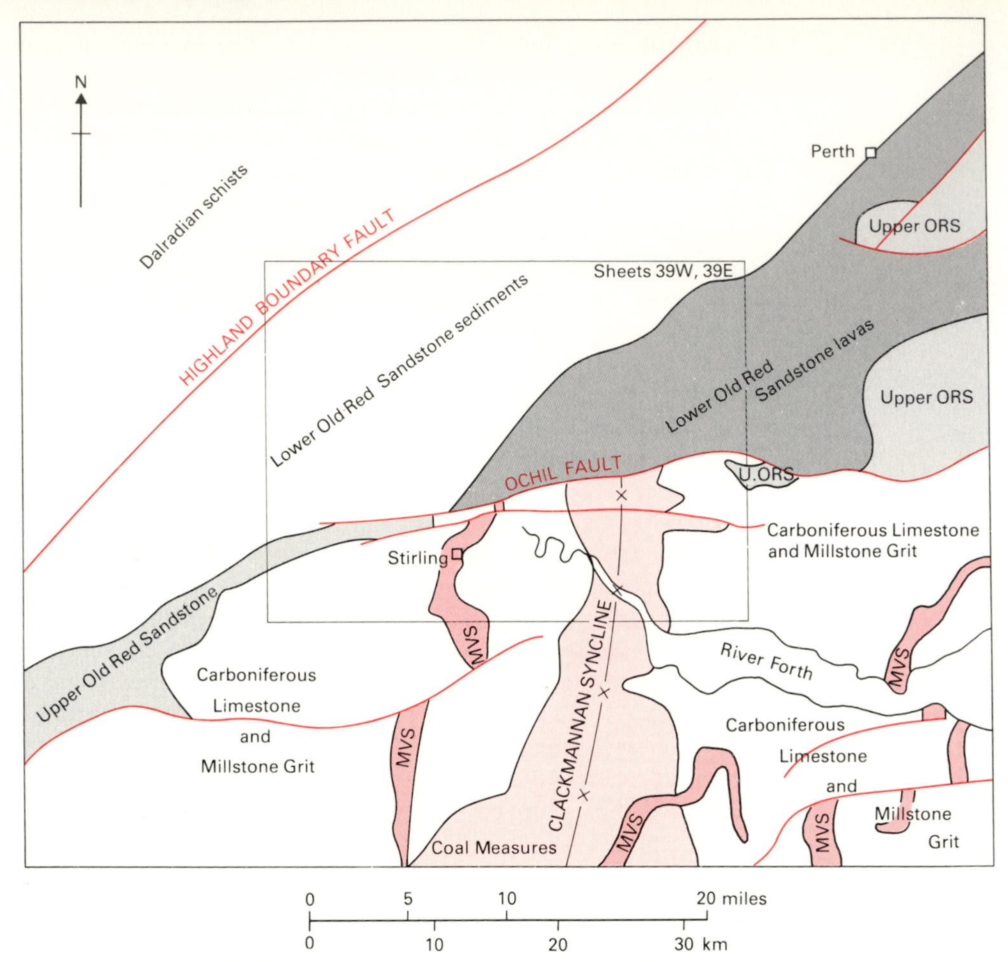

Fig. 9.2 Simplified geological map of part of the Midland Valley of Scotland, based on part of the 1 : 625 000 Geological Map of the United Kingdom (North Sheet) showing the location of Sheets 39W and 39E (Fig. 9.3 *et seq.*). The principal structural and lithological features recognisable at this scale are labelled. The Highland Boundary Fault separates Dalradian schists metamorphosed during the Caledonian orogeny from the Old Red Sandstone (approximately Devonian) and Carboniferous sediments and volcanic rocks of the Midland Valley. The Ochil Fault is the largest of a series of east–west faults, and forms the northern boundary of the Clackmannan Syncline of Upper Old Red Sandstone (Upper ORS) and Carboniferous rocks. The Midland Valley Sill (MVS) intrudes the Carboniferous sediments at various levels in the Clackmannan Syncline.

Fig. 9.3 Simplified geological map of the area of B.G.S. Sheets 39W and 39E, showing the principal features, with representative dips. The Highland Boundary Fault (HBF) separates Dalradian schists (probably Cambrian) from Lower Old Red Sandstone (L. ORS, Devonian) in the major northeast–southwest Strathmore Syncline. Upper Old Red Sandstone (U. ORS, Devonian) lies with angular unconformity on Lower Old Red Sandstone in the southwest. Upper Old Red Sandstone and Carboniferous rocks (Calciferous Sandstone Measures to Coal Measures) occur in the major north–south Clackmannan Syncline, separated from the older rocks of the Strathmore Syncline by the Ochil Fault; four sub-units of the Carboniferous have been selected for this sketch map in order to emphasise the shape of the Clackmannan Syncline. The transgressive quartz-dolerite Stirling Sill forms the northwest part of the Midland Valley Sill (cf. Fig. 9.2).

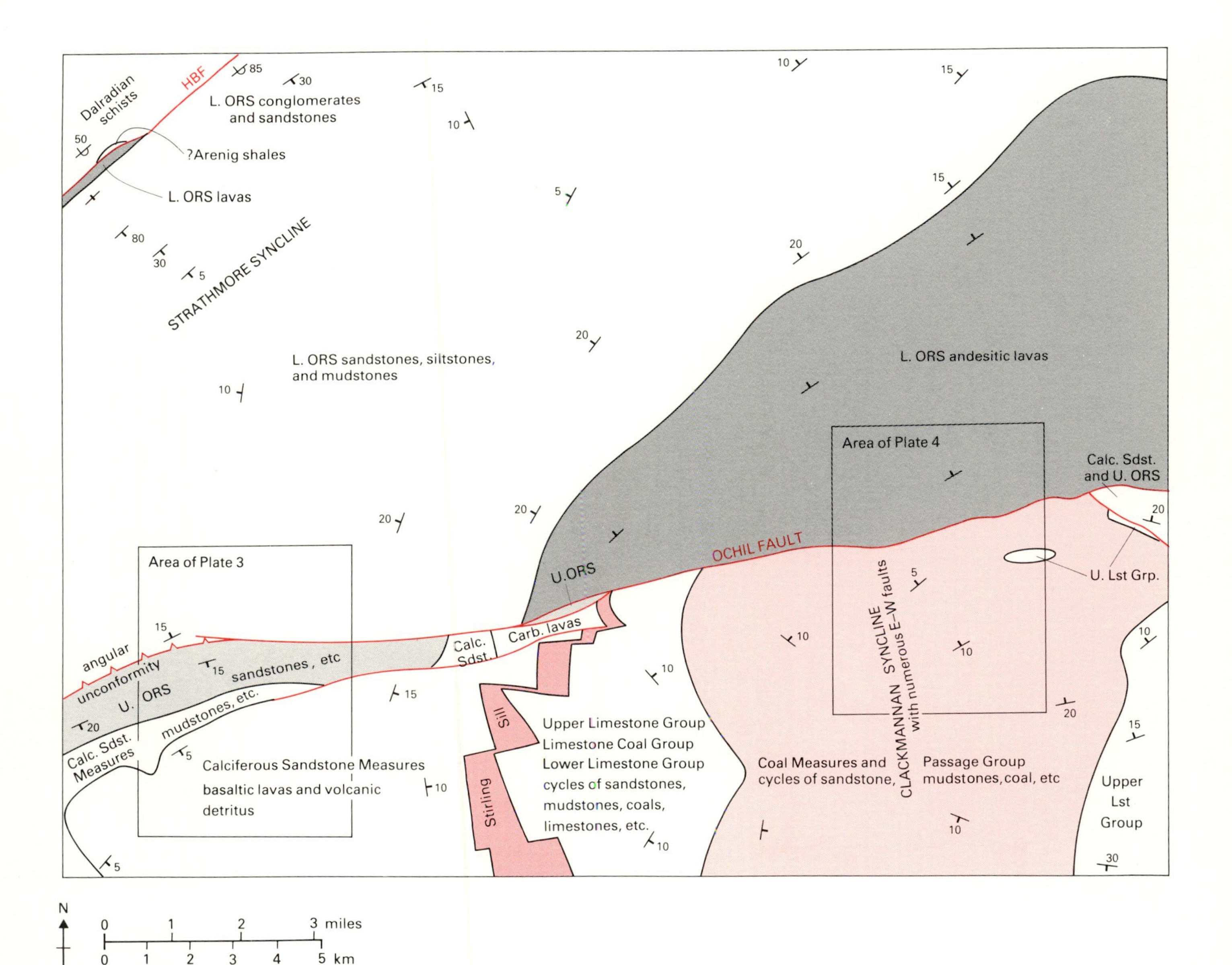
Dalradian schists
HBF
L. ORS conglomerates and sandstones
?Arenig shales
L. ORS lavas
STRATHMORE SYNCLINE
L. ORS sandstones, siltstones, and mudstones
L. ORS andesitic lavas
Area of Plate 4
Calc. Sdst. and U. ORS
OCHIL FAULT
U.ORS
U. Lst Grp.
Area of Plate 3
angular unconformity
U. ORS
sandstones, etc
mudstones, etc.
Calc. Sdst. Measures
Calciferous Sandstone Measures basaltic lavas and volcanic detritus
Calc. Sdst.
Carb. lavas
Sill
Stirling
Upper Limestone Group
Limestone Coal Group
Lower Limestone Group
cycles of sandstones, mudstones, coals, limestones, etc.
CLACKMANNAN SYNCLINE with numerous E–W faults
Coal Measures and cycles of sandstone,
Passage Group mudstones, coal, etc
Upper Lst Group
N
3 miles
5 km

structural units (Sect. 6.G) or groups of rocks which have different structural histories. Major faults, igneous intrusions, and boundaries between stratigraphical systems should be shown; a few additional boundaries may be delineated because they help to illustrate the distribution of the rock-types and the structure of the area. Superficial deposits may also be shown if appropriate. *Fig. 9.3*

Determine the thicknesses of lithological units, using the methods of Section 2.C and information printed on the map or in additional explanatory material. *Fig. 9.4*

Table 9.1 Identification of lithological units on Figs 9.4–9.8 and 10.1–10.8 and on Plates 3 and 4. Based on B.G.S. Sheets 39W (Stirling) and 39E (Alloa)

Figures	*Plates*	*Lithological unit*	*Composition*	*Age*
Q^D	Q^D, Q^T	Intrusive rocks	Quartz-dolerite and tholeiite	Probably Permo-Carboniferous
D	K, K^D, D, X	Intrusive rocks	Dolerite, tholeiite, olivine basalt, trachybasalt	Probably Carboniferous
V	—	Volcanic vents	Agglomerate	Carboniferous
d^{C3}	d^{C3}	Upper Coal Measures	Sandstones, siltstones, mudstones, and seatearths	Westphalian
d^{C2}	d^{C2}	Middle Coal Measures	Cyclic sequence of sandstones, siltstones, mudstones, coals, and seatearths	Westphalian
d^{C1}	d^{C1}	Lower Coal Measures	Cyclic sequence of sandstones, siltstones, mudstones, coals, and seatearths	Westphalian
d^{MC}	d^{MC}	Passage Group	Sandstones with clay-rocks, seatearths, thin mudstones, coals, and three thin limestones	Westphalian–Namurian
d^{M2}	d^{M2}	Upper Limestone Group	Cyclic sequence of sandstones, siltstones, and mudstones, with marine limestones, thin coals, seatearths, and (in the east) tuffs	Namurian
d^{M1}	d^{M1}	Limestone Coal Group	Cyclic sequence of sandstones, siltstones, mudstones, coals, and seatearths and (in the east) tuffs	Namurian
d^{L4}	d^{L4}	Lower Limestone Group	Sandstones, siltstones, and mudstones, with marine limestones, coals, and seatearths	Dinantian
d^{L1-3}	d^{L1-3}	Calciferous Sandstone Measures	Mudstones, siltstones, sandstones, limestones; argillaceous dolomite (cementstone) near base. Red dots on Plate 3 denote volcanic detritus	Dinantian
d^{LV}	W, W^M BW, B, miB, B^J, B^M	Extrusive rocks within Calciferous Sandstone Measures	Trachybasalt, mugearite, basalt, olivine basalt	Dinantian

Table 9.1 (cont.)

Figures	*Plates*	*Lithological unit*	*Composition*	*Age*
c^{3C}	c^{3C}	Cornstone Beds	Sandstones, siltstones, mudstones, and conglomerates, with cornstones (calcitic and dolomitic concretionary and conglomeratic siltstones – CS on Plate 3)	Upper Old Red Sandstone
c^{3}	c^{3}	Gargunnock Sandstones	Red sandstones with marls; pebbly sandstones and conglomerates in lower part	
c^{1CG}	—		Conglomerates	Lower Old Red Sandstone
c^{1T}	c^{1T}	Teith Formation	Sandstones, with siltstones, mudstones, and pebbly sandstones	
c^{1C}	c^{1C}	Cromlix Formation	Mudstones and siltstones	
c^{1D}	c^{1D}	Dunblane Formation	Sandstones with subsidiary mudstones	
c^{1R}	c^{1R}	Ruchill Formation (in NW) Buttergask Formation (in SE)	Sandstones, siltstones, and shales, with rare pebbly bands	
c^{1S}	c^{1S}	Sherriffmuir Formation	Sandstones with subsidiary shales; andesite and basalt lavas near base	
c^{1VC}	—	Volcanic conglomerate	Conglomerate with pebbles and boulders of lava, and intercalations of andesite and basalt lavas	
c^{1V}	N, hA, pA, AB, B, Z	Extrusive rocks	Rhyodacite, trachyandesite, andesite, basalt, tuff, agglomerate	
P	F^{P}, qF^{P}, P, P^{P}, K^{D}	Intrusive rocks	Felsite, porphyry, andesite, dolerite, basalt, tholeiite	
H	H	Intrusive rocks	Diorite	
b^{1}	—	Highland Border Series	Black shale with chert	?Ordovician
μ	—		Serpentinite	?Arenig
Dal	—	Dalradian Schists	Graphitic shales, slates, grits, and thin limestones	Lower Cambrian and ?Cambrian

Ag, Ba, Co, Cu Mineral veins
Individually mapped coal seams and limestones are labelled on Plates 3 and 4 (e.g. A CH, A SP, CN M, . . . CC, PL No. 1, CM, . . .)

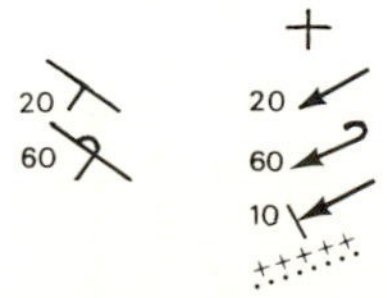

Figures	Plates	
	+	Horizontal strata
20	20	Inclined strata, dip in degrees
60	60	Overturned strata, dip in degrees
	10	Inclined strata underground, dip in degrees
	+++++	Margin of metamorphic aureole

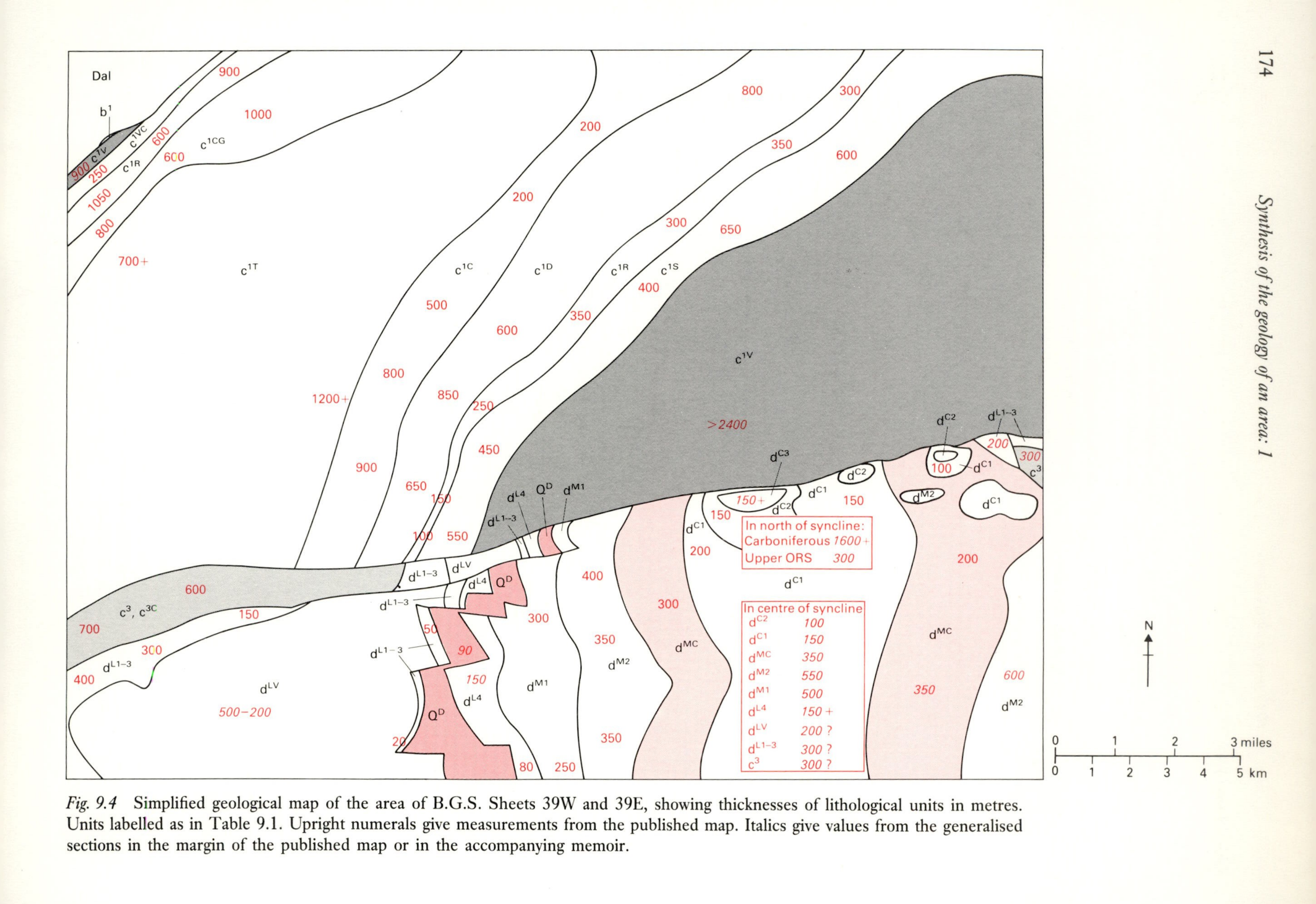

Fig. 9.4 Simplified geological map of the area of B.G.S. Sheets 39W and 39E, showing thicknesses of lithological units in metres. Units labelled as in Table 9.1. Upright numerals give measurements from the published map. Italics give values from the generalised sections in the margin of the published map or in the accompanying memoir.

9.B TO CONSTRUCT A STRUCTURAL OR TECTONIC MAP OF AN AREA

The following procedure is appropriate for most areas:

1. Study the map to decide which features to illustrate.
2. Select an appropriate scale (see Sect. 9.A.2).
3. Using a distinctive colour or line-symbol, delineate major faults and label their directions of movement (Sect. 4.A, 4.C.2, 4.G).

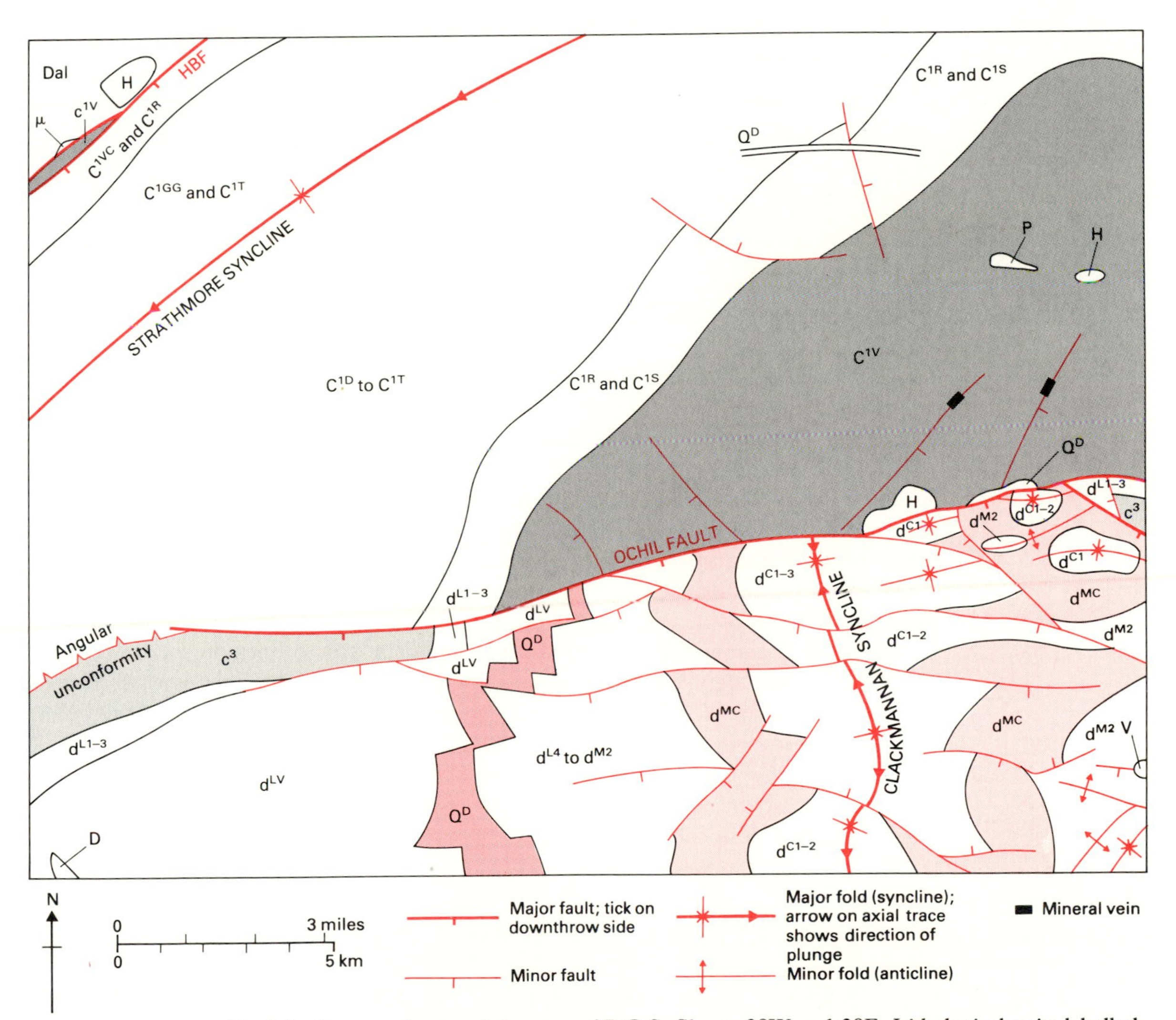

Fig. 9.5 Structural map of the area of B.G.S. Sheets 39W and 39E. Lithological units labelled as in Table 9.1. The map shows the different structural styles of the areas each side of the Ochil Fault – the northeast–southwest Strathmore Syncline with predominantly northwest–southeast faulting on the north side and the north–south Clackmannan Syncline with predominantly east–west faulting on the south side.

4. Using another distinctive colour or line-symbol, delineate unconformities (Sect. 6.A).
5. Using both the stratigraphical succession and dip information on the original map determine the positions of fold axial traces and delineate them distinctively (Sect. 5.C.1). Determine and label directions of plunge (Sect. 5.C.3).
6. Select igneous intrusions and add them to the map, simplifying the outcrop shapes (Sect. 3.D).
7. Indicate the positions of mineral veins or any other special rock types.
8. Mark the boundaries of stratigraphical systems.
9. Select any additional stratigraphical boundaries that illustrate the effects of folds, faults, unconformities, and intrusions and add them to the map, simplifying their outcrop shapes.
10. Label the lithological units and the structures and/or make an explanatory legend. Add a scale-bar and north arrow.

Fig. 9.5

9.C TO CONSTRUCT A VERTICAL CROSS-SECTION THROUGH AN AREA

The construction of cross-sections through individual structures has been described in Sections 4.D and 5.B. Here we shall describe the integration of all the constructions to produce an interpretation of the structure of a whole area.

The vertical and horizontal scales of the cross-section should be the same, so that observed or calculated dips of lithological boundaries and thicknesses of lithological units can be plotted directly on the section, and so that the completed section is a true-scale representation of the geology (Sect. 5.B.2).

As the construction of the cross-section progresses a clearer appreciation of the three-dimensional structure of the area begins to emerge. In the light of the growing knowledge it is commonly necessary to make adjustments to lines drawn at an earlier stage; in general, start with lightly-drawn constructions so that amendments are easy to make. It is usually convenient to establish most of the near-surface geology before extending the construction to deeper levels.

The following sequence of operations is suitable for the construction of most sections, but for the reasons given above it is best to keep a flexible approach rather than to apply a rigid routine.

9.C.1 SELECT A SUITABLE LINE ACROSS THE MAP (Sect. 5.B.1)

9.C.2 CONSTRUCT THE TOPOGRAPHICAL PROFILE ALONG THE LINE (Appendix 3)

9.C.3 MARK ON THE TOPOGRAPHICAL PROFILE THE OUTCROPS OF LITHOLOGICAL BOUNDARIES AND FAULTS

9.C.4 MARK THE LOCATIONS OF SUPERFICIAL DEPOSITS (Ch. 7)

Note that many superficial deposits are very shallow (only a few tens of metres) and can frequently be omitted, or their presence indicated by a symbol placed above the representation of the solid geology.

9.C.5 CONSTRUCT LINES TO REPRESENT FAULTS (Sect. 4.D)

Initially extend each fault line to only a shallow depth (say, a few hundred metres on the scale of the map) as it may later be found that the fault dies out, changes direction in depth, or is intersected by another structure.

9.C.6 CONSTRUCT LINES TO REPRESENT DISCORDANT IGNEOUS INTRUSIONS (Sect. 3.D)

If the intrusion is known or can be reasonably inferred to have vertical boundaries (e.g. many dykes, vents, and plutons), these can be constructed immediately and extended (as with faults, see above) to a small depth. Intrusions of unknown shape may have to be left till a later stage of construction; sills are best constructed at the same time as the lithological units they intrude (see Sect. 9.C.8), bearing in mind that they may change from one stratigraphical level to another (see Sect. 3.D, 9.C.10).

9.C.7 CONSTRUCT LINES TO REPRESENT UNCONFORMITIES (Sect. 6.A)

If the dip of the overlying lithological units is known or can be assumed to be the same as that of the unconformity, construct a short line at the angle of dip; in further constructions (Sect. 9.C.8) use the same procedure as for the *overlying* units. Use a distinctive line-symbol to represent the unconformity.

If the unconformity marks a buried landscape (Sect. 6.B.1) or if there is overlap (onlap) (Sect. 6.D), construct a short line at the same dip as the overlying units, but defer further construction until more is established about the geometry of the unconformity.

9.C.8 CONSTRUCT LINES TO REPRESENT STRATIGRAPHICAL BOUNDARIES

This is the main part of the construction of the cross-section and follows the procedure set out in Section 5.B, which is briefly summarised below:

1. Mark on the topographical profile the positions of observed dips, with their directions and amounts.
2. Construct dip-lines at the angle of dip and guide-lines perpendicular to the dip. If faults, unconformities, or discordant igneous rocks are present, guide-lines must terminate at the point where they intersect such structural discontinuities.
3. Mark the positions of lithological boundaries on the topographical profile.
4. Mark the positions of fold axial traces (Sect. 5.C.1).
5. Construct lines to represent the lithological boundaries, parallel to the dip-lines.

6. Check the thicknesses of lithological units, as revealed by the section at this stage, by comparison with other data on the map.
7. Make any necessary adjustments to the topographical profile or to the locations of changes of dip relative to the positions of guide-lines.

9.C.9 INTEGRATE THE SEPARATE STRUCTURES

By this stage the separate parts of the cross-section are beginning to impinge upon each other. For the completion of the near-surface interpretation and extension in depth proceed as follows:

1. Determine from the surface outcrop patterns the displacement of each fault (Sect. 4.F, 4.H). Construct the lithological boundaries that are cut by a fault so that they show the same displacement.
2. Determine the probable sub-crop pattern of the older rocks below an unconformity (Sect. 6.C.3). If possible use accurate constructions to determine the positions and dips of the lithological boundaries of the older units on the plane of the section (e.g. Sect. 9.D). If accurate construction is not possible, indicate the probable attitude of the older rocks schematically, for example with a dashed line.
3. Determine from the surface outcrop patterns the ages of igneous intrusions relative to faults, folds, and unconformities (Sect. 3.K). Construct the cross-section so that intersections of igneous bodies with these structures show the correct relative ages.

9.C.10 INSPECT THE MAP FOR ADDITIONAL INFORMATION

Examine the areas of the map adjacent to the line of section to see whether there are lithological units or structures which are covered over by superficial deposits or which are laterally discontinuous (lenticular sedimentary units, igneous intrusions, mineral veins, etc.):

(a) if they appear on the line of section they may not continue in depth;
(b) if they do not appear on the line of section at the surface they may be present at depth.

Examine the whole map (including any accompanying descriptive material) for additional information about lithological units or structures to which statements (a) or (b) above may apply (transgressive sills, concealed igneous intrusions and their metamorphic aureoles, lithological units below an unconformity or below the oldest rocks on the line of section, etc.). Make use of data (if available) from boreholes, underground workings, or geophysical mapping.

In all cases representation of such lithological units or structures may have to be schematic and conjectural (and indicated accordingly), but their recognition is essential to the interpretation of the cross-section.

9.C.11 CONTINUE THE CONSTRUCTION TO GREATER DEPTH

The construction of the cross-section should be continued in depth (and it can also be extended above the present-day land surface using the same geometrical principles).

The limit to which the construction should be carried is usually set by one of the following criteria:

(a) Economic objectives have been attained (e.g. the determination of the configuration of mineral-bearing rock units or of potential hydrocarbon reservoir rocks).
(b) The cross-section adequately illustrates the lithology and structure of the area.
(c) The limit of reasonable extrapolation of the surface information has been reached.

It sometimes happens that (c) intervenes before objectives (a) or (b) are attained. The probable or possible deeper extension of the stratigraphy and structure should then be indicated by the use of dashed lines, question marks, or other symbols to indicate the lower certainty of the relevant parts of the cross-section. In some parts of the section there may be no immediate way of determining what lies below the surface rocks and structures – for example, below an unconformity or a thrust, or in the core of an antiform, or the deeper extension of an igneous intrusion. Nonetheless, information should be sought from elsewhere on the map, from adjacent or smaller-scale maps, or from regional stratigraphical or structural studies in order to make the interpretation of the section as complete as possible.

9.C.12 CHECK THE WHOLE CROSS-SECTION FOR GEOLOGICAL CONSISTENCY AND PLAUSIBILITY

Up to this stage the interpretation of the cross-section has been based on geometrical construction from the data available on the map. In some cases, however, rigorous geometrical principles may conflict with geological realities. The following questions are useful for calling attention to particular aspects of the cross-section – they are based in part on the concept of reversing structural processes to restore sedimentary rocks to their original orientations as depositional layers (cf. Sect. 6.G):

1. Do the thicknesses of lithological units match the thicknesses determined from the map or shown in stratigraphical sections?
2. Are the thicknesses of stratigraphical units constant across the section (most sedimentary rock units), or do they vary in a manner consistent with their composition (cf. Ch. 3), both in total thickness and in the angle between top and bottom surfaces?
3. Are the thicknesses of a rock unit each side of a fault the same? Do its geological boundaries make the same angle with the fault plane? If the original positions of the rocks each side of the fault were restored by reversing the movement of the fault, would the lithology and structure be continuous from one side to the other?
4. Is there a topological similarity between the rock relationships shown on the map and on the cross-section? As examples: the shape of a fold (angular, gently curved, box-shaped, etc.) is usually comparable on the horizontal surface of the map and in the vertical cross-section – cf. Section 5.C.3; the map and the cross-section of an unconformity are often directly comparable – cf. Fig. 6.4.
5. Are the rock relationships consistent with what is known of the geological history of the area (to be determined in more detail as described in Section 10.A)?

Make any necessary adjustments to the section in the light of these questions.

9.C.13 COMPLETION OF THE CROSS-SECTION

Ink over the final version of the section. Erase (or make fainter) all construction lines and temporary marks and comments. Add a scale. Label all lithological units and structures, geographical locations, and the locations of the ends of the section. If necessary, make an explanatory key of all the symbols and abbreviations used. The appearance and intelligibility of the section is often improved if one or two significant lithological units are coloured or shaded so that the interpretation of the geological structure and the relationships of the different parts of the section are made as clear as possible.

Figs 9.6 and 9.7

It may be necessary to construct more than one cross-section in order to illustrate and understand the three-dimensional geometry of an area.

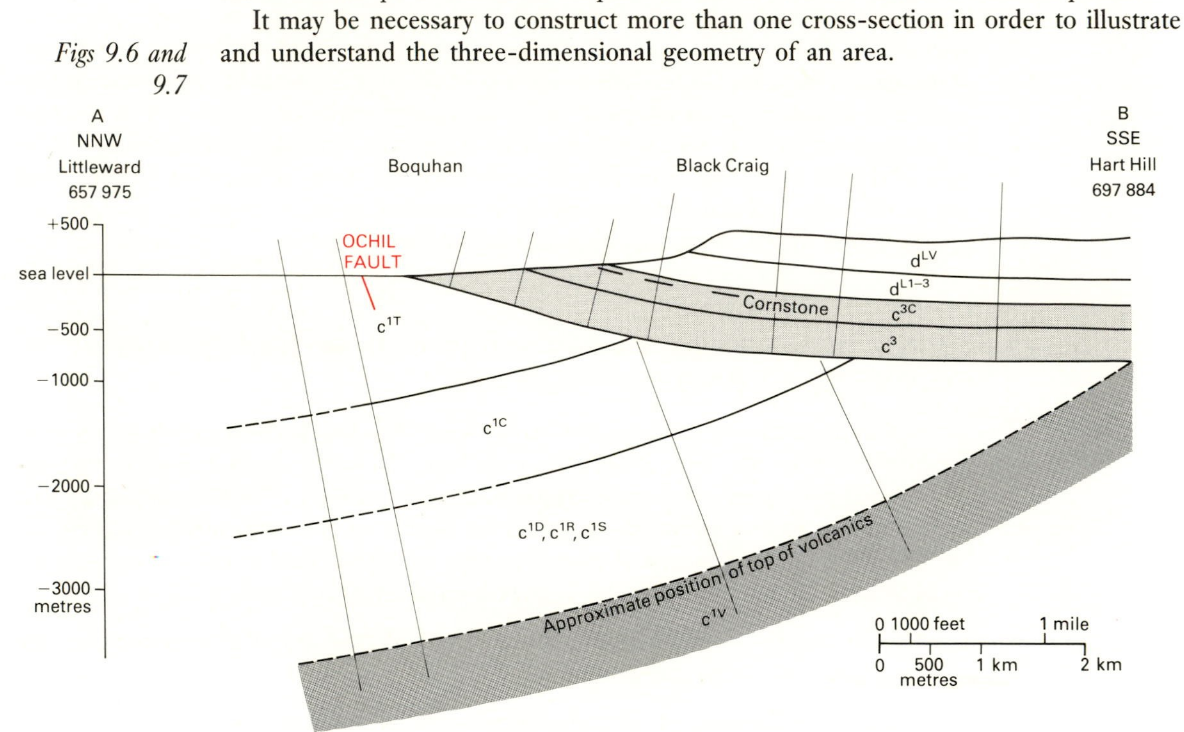

Fig. 9.6 Vertical cross-section along the line from Grid Reference 657 975 (Littleward) to 697 884 (Hart Hill) on Plate 3, showing the stratigraphy and structure of the Old Red Sandstone and Carboniferous rocks. Units labelled as in Table 9.1. Some guide-lines for dips have been retained.

The following observations were taken into consideration in constructing the section. The reader is recommended to review the evidence and to construct his or her own cross-section.

1. Dips in the Upper Old Red Sandstone and Carboniferous are consistent over wide areas and can be interpolated where necessary onto the line of section.
2. The Upper Old Red Sandstone and Calciferous Sandstone Measures thin towards the east (Fig. 9.4).
3. There are intermittent outcrops of cornstone in the Upper Old Red Sandstone.
4. There are numerous small dykes, but their orientation and distribution are too irregular to allow meaningful interpolation onto the section.
5. The probable subcrop of the Lower Old Red Sandstone rocks below the unconformity can be determined by construction of a structure contour map (Sect. 9.D, Fig. 9.8).
6. The probable thicknesses of lithological units of the Lower Old Red Sandstone can be determined by extrapolation from measurements of their exposed thicknesses (Fig. 9.4).
7. The possible structure of the unexposed Lower Old Red Sandstone units can be guessed from observations of dips and their relative steepness in the area north of the Ochil Fault.
8. The section crosses the Ochil Fault close to its west end.

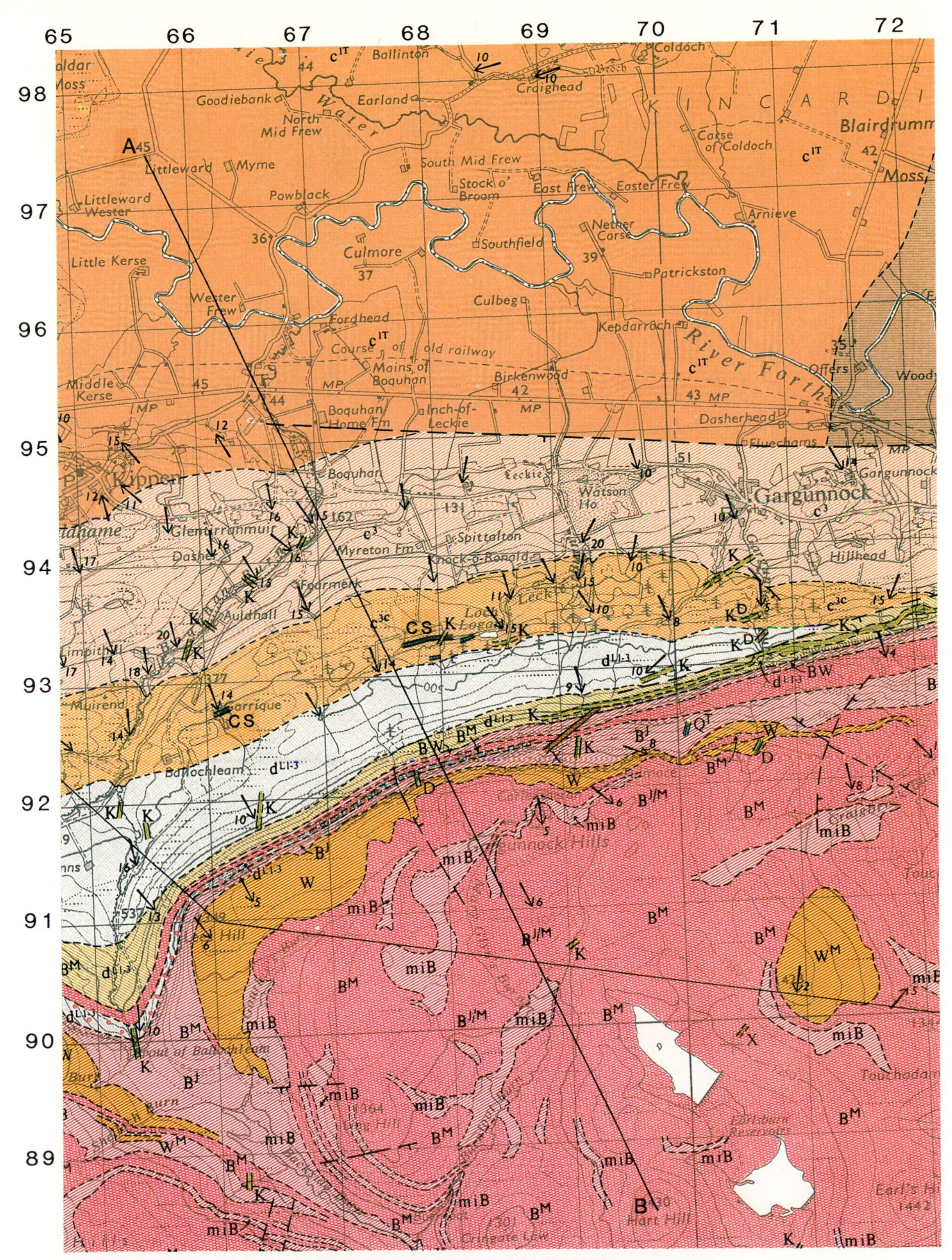

Plate 3. Extract from B.G.S. Sheet 39W (Stirling). Scale 1 : 50 000. Caption on page xi.

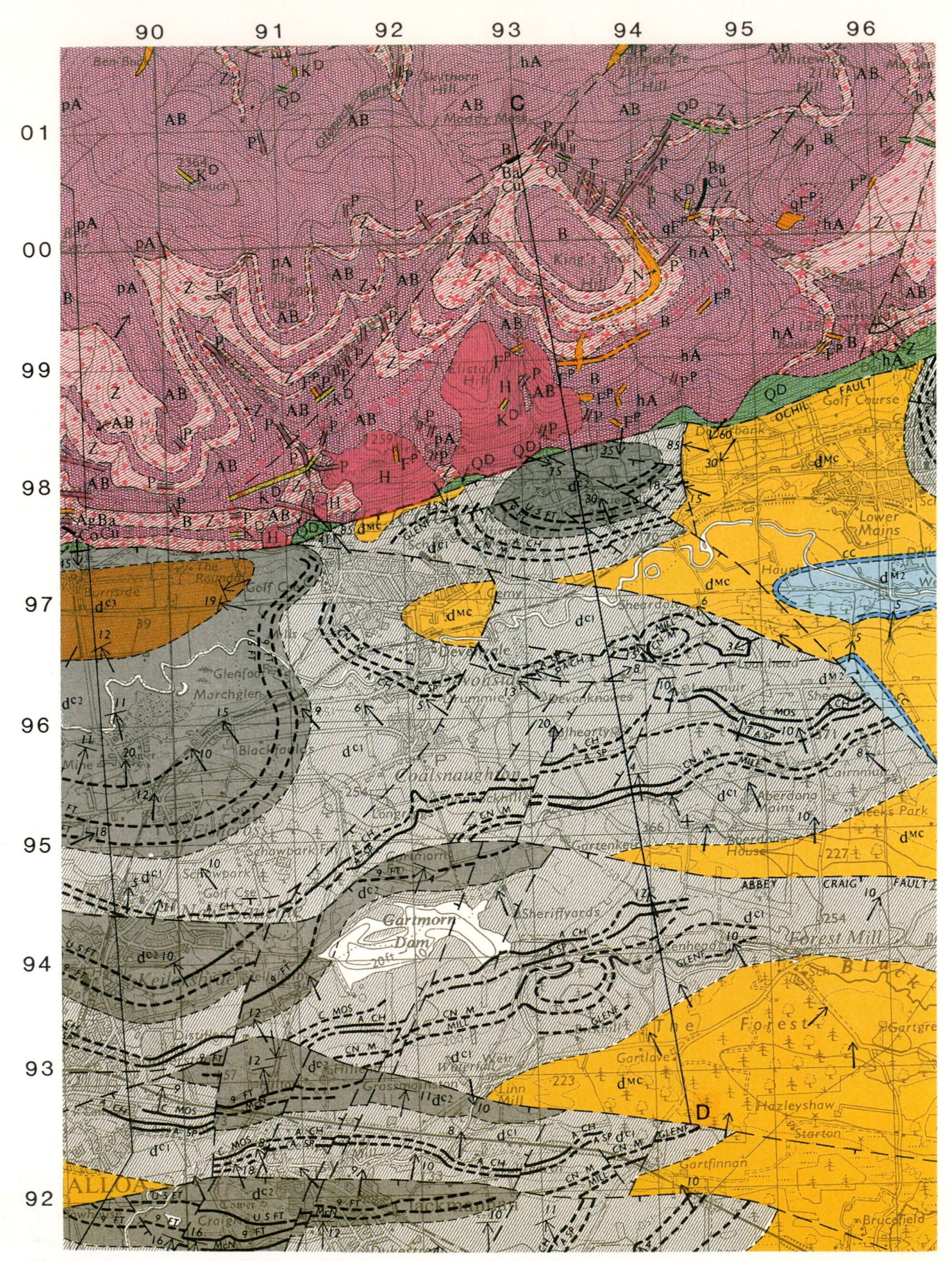

Plate 4. Extract from B.G.S. Sheet 39E (Alloa). Scale 1 : 50 000. Caption on page xi.

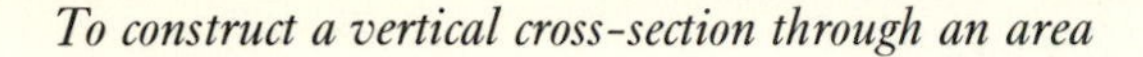

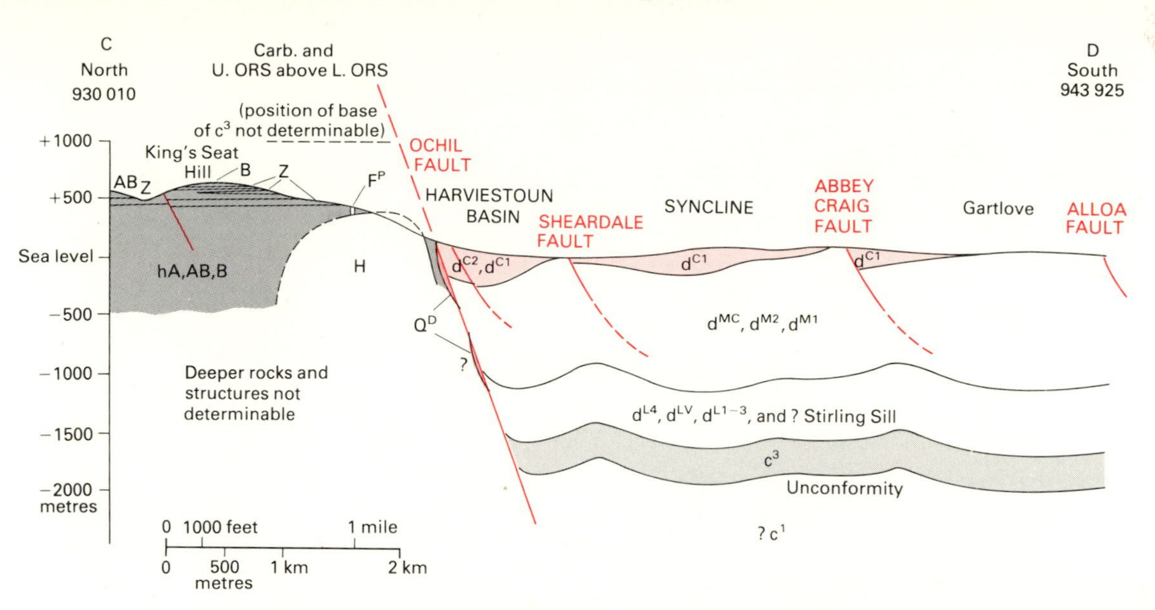

Fig. 9.7 Vertical cross-section along the line from Grid Reference 930 010 (near Maddy Moss) to 943 925 (the fault near Gartfinnan) on Plate 4, showing an interpretation of the stratigraphy and structure of the Old Red Sandstone and Carboniferous rocks. Units labelled as in Table 9.1.

The following observations were taken into account in constructing the section. The reader will see that the interpretation shown in the section is necessarily more speculative than that in Fig. 9.6.

1. The dip of the Lower Old Red Sandstone volcanics can be determined by measurement from structure contours on King's Seat Hill (9399 and 9300).
2. The shape of the diorite (H) is suggested by the shape of outcrop of its metamorphic aureole.
3. The relative ages of the porphyry dyke (F^P) and the diorite (H) are indicated by their outcrop relationships, though the evidence is equivocal.
4. The base of the Upper Old Red Sandstone outcrops on the north side of the Ochil Fault 8–10 km east of the east edge of Sheet 39E (on the adjoining Sheet 40, Kinross). It must rise over the highest parts of the Ochil Hills (up to 720 m above sea-level).
5. The Ochil Fault dips at about 70° to the south (data from the Sheet memoir, Francis *et al.* 1970, p. 248), and is steeper than the dip of the lower contact of the Q^D intrusion, which is determinable at Bank Hill (9498 and 9598). Other faults dip at about 60° (based on Dinham & Haldane 1932, p. 174).
6. The displacements of the fault at 935 981 (see Plate 4), the Sheardale Fault, and the Abbey Craig Fault can be approximately determined from the stratigraphical displacements at their outcrops. Other faults have displacements which are too small to show on the cross-section.
7. The location of the axis of the syncline near 938 961 can be determined from the outcrops of the A SP and A CH coal seams.
8. Dips in the region of this syncline are probably local, due to minor folding, and not representative of the structure as a whole. Stratigraphical thicknesses of units can be used to determine the shape of the fold.
9. The deeper extrapolation of the Carboniferous and Upper Old Red Sandstone rocks can be determined from inferred thickness variations (Francis *et al.* 1970, p. 247; Fig. 9.4). Guide-lines and dip-lines can be used only for the near-surface structure; the positions of deeper boundaries must be constructed by measurement of thicknesses of lithological units. Thicknesses have been assumed to vary linearly with distance.
10. The shape and the downward extent of the faults in the Carboniferous rocks are based on the assumption that they are listric extensional faults (cf. Fig. 4.3).

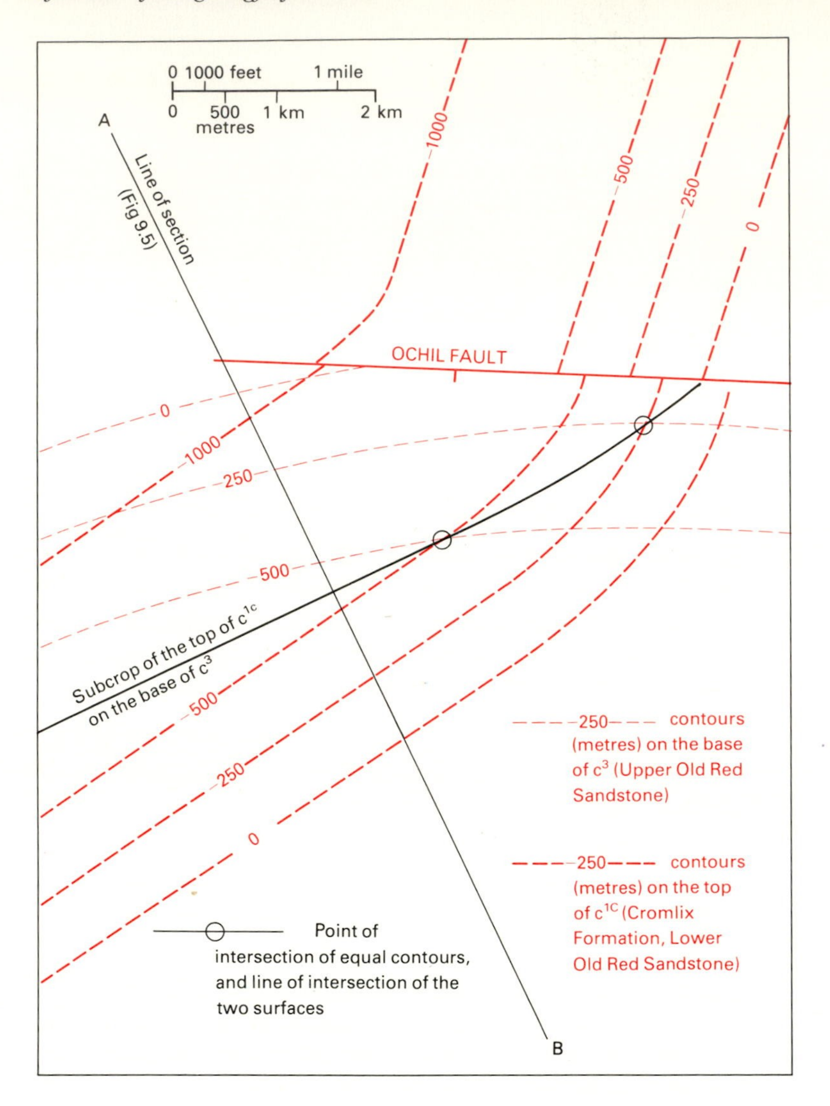

Fig. 9.8 Structure contour map of the area of Plate 3, showing contours on the top of c^{1C} (Cromlix Formation, Lower Old Red Sandstone) and on the base of c^3 (Upper Old Red Sandstone). The top of c^{1C} dips to the west and northwest; the base of c^3 dips to the south. The contours of the two surfaces are shown as intersecting, so that the subcrop of the top of c^{1C} on the angular unconformity at the base of c^3 can be located; in fact the top of c^{1C} was eroded away, to the south of the line of its subcrop; so also were progressively lower levels of the Lower Old Red Sandstone in areas further south (cf. Fig. 9.6).

The following observations and assumptions were taken into consideration in constructing the map. The reader is recommended to review the evidence and to construct his or her own structure contour map.

1. Thicknesses of lithological units vary as shown in Fig. 9.4.
2. Northwestward dips in the area of Kippen (652 948) influence the southward continuation of the Lower Old Red Sandstone units below the unconformity. The location of the change of strike of the Lower Old Red Sandstone units is constrained by these northwestward dips and by the NNE–SSW strike of the outcrops north of the Ochil Fault.
3. The Ochil Fault displaces the rocks of the Lower and Upper Old Red Sandstone.

9.D TO CONSTRUCT A STRUCTURE CONTOUR MAP (see also Sect. 2.H)

1. Select a lithological boundary of interest whose depth (elevation) can be determined over a wide area of the map.
2. Draw accurate cross-sections (Sect. 9.C) to determine the shape of the lithological boundary. Alternatively, provided the dips are uniform over wide areas, the *depth* (Sect. 2.C) of a lithological boundary can be determined from stratigraphical thicknesses.
3. Mark the positions on each cross-section where the lithological boundary is at particular depths (e.g. every 100 m, 500 m, or 1000 m below sea-level).
4. Construct a skeleton map of the area at a suitable scale. Mark on it the locations of the determinations of depth from stages 2 and 3.
5. Construct smooth contour lines to achieve a best fit among the locations of known depth. Take into account the positions of faults and axial traces of folds that interrupt or reverse the contour pattern. Similarly, show the effects of igneous intrusions or unconformities that locally remove the lithological boundary by stoping or by erosion.

The same procedure can be applied to determine elevations above ground level – that is, to determine the presumed position of the lithological boundary in the past before the rocks were removed by erosion.

Fig. 9.8

9.E ASSESSMENT OF MINERAL RESOURCES

Mineral resources occur in a wide variety of forms (Table 3.4). Different approaches to their assessment are needed for each raw material (coal, oil and gas, metalliferous ores, industrial minerals, building stone, water) and each kind of geological situation (regular or irregular-shaped bodies of rock, massive or disseminated ores, high- or low-value materials, etc.). Many factors – geological, economic, social, political, and environmental – determine whether a particular resource is of economic value (Link 1982; Ward 1984; Selley 1985a; Hobson & Tiratsoo 1985; North 1985).

There is only space here to give a very simple and generalised description of the geological factors that are used in the assessment of mineral resources. These are:

(a) The size and shape of the resource.
(b) The depth of the resource in relation to extraction by quarrying, mining, or drilling.
(c) The richness of the resource (grade of metalliferous ores, thickness of coal seams and other bedded ores, probability of occurrence of hydrocarbons, etc.).
(d) The volume or tonnage of the resource that can be extracted.

1. Use vertical cross-sections (Sect. 9.C), thickness contour maps (Sect. 2.E), structure contour maps (Sect. 9.D), and any other information to determine the three-dimensional shape of the resource of interest. Record the information on a map or cross-sections as appropriate.

2. Demarcate areas where the resource is absent because of erosion, faulting, unconformities, or igneous intrusions.
3. Apply any known constraints (e.g. maximum workable depth, geographical boundaries of permitted operations, etc.) to determine the limits to which the resource can be extracted.
4. Determine the volume of the resource and its probable value.

Figs. 9.9 and 9.10

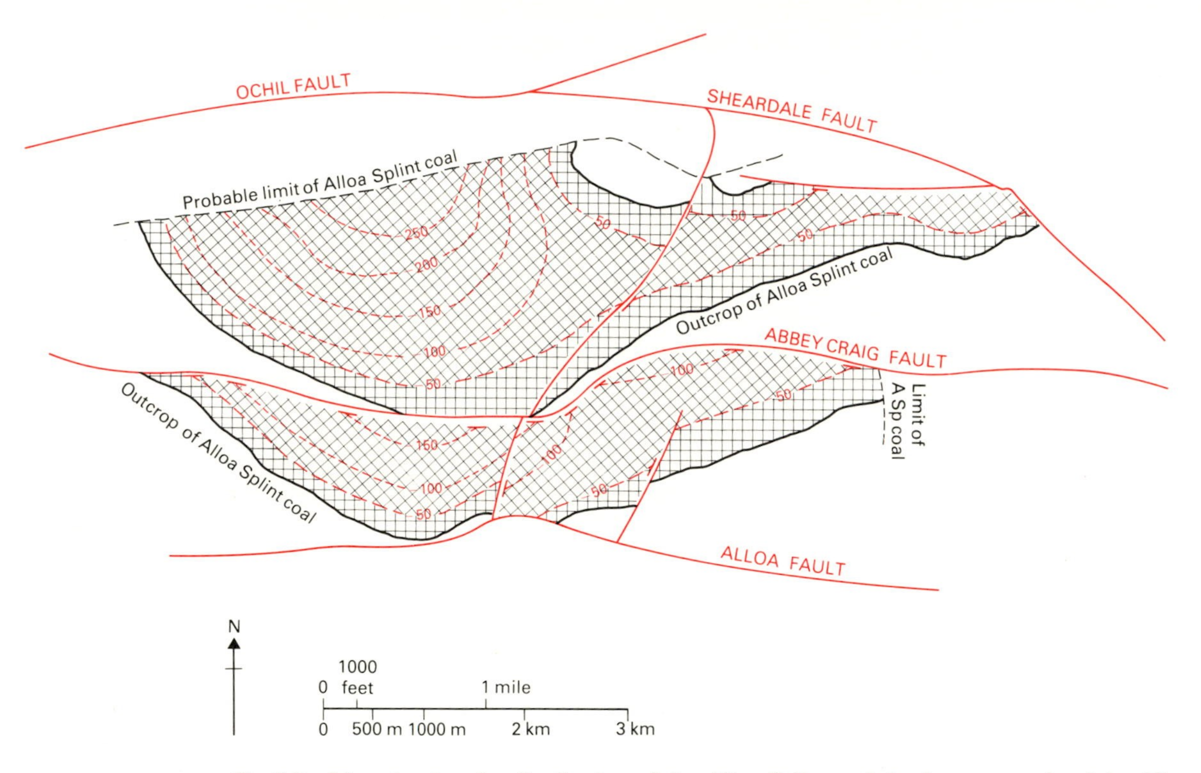

Fig. 9.9 Map showing the distribution of the Alloa Splint coal, in the area north of the Alloa Fault, derived from B.G.S. Sheet 39E (Alloa). Solid line shows the outcrop of the coal, dashed line shows its probable northern and eastern limits. The map has been slightly simplified by omitting faults with throws of less than about ten metres.

Contours are *depths (in metres) below ground level* (not structure contours) calculated from cross-sections (Fig. 9.7 and Section B – B on the published map) and by estimating the depth of the coal below other lithological boundaries and coal seams using the generalised stratigraphical section in the margin of the published map. Short lines adjacent to faults mark the subsurface positions where they intersect the coal seam (assuming that the dip of the fault planes is 60°).

The diagonal shaded areas show where the Alloa Splint coal is at a depth greater than 50 metres, and therefore accessible by deep-mining methods. The area occupied by the coal, determined by counting the small squares (100 metres each side), is 12.4×10^6 m^2; the actual area of the coal seam is slightly greater because of the dip (10–15° over most of the area). Given that the thickness of the coal is 0.76 to 0.84 m, and assuming that its density is 1.4×10^3 kg m^{-3}, the total weight of the coal below 50 m from the surface is approximately 13.9 million tonnes. About 40 per cent of this would have been recoverable.

In a similar way the amount of coal within 50 m of the surface (shown by horizontal/vertical squares) can be estimated as 6.0×10^6 tonnes, accessible by open-cast methods, with a recoverability of about 90 per cent.

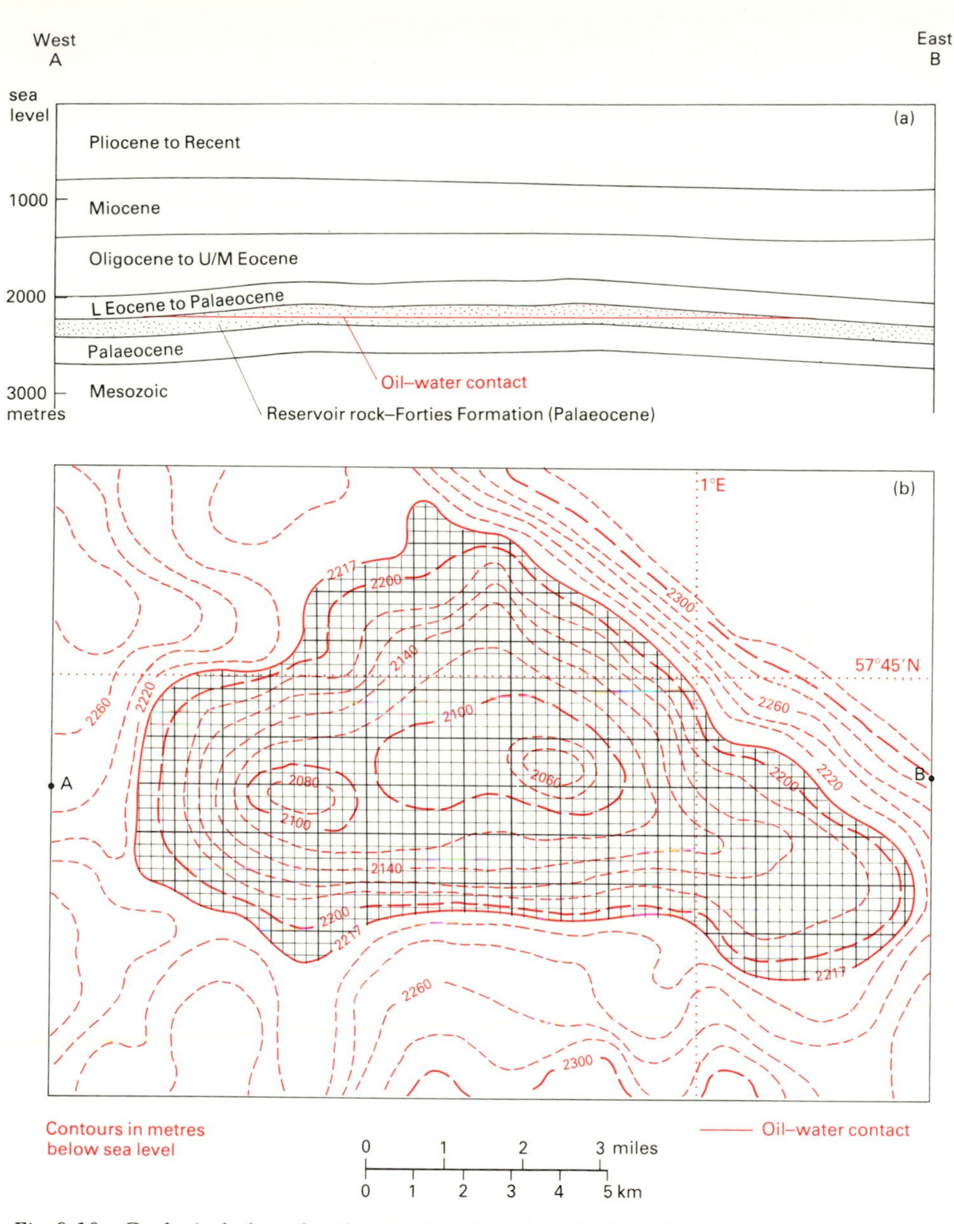

Fig. 9.10 Geological data for the Forties oil-field, Northern North Sea, 150 km ENE of Aberdeen, Scotland, from P. J. Walmsley, Ch. 37, 'The Forties Field', in A. W. Woodland (ed.). 1975, *Petroleum and the Continental Shelf of North-West Europe*, John Wiley & Sons, New York.

(A) East–west cross-section through the field, showing the reservoir rock – the Forties Formation (Palaeocene) – and the oil–water contact.

(B) Structure contour map on the top of the reservoir horizon; the contours are the depths in metres below sea-level. The superimposed grid shows 1 km and 0.25 km squares.

The geology of the oil-field is more complex than is shown here, in particular because the reservoir rock is composed of several sandstone units (with intervening shales) which are both laterally and vertically discontinuous. Nonetheless the size of the oil-field can be simply and meaningfully estimated as follows:

Determine the area enclosed by each contour by the method of counting squares. Calculate the volume of each horizontal layer defined by adjacent contours by taking the mean area and

(*Caption continues on p. 186*)

(*Caption of Fig. 9.10 cont.*)

multiplying by the contour interval. Sum the volumes of the layers to obtain the total volume of reservoir rock above the oil–water contact:

For example, the areas enclosed by the -2140 m and -2160 m contours are 35.5 km^2 and 46.6 km^2 respectively. The volume between the contours is then $(35.5 + 46.6)/2\ km^2 \times 20\ m = 822 \times 10^6\ m^3$. The total volume of the reservoir rock above the oil–water contact is $5820 \times 10^6\ m^3$.

Given that 60 per cent of the reservoir horizon in the Forties field is sandstone, with average porosity 27.5 per cent, filled with 70 per cent hydrocarbon + 30 per cent water, the volume of hydrocarbon-in-place is calculated as $5820 \times 10^6 \times 0.6 \times 0.275 \times 0.7 = 672 \times 10^6\ m^3$ or 4230 million barrels. With existing technology about 40 per cent or 1690 million barrels would be recoverable.

10 SYNTHESIS OF THE GEOLOGY OF AN AREA: 2. GEOLOGICAL INTERPRETATION

A geological map is a cartographical representation of real rocks; its geological interpretation must relate to the geological environments and processes that created the rocks and structures that are now seen on the surface of the Earth and are recorded on the map in symbolic form. Geometrical construction governed by geological principles (Ch. 9), is a necessary preliminary to understanding the geology, but creates an essentially static view of the present-day three-dimensional shapes of the rocks and structures. In order to convert the geometry into a dynamic geological interpretation – to make the map 'come alive' – it is necessary to consider the lithologies of the rocks, the shapes, orientations, and relationships of the structures, and the dates and intervals of geological time during which the rocks were formed and deformed. These concepts were introduced in earlier chapters in relation to individual structures; we must now show how they can be integrated to produce a dynamic account of the evolution of an area of varied rocks and structures produced over an extended period of geological time.

As in any scientific investigation, it is necessary to work towards the most complete understanding that is possible *within the limits of the evidence that is available.* The detail and depth of interpretation of a map depends on the amount of information on the map, together with any additional material published in map explanations, memoirs, or other reports. Sometimes the evidence is insufficient for a complete and precise interpretation – there may be several possible interpretations of a particular lithology or structure, or there may be no rocks or structures that relate to the development of the area during part of its history. In such cases the interpretation of the map must set out the alternatives, or if necessary state that the evidence does not allow a complete interpretation.

There are four kinds of information contained in each rock unit or structure (fold or fault) on a geological map:

1. The lithology (composition) of the unit or the geometry of the structure.
2. The spatial relationships of the unit or structure to contemporaneous units and structures elsewhere in the area.
3. The dimensions (thickness, width, and volume) of the unit or the amount of movement produced by the structure.
4. The time interval during which the rock unit or structure was formed.

1 and 2 provide evidence about the environment of formation of the rock unit or structure (sedimentary facies, igneous province, or conditions of metamorphism or of deformation). 4 defines the length of time during which the conditions identified by

(*Text continues on p. 194*)

Table 10.1 Characteristics of principal plate tectonic settings

This table presents an outline of the characteristic lithological and structural features of the principal plate tectonic settings. Like all the similar tables in this book, it is a summary of the common features of individual instances of each setting, and is therefore necessarily a simplification and generalisation of complex and variable geological situations. In attempting to identify the former plate environments of ancient rocks geologists make use of all the available data, including geophysical, geochemical, geochronological, and palaeontological evidence which is not usually recorded on geological maps. Table 10.1 provides a guide to possible settings; identification of plate environments from the evidence only of rock-types and structures recorded on maps must be regarded as provisional and subject to confirmation with additional evidence.

Identification of former plate environments is made more complex by the relatively rapid rates of plate movements. Rocks formed in one environment (e.g. a forearc) at one time in the geological history of an area may subsequently be incorporated into another environment (e.g. a continental collision zone) a few tens of millions of years later. It is necessary to study the rocks of an area stage by stage in order to recognise possible successive changes of environment.

	Oceanic plate	*Forearc*	*Island arc and active continental margin*
Sediments (see Table 2.1)	Thin pelagic sediments (mudstones, turbidites, cherts, calcareous rocks, evaporites, metalliferous sediments)	Thin pelagic sediments from the oceanic plate, and thick clastic shelf and oceanic sediments from the adjacent island arc or active continental margin. Sedimentary and tectonic mélanges. Lateral and vertical variation of sedimentary facies, with unconformities	Thick clastic shelf and oceanic sediments and/or volcaniclastic sediments. Lateral and vertical variations of sedimentary facies, with unconformities
Igneous rocks (see Table 3.2)	Ultrabasic and tholeiitic basic rocks of the ophiolite sequence: peridotite (or serpentinite), gabbro (commonly layered), sheeted dolerite dykes, basalt lavas (in part pillowed; spilites). Na-rich intermediate and acid igneous rocks. Tholeiitic and alkalic shield volcanoes at hot-spots	Generally little igneous activity, but ophiolite rocks from oceanic crust may be tectonically emplaced; calc-alkaline and other volcanic rocks from the adjacent island arc or active continental margin may extend to the forearc	Abundant calc-alkaline and tholeiitic pyroclastics and lavas (basalts, andesites, dacites, and rhyolites) – may be submarine in island arcs. Erosion to deeper levels reveals calc-alkaline stocks and batholiths, with diorites and granodiorites predominant, gabbros and granites less common

Continental collision zone	*Transform plate margin*	*Passive continental margin*	*Continental plate (stable craton)*
Sediments of island-arc or active-continental-margin type (q.v.), with or without sediments of passive-continental-margin type (q.v.). After collision, subsiding basins filled with thick continental or marine sediments may form adjacent to the collision zone. Lateral and vertical variations of sedimentary facies, with unconformities	Sedimentary sequences of variable facies and variable (but usually great) thickness in fault-bounded basins. Rapid lateral and vertical variations of facies, with unconformities	Coastal and shelf sediments deposited in block-faulted basins, with a general ocean-ward trend from shallow to deep marine facies. Pelagic sediments on adjacent ocean floor	High-grade metamorphic rocks, mostly of Precambrian age, either exposed or with an unconformable cover of a thin sequence of predominantly shelf or continental sediments. Thick continental or coastal sediments in rift valleys. Crustal extension to produce 'failed rifts' (aulacogens) may create downwarping basins with thick continental or marine sediments unconformably overlying block-faulted older rocks
Calc-alkaline lavas and pyroclastics (andesites, dacites, and rhyolites). Erosion to deeper levels reveals calc-alkaline batholiths, with granites predominant, diorites and granodiorites less common	Sparse calc-alkaline or alkalic extrusive and intrusive rocks	Tholeiitic or alkalic basalt lavas, dolerite dyke swarms and sills, layered and ring intrusions. Ophiolite sequence in adjacent ocean floor	Granitic gneisses and greenstone belts (Archaean) Anorthosites (Proterozoic) Alkalic and potassic extrusive and intrusive rocks; kimberlites and carbonatites. Tholeiitic and alkalic flood basalts associated with hot-spots and early stages of continental rifting

(*Table 10.1 cont.*)

	Oceanic plate	*Forearc*	*Island arc and active continental margin*
Mineralisation (see Table 3.4); coal and hydrocarbons	Chromitite and cupriferous pyrite (Cyprus-type ores) in ophiolite rocks; metalliferous sediments and Mn nodules in pelagic sediments	Generally little mineralisation and low potential for hydrocarbons	Massive stratiform Zn-Pb-Cu ores in rhyolitic pyroclastics (Kuroko ores) or porphyry Cu and Mo ores in calc-alkaline intrusive complexes. Sn, W in acid volcanics and intrusive rocks. Au and Ag in igneous rocks. Au, Sn in fluviatile placer deposits. Hydrocarbons in fold belts and basins adjacent to the island arc or active continental margin
Structural style (see sections 4.K, 5.I)	Simple extensional structures in oceanic crust as initially formed; may become structurally complex as a result of thrusting (obduction) over another plate in a convergent plate margin	Compressional structures, with thrusting and crustal shortening. Oblique subduction may produce strike-slip movement. Accretionary prism, with distinct sedimentary sequences now juxtaposed, formed by repeated thrusting as sediments are scraped off a subducting oceanic plate	Compressional structures with crustal shortening – folds, nappes, and thrusts; intense deformation in axis of the arc or active margin, diminishing towards the overriding plate. Commonly a major fault (boundary fault) between the arc or active margin and the forearc. Older continental crustal rocks ('basement') may become involved in the deformation

Continental collision zone	*Transform plate margin*	*Passive continental margin*	*Continental plate (stable craton)*
Hydrothermal Sn-W-U mineralisation in granitic rocks. Hydrocarbons in fold belts and basins adjacent to the collision zone	Sparse mineralisation. Hydrocarbons in stratigraphical traps and faulted structures	Stratabound Pb-Zn (Mississippi-Valley-type) mineral deposits in calcareous rocks. Phosphorites, evaporites. Banded iron formations (Precambrian). Sedimentary ironstones (Phanerozoic). Deposition of future coal beds on continental margins. Hydrocarbons in block-faulted structures and basins	A. In sedimentary cover and younger igneous rocks: Chromite, Ni, Pt in basic/ultrabasic layered intrusions. Nb, Ta, and phosphates in carbonatites and nepheline syenite complexes. Diamond in kimberlites. Stratiform massive sulphide ores of Cu, Pb, Zn, Ag. Stratabound ores of U, V, Cu. Diamonds, Au, U, Sn in placers. Bauxite. Deposition of future coal beds on continental margins and inland basins. Hydrocarbons in sediments of rifts and aulacogens. B. In basement rocks: Fe, Ni, Au, base metal sulphides, U.
Compressional structures with crustal shortening – folds, nappes, and thrusts; decrease in structural complexity away from the collision zone. The original boundary between the two formerly separated continents may be marked by a suture zone – a major fault zone with tectonic slices of oceanic crustal rocks	Contemporaneous extensional and compressional structures in adjacent areas. Strike-slip faulting and *en échelon* folds	Simple extensional structures with block-faulting and tilting	Generally horizontal or gently-dipping sediments and/or flood basalts unconformably overlying intensely deformed ancient schists and gneisses of the older continental crust. Extensional block-faulting in rifts and aulacogens

(*Table 10.1 cont.*)	*Oceanic plate*	*Forearc*	*Island arc and active continental margin*
Metamorphic rocks (see Table 3.3)	Non-metamorphic or zeolite, greenschist, or amphibolite facies, with hydrothermal metamorphism from circulating sea-water during initial cooling of oceanic crust. Later involvement in convergent plate margin processes may produce blueschist facies metamorphism	Non-metamorphic or blueschist facies (high-P, low-T) metamorphism, which may be paired with medium-P metamorphism in the adjacent island arc or active continental margin	Medium-P metamorphism – metaclaystones contain Al silicates and cordierite or garnet, metabasics are greenschists and amphibolites. Metamorphic zonation, with higher P and T in the axial zone of the island arc or active continental margin
Tectonic situation	Present-day ocean floor. Ancient oceanic crust may be tectonically emplaced in forearcs or active continental margins or in suture zones in continental collision zones	Linear belt at a convergent margin between an oceanic plate and either an oceanic or a continental plate. Present-day examples are on the convex (oceanic) sides of island arcs or active continental margins. Ancient examples may become incorporated in continental collision zones	Linear belt at a convergent margin between an oceanic plate and either an oceanic or a continental plate. Ancient examples may become incorporated in continental collision zones
Distinctive features	Ophiolite sequence. Metalliferous sediments	Tectonic situation between an oceanic plate and an island arc or active continental margin. Repeated thrusting of marine sediments (accretionary prism). Blueschist facies metamorphism (Classical eugeosyncline)	Calc-alkaline igneous rocks. Strongly compressional structures. Medium-P metamorphism. Kuroko and porphyry Cu ores (Classical orogenic belt)
Present-day examples	Ocean floor (60% of Earth's surface)	Mentawai islands and strait, SW of Sumatra.	Island arcs and continental margins of the Pacific Ocean
Ancient examples	Oman ophiolite (Cretaceous). Ophiolite belts of Appalachians, eastern USA and Canada (early Palaeozoic)	Upper Jurassic – Cretaceous Franciscan Complex and Great Valley Sequence of California. Ordovician – Silurian of Southern Uplands of Scotland	Mesozoic of Rocky Mountains, Canada and USA. Devonian of Sidlaw Hills, Borrowdale Volcanic Group of Lake District, Britain. Cenozoic of Japan

Continental collision zone	*Transform plate margin*	*Passive continental margin*	*Continental plate (stable craton)*
Medium-P metamorphism – metaclaystones contain Al silicates and cordierite or garnet, metabasics are greenschists and amphibolites. Metamorphic zonation, with higher P and T in the axial zone of the collision belt. Pre-orogenic continental plate (cratonic) rocks may be high-grade schists and gneisses	Little or no metamorphism of the sediments formed at the time of the transform movement. Older rocks of the continental plate may be high-grade schists and gneisses	Little or no metamorphism of the sedimentary cover. Older rocks of the continental plate may be high-grade schists and gneisses	High-grade metamorphic rocks (amphibolite and granulite facies) in the older cratonic rocks. Non-metamorphic sedimentary cover
Linear belt at a convergent margin between two continental plates. One or both of the edges of the formerly separated plates must necessarily have been an active continental margin before collision. Ancient examples now lie within continental plates and may be cut across by later passive continental margins as a result of subsequent plate movements	Linear belt (but shorter than those at convergent and divergent margins) at a transform margin between a continental plate and either a continental or an oceanic plate. Ancient examples lie at present or former continental margins which trend parallel to spreading directions of plate movement	Linear belt between continental and oceanic crust of the same plate. The continental margin and adjacent oceanic crust were created at a former divergent margin at the date of continental rifting. Ancient examples may become incorporated in continental collision zones	Interiors of continents, or peripheries of continents whose margins have been created by geologically recent spreading
Tectonic situation between two continental plates. Calc-alkaline igneous rocks. Strongly compressional structures. Medium-P metamorphism (Classical orogenic belt)	Rapid variation of sedimentary facies and thicknesses. Contemporaneous compressional and extensional structures. Strike-slip faulting	Extensional structures with block faulting and sedimentary basins (Classical miogeosyncline)	Thin or absent sedimentary cover with simple structures over high-grade (commonly Precambrian) schists and gneisses. Thick sedimentary sequences in extensional basins unconformably overlying continental crust (Classical continental shield)
Alpine – Himalayan mountain belt (Tertiary to Recent)	Sedimentary basins associated with the San Andreas Fault, California	Gulf of California, Red Sea. Atlantic margins of America, Europe, Africa	Canadian Shield, Russian and Siberian Platforms, peninsular India, Africa
Late Palaeozoic of Urals, USSR. Palaeozoic of Appalachians and Caledonides, NW Europe and eastern North America.	Late Carboniferous of Cantabrian Mountains, Spain	Upper Jurassic – Cretaceous of Mediterranean	Lewisian of Scottish Highlands

1 and 2 persisted. 3 and 4 taken together can be used to calculate the rate of formation of the unit or structure (rate of sedimentation, rate of deformation, etc.). Metamorphic rocks contain information both about the original rock-type (protolith) and its environment of formation and about the conditions of metamorphism. Unconformities are the product of several successive processes (Sect. 6.G) and chronostructural units bounded by unconformities are usually of key importance in the interpretation of the evolution of an area. Faults commonly have an extended history, with repeated movement during further periods of stress.

The determination of the **time-sequence** of rocks and structures of an area (Sect. 10.A) follows the logic of the determination of relative ages set out in Section 3.K. The time-sequence can be related to the **geological time-scale** (Sect. 10.B) by using the known dates of stratigraphical boundaries (Appendix 2) and radiometric determinations of the ages of individual rocks from the area.

The associations of rock-types and styles of deformation that characterise different geographical and tectonic regions of the Earth at the present day were set out in general terms in Table 1.1 and in more detail in the appropriate parts of Chapters 2 to 8. Comparison of the geology of an area with present-day environments allows the **identification of the environments of formation** of rocks and structures of the past (Sect. 10.C).

The **rates of geological processes** are determined from geometrical measurement of the dimensions of rock units and structures in relation to the interval of geological time during which the formational process operated. Such calculations reveal and quantify episodes of relative mobility or stability of the area during the course of its development, and can be matched to the rates of comparable processes at the present day (cf. Table 1.2).

The extent of further interpretation of an area depends on the degree of insight, on the part of the interpreter, into the significance of the evidence. While insight comes more from acquired experience than from formal learning (and therefore cannot be adequately taught from a textbook), it is possible to apply simple principles to form a picture of the probable palaeogeography of the area at different stages of its history, based on analogies with comparable present-day Earth environments (Sect. 10.E). The ultimate objective of this **actualistic interpretation** is to link together the successive 'bird's-eye views' of the area and create a model of the geology of the area as the product of a continuously evolving pattern of environments and processes (cf. Sect. 1.C).

On the widest scale the palaeogeography of an area can be interpreted in relation to the former distribution of continents and oceans and the locations of plate boundaries and continental margins (Table 10.1; see also Ch. 8). This **plate-tectonic interpretation** (Sect. 10.F) requires a wider range of information, both regional and in the extent of geological and geophysical detail, than is available on most published maps. However, it is usually possible to make a limited but meaningful interpretation of the plate-tectonic situation of an area, though with an appropriate degree of caution in the recognition that this is necessarily more speculative than the preceding levels of interpretation.

As in Chapter 9, the illustrative examples are from the British Geological Survey 1 : 50 000 Sheets 39W and 39E (Stirling and Alloa). Figure 10.1 presents a simplified summary of the three-dimensional geometry of the area southeast of the Highland Boundary Fault, based on the results of the graphical techniques used in Chapter 9; it will be shown that this simple geometrical interpretation must be substantially modified in the light of additional evidence (see the footnotes for Table 10.3 and Sect. 10.F). Table 10.2 presents radiometric ages and some other data not printed on the published maps but of a kind commonly included in the most recent geological maps and essential for the interpretation of an area.

Table 10.2 Radiometric ages, temperatures of metamorphism, and dips of faults in the area of B.G.S. Sheets 39W and 39E

Rocks	*Age*	*Reference*
Stirling Sill and quartz-dolerite dykes	295 Ma	(1)
Carboniferous lavas	345–320 Ma	(1)
Diorite similar to those that cut the Lower Old Red Sandstone lavas (cf. Plate 4)	407 Ma	(2)
Intrusive igneous rocks similar to those that cut the Dalradian schists	435–390 Ma (mostly 410–400 Ma)	(3), (4)
Lower Old Red Sandstone lavas	~ 410 Ma	(2)
Regional uplift of the Dalradian	460–440 Ma	(5)
Peak of metamorphism of Dalradian	520–490 Ma	(5)
Temperature of metamorphism of Dalradian schists in the northwest of B.G.S. Sheet 39W	~ 300 °C	(6)
Probable geothermal gradient at the time of the Dalradian metamorphism	~ 30 °C/km	(6)
(These two pieces of data can be combined to infer that the depth of the Dalradian schists in this area at the time of metamorphism was 10 km below the surface.)		
Dip of the Highland Boundary Fault	54° to N. 40° W.	(7)
Dip of Ochil Fault	about 70° to S.	(7)

References:
(1) E. H. Francis, Chapter 10 in Craig (1983).
(2) B. J. Bluck, pp. 275–295 in Bluck *et al.* (1984).
(3) Chapter 3 in Francis *et al.* (1970).
(4) P. E. Brown, Chapter 7 in Craig (1983).
(5) B. Harte *et al.*, pp. 151–163 in Bluck *et al.* (1984).
(6) M. R. W. Johnson, Chapter 4 in Craig (1983).
(7) Chapter 16 in Francis *et al.* (1970).

The first geological survey map of the Stirling area was published just over a hundred years ago (1882). The geology of Scotland has been studied for nearly two centuries and continues to be the subject of vigorous geological research, summarised in recent publications (Craig 1983; Bluck *et al.* 1984). Some of the ideas, especially those relating to plate-tectonic interpretations of the regional geology, are in process of development and are still controversial. In writing this chapter, and having in mind the overall purpose of the book and its intended readership, we have decided to keep closely to the evidence available on the published maps themselves. For a more extended account of the geology of the area, the reader is recommended to the publications listed above.

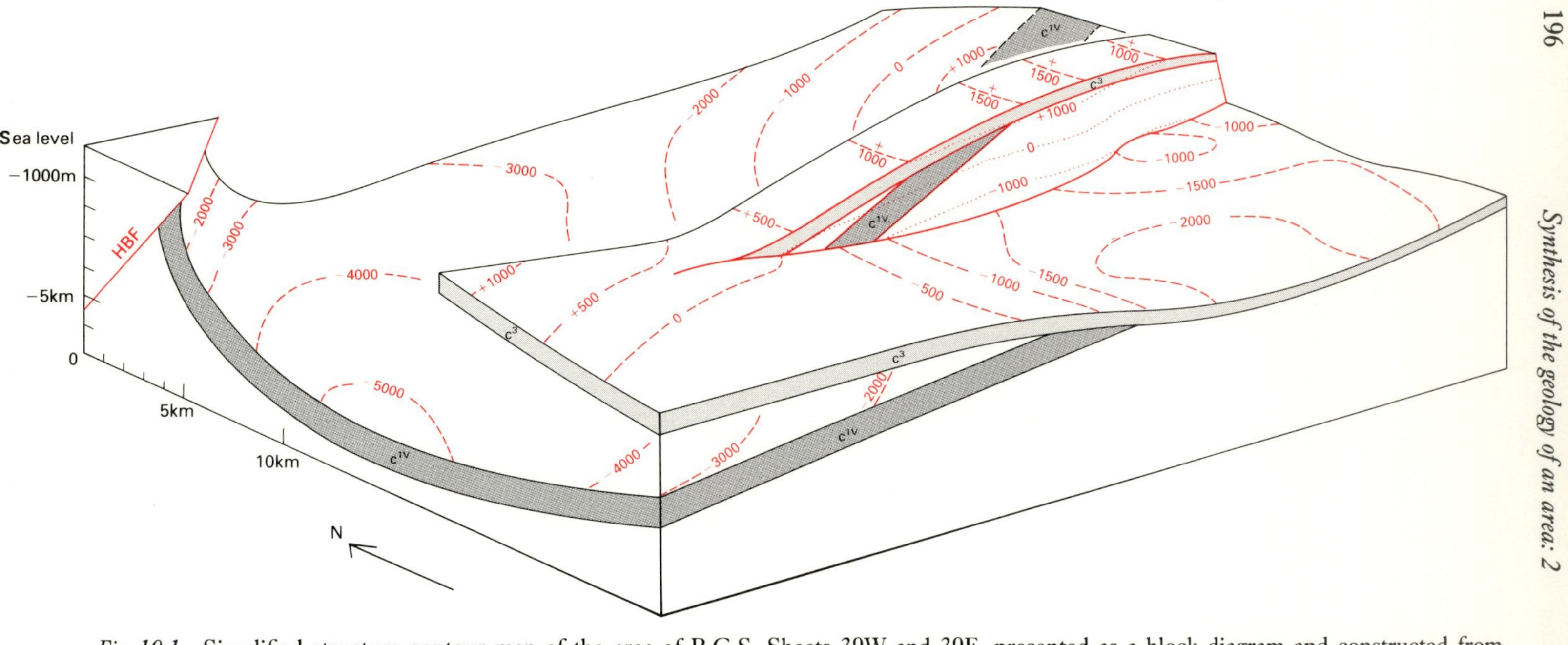

Fig. 10.1 Simplified structure contour map of the area of B.G.S. Sheets 39W and 39E, presented as a block diagram and constructed from numerous vertical cross-sections such as Figs 9.6 and 9.7.

The construction of the diagram is based on the assumption that the units of both the Lower Old Red Sandstone and the Upper Old Red Sandstone and Carboniferous originally extended over the whole area southeast of the Highland Boundary Fault. This geometrical interpretation makes the structure of the area conceptually easier to understand, and serves as a simple introduction to the geology, but it will be shown in the course of this chapter that it is likely to be an oversimplified and incorrect interpretation of the real geology of the area.

Two layers are shown:

1. A layer within the volcanic sequence in the Lower Old Red Sandstone (c^{1V}), with structure contours (heights above and below sea-level in metres) on its upper surface, which is 1000 metres below the top of the volcanic sequence.
2. A layer to represent the Upper Old Red Sandstone sediments (c^{3}), with structure contours on its upper surface, which is also the base of the Carboniferous. This layer has been cut off along a line 3 km north of the Ochil Fault so as to reveal the structure in the Lower Old Red Sandstone rocks below. The Upper Old Red Sandstone rests with angular unconformity on the Lower Old Red Sandstone; the subcrop of the lower layer on the plane of the unconformity is shown in the eastern part of the area north of the Ochil Fault.

No structure is shown in the Dalradian schists northwest of the Highland Boundary Fault (HBF).

To simplify the diagram only the Highland Boundary Fault and the Ochil Fault are shown. All other faults, including the major Arndean Fault in the northeast corner of the block south of the Ochil Fault, have been omitted, though their effects have been allowed for in the construction of the structure contours. Igneous intrusions, including the Stirling Sill, have also been omitted.

10.A TO DETERMINE THE SEQUENCE OF ROCKS AND STRUCTURES IN GEOLOGICAL TIME

The sequence can be set out in any convenient format – for example, as a description starting with the oldest rocks and continuing towards the youngest, or in an appropriate tabulated form. It will be found convenient to leave space within or beside the description or tabulation for the addition of further detail from subsequent stages of interpretation (Sect. 10.B to 10.F).

1. Rank the stratigraphical units (both sedimentary and volcanic) in order from oldest to youngest (Sect. 2.D). Leave space for the insertion of episodes of deformation, igneous and metamorphic activity, and erosion (see below). Label the stratigraphical periods, epochs, and ages of the units in as much detail as is available.
2. Determine the relative dates of episodes of faulting (Sect. 4.I), folding (Sect. 5.F), igneous intrusion and metamorphism (Sect. 3.K), and place them in the correct sequence in relation to the stratigraphical succession.
3. Determine the time-spans of unconformities, from the youngest pre-unconformity rock or structure to the oldest post-unconformity rock or structure (Sect. 6.F), and add them to the sequence.
4. Identify episodes of erosion, including those associated with unconformities, and add them to the sequence.
5. Determine the sequence of recent erosional and depositional episodes, making use of information from the topography and from superficial deposits (Ch. 7), and add them to the sequence.

Example See Table 10.3 (left half).
Note: Table 10.3 and its footnotes are a compilation of numerous observations and interpretations and are the end-result of a lengthy and detailed study of the map. The reader is recommended to examine each individual item in the Table, to find the evidence on which the interpretations are based, and to come to his or her own conclusions about the geological history that is recorded on the map.

10.B TO RELATE THE SEQUENCE OF ROCKS AND STRUCTURES TO THE GEOLOGICAL TIME-SCALE

1. Determine the chronometric dates of stratigraphical boundaries and of structural episodes so far as possible, using data from Appendix 2.
2. Make use of radiometric dates of igneous and metamorphic rocks, as printed in the margin of the map or from other sources.

Add the data to the sequence determined in Section 10.A.

Example See Table 10.3 (centre).

Table 10.3 Sequence of rocks and structures, dates, and interpretation of the environments in the area of B.G.S. Sheets 39W and 39E

	Stratigraphical column	*Sequence of rocks and structures (Sect. 10.A)*	
	Quaternary	Erosion to produce present-day topography and outcrop pattern	
	Tertiary		
	Mesozoic		
	Permian	Mineralisation[2]	
		Faulting [3] Intrusion of quartz-dolerite Stirling Sill and east–west dykes[3] (Movement of Highland Boundary Fault completed before this time)	
Carboniferous	Stephanian	Movement of Ochil Fault and of other east–west and northeast–southwest faults in southern part of the area[4] Formation of major north–south Clackmannan Syncline, folding the Carboniferous rocks, with the Ochil Fault as its northern boundary; presumed contemporaneous folding of the L ORS rocks north of the Ochil Fault[4]. Minor east–west domes and basins	
Carboniferous	Westphalian	Coal Measures	Cyclic sequence of sandstones, siltstones, mudstones, and coals. Sandstones predominate in the Passage Group. Some limestones in the lower part of the succession
Carboniferous		Passage Group	
Carboniferous	Namurian	Upper Limestone Group	Contemporaneous differential subsidence (folding) and faulting[5]. Emplacement of volcanic vents and olivine-dolerite sills (in southwest and southeast)[6]
Carboniferous		Limestone Coal Group	
Carboniferous		Lower Limestone Group	Sandstones, siltstones, mudstones, coals, limestones
Carboniferous		Calciferous Sandstone Measures (upper part)	
Carboniferous	Dinantian (Viséan and Tournaisian)	Volcanic rocks	Olivine basalts, mugearites, and trachybasalts; volcanic detritus
Carboniferous		Calciferous Sandstone Measures (lower part)	Mudstones, shales, sandstones, cementstones (argillaceous dolomites)
	Late Devonian	Upper Old Red Sandstone	Sandstones, siltstones, mudstones, conglomerates, cornstones (calcitic and dolomitic siltstones), marls

Ages of rocks and structures (Sect. 10.B)[1]		
Chronometric age (App. 2)	*Radiometric age (Table 10.2)*	*Environments of formation of rocks and structures (Sect. 10.C)*
0 Ma		No evidence about earth movements or deposition later than Carboniferous. Net erosion. (In adjacent areas: deposition of desert sediments in Permian and Triassic, marine sediments in Mesozoic; in the Cenozoic deposition of thick marine sediments in the North Sea and volcanic activity in western Scotland – Ziegler 1982; Craig 1983.)
290 Ma	295 Ma	North–south crustal extension with emplacement of tholeiitic dykes and Stirling Sill; possibly related to continental hot-spot
300 Ma		North–south extension and subsidiary northwest–southeast extension. East–west compression and minor north–south compression. Development of the Ochil Fault as a structural boundary between an area of uplift on the north and an area of downwarp on the south
310 Ma		Cyclic sedimentation in tropical, deltaic or river [or lake or clastic shoreline][7] environment. Subsidence greater in the central part of the basin (approximately coinciding with the southern end of the Clackmannan Syncline in the area of the map), with less subsidence and thinner sediments in the north and west. Deposition gradually gaining on subsidence towards the top of the succession. Minor volcanic activity
325 Ma		Sedimentation in semi-arid deltaic [or fluviatile, lake, or clastic shoreline] environment. Rate of subsidence approximately equal to rate of sedimentation.
	320–345 Ma	Volcanic activity of alkalic series composition; possibly related to continental hot-spot
		Sedimentation in arid or semi-arid shoreline or lake [or desert] environment. Rate of subsidence approximately equal to rate of sedimentation
355 Ma		Sedimentation in semi-arid river or lake [or desert or shoreline] environment. Rate of subsidence approximately equal to rate of sedimentation, with higher rates and thicker sediments in the west

(*Table 10.3 cont.*)

Stratigraphical column	*Sequence of rocks and structures (Sect. 10.A)*	
Middle Devonian	Erosion. Major movement on Highland Boundary Fault?[8] Faulting[9] Formation of broad major northeast-southwest Strathmore Syncline	Formation of angular unconformity at base of Upper Old Red Sandstone
Early Devonian	Lower Old Red Sandstone (upper part)	Sandstones, siltstones, and mudstones; thick conglomerates in the northwest, wedging out to the southeast Possible movement on the line of the Highland Boundary Fault
	Intrusion of diorites and dykes into the Lower Old Red Sandstone lavas[10], thermal metamorphism and formation of a radial dyke swarm around the diorites. Intrusion of diorites and dykes into Dalradian schists[10]	
Silurian	Lower Old Red Sandstone (lower part)	Basalt, andesite, and rhyodacite lavas, with tuffs and agglomerates
	Uplift of Caledonian orogenic belt	
Ordovician	?Arenig	Black shale and chert; emplacement of serpentinite or of peridotite with subsequent serpentinisation[11]
	Intense folding of Dalradian sediments on northwest–southeast axes, with formation of cleavage and schistosity[11]	
Cambrian	Lower Cambrian and probably Cambrian	Mudstones (in part carbonaceous), grits, and thin limestones – Dalradian sediments
Precambrian		

Ages of rocks and structures (Sect. 10.B)[1]		
Chronometric age (App. 2)	*Radiometric age (Table 10.2)*	*Environments of formation of rocks and structures (Sect. 10.C)*
375 Ma		Northwest–southeast compression producing a large fold with Caledonoid trend, and contemporaneous erosion of the uplifted L ORS sediments and lavas. The direction of movement of the Highland Boundary Fault is not determinable (strike-slip movement is possible), but there may have been further uplift of the former Caledonian orogenic belt
390 Ma		Sedimentation in semi-arid, river or lake [or desert] environment with alluvial fans of conglomerate in the northwest. Erosion products of the Caledonian orogenic belt were deposited in a uniformly subsiding basin. Possibly the Highland Boundary Fault was active (downthrow to the southeast) creating the subsiding basin in which the L ORS was deposited
405 Ma	407 Ma 390–435 Ma (mostly 400–410 Ma) ~ 410 Ma	Post-orogenic calc-alkaline igneous activity; probably the area of the Midland Valley of Scotland was the site of a convergent plate margin
435 Ma	440–460 Ma	End of Caledonian orogenic activity; possible cessation of relative plate movements in the area
~ 480 Ma		Pelagic oceanic sediments and ultrabasic rocks; possibly oceanic lithosphere
⩾510 Ma	490–520 Ma	Orogeny, with northwest–southeast compression and regional metamorphism; convergent plate margin processes
		Sedimentation in shelf or oceanic environment
⩽570 Ma		

Notes: Table 10.3

1. The chronometric dates of some stratigraphical units are not directly determinable – notably the Upper Old Red Sandstone sediments and the Lower Old Red Sandstone sediments and volcanics, which are commonly correlated with Late and Early Devonian, but which may be in part early Carboniferous and late Silurian, respectively.
2. The dates of the mineral veins in the Ochil Hills can only be determined from the map as later than the faulting of the Lower Old Red Sandstone volcanics; they are believed to be in part middle Carboniferous to Permian and in part Tertiary (Francis *et al.*, 1970, p. 293).
3. The evidence on the map for the relative ages of the Stirling Sill and the faulting is equivocal. Fault movement probably occurred throughout the Carboniferous (see notes 4 and 5); the emplacement of the sill occurred towards the end of the history of faulting. The sill was emplaced soon after the Carboniferous sedimentation and the folding, and it increases in thickness towards the centre of the Kincardine Basin (note 5) and the axis of the Clackmannan Syncline (Francis, 1982). The east–west quartz-dolerite dykes were probably contemporaneous with the Stirling Sill.
4. The relative ages of the faults and of the folding are not determinable from the evidence on the map. The interpretation of the structural significance of the Ochil Fault given in the final column requires at least part of its movement to be contemporaneous with the folding.
5. The thickness of the Carboniferous sediments is greatest in the centre of the Kincardine Basin, which approximately coincides with the axis of the Clackmannan Syncline (Francis *et al.*, 1970 – cf. Fig. 9.4), so some of the folding is contemporaneous with sedimentation. Thickness and facies changes across fault lines (not seen on the Stirling and Alloa maps, but demonstrable elsewhere in the Midland Valley of Scotland) indicate syn-sedimentary (growth) faulting. The Ochil Fault may have begun to form at about this time and to have influenced or limited Carboniferous sedimentation on the northern side of the Kincardine Basin.
6. The dates of emplacement of the vents and olivine–dolerite sills are only determinable from the evidence on the map as later than the Upper Limestone Group; they are Namurian in age (E. H. Francis, personal communication).
7. The environmental interpretations are based on additional evidence (Craig, 1983). Possible alternative (though incorrect) interpretations based on the generalised correlations in Table 2.1 are given in square brackets.
8. From the evidence on the map the date of movement of the Highland Boundary Fault can only be determined as later than the Lower Old Red Sandstone sediments (and probably later than the folding of the Lower Old Red Sandstone) and before the intrusion of the quartz-dolerite dykes.
9. Faulting of the Lower Old Red Sandstone rocks in the Ochil Hills and north-centre of the area is largely parallel and perpendicular to the trend of the Strathmore Syncline (cf. Fig. 9.5) and is probably related to that structure rather than to the faulting that occurred during the Carboniferous.
10. The ages of the diorites (and dykes) that intrude the Lower Old Red Sandstone volcanics relative to those that intrude the Dalradian schists are not determinable from the evidence on the map; their radiometric dates (Table 10.2) suggest contemporaneity.
11. The ages of the ?Arenig shales and serpentinite relative to the folding and metamorphism of the Dalradian sediments are not determinable from the evidence on the map. Chronometric and radiometric dates indicate the sequence given.

10.C TO IDENTIFY THE ENVIRONMENTS OF FORMATION AND OF DEFORMATION OF ROCKS

Tables 2.1, 3.2, 3.3, 3.4, and 7.1 show how the lithologies of rock units are related to environments of formation. Figure 4.3 and Sections 4.K and 5.I describe and illustrate the stress systems for the formation of faults and folds.

Note that it is possible for different environments to co-exist in different areas of the map – for example, there may be variations of facies, erosion in one area and deposition in another, or variations in structural style during a single deformational episode.

It is most important to recognise the limits of valid interpretation of geological environments from evidence on the map, and not to attempt interpretations beyond the available evidence. The tables, figures, and sections cited above are necessarily based on generalisations, and must be used with discrimination and with due concern for alternative interpretations. Whenever possible, additional information should be sought in more detailed descriptions of the rocks and structures in books and papers. However, even in the absence of such additional material, it should be possible to make a broad-scale interpretation of the environments of formation of the rocks and structures.

1. Use descriptions of lithological units in the margin of the map, or make use of Tables 2.1, 3.2, 3.3, 3.4, and 7.1 to identify the probable environments of deposition of sedimentary rocks and of formation of igneous and metamorphic rocks. Add these environmental interpretations to the historical sequence of Section 10.A.

2. Use descriptions of the faults and folds in the margin of the map, together with information from cross-sections or other constructions (Fig. 4.3, Sect. 4.K and 5.I) to identify the probable environments of formation of the structures, including the orientations of the stresses that produced them. Refer to small-scale maps covering the area in order to see the structures in their regional setting. Add the structural interpretation to the historical sequence of Section 10.A.

3. Identify from Section 10.A any significant intervals of geological time during which there is no record of rocks or structures preserved in the area. Clearly, lacking any direct evidence, it is impossible to determine with certainty what was happening in the area (for example, whether there was continuous erosion, or whether rocks were deposited during part of the time and subsequently eroded away). Refer to adjacent maps or to small-scale maps to see whether there is indirect evidence about probable environments from nearby areas with a more complete geological record. Add interpretations about such intervals to the historical sequence in Section 10.A.

Example See Table 10.3 (right half).

10.D TO DETERMINE THE RATES OF GEOLOGICAL PROCESSES

If the time interval during which a rock unit or structure was formed is known (Appendix 2), the rate of the process of formation can be calculated from geometrical dimensions measured from the geological map, stratigraphical sections, vertical cross-sections, or structure contour maps (Sect. 2.G, 4.J, 5.H, 6.D). Measured or assumed densities can be used to calculate the mass of a unit of rock. If only limiting dates are known, a minimum rate for the process can be calculated.

The calculated rates can be compared with the rates of similar processes at the present day (Table 1.2). As a useful generalisation, rates less than about 0.01 mm/y may be characterised as slow, indicating periods of relative stability of the crust of the Earth in the area; rates greater than about 0.1 mm/y are geologically rapid, indicating periods of mobility commonly associated with mountain-building; rates greater than about 10 mm/y are very rapid, and possibly indicate processes associated with relative plate movements.

Example Table 10.4.

Table 10.4 Rates of geological processes within the area of B.G.S. Sheets 39W and 39E

	Duration of process (or limiting dates)		
Stratigraphical interval	*Start*	*Finish*	*Duration*
After quartz-dolerite intrusions to present day	295 Ma	0 Ma	295 m.y.
At time of emplacement of quartz-dolerite intrusions	< 300 Ma	295 Ma	< 5 m.y.
Post-Westphalian, pre-quartz-dolerite intrusions	300 Ma	295 Ma	5 m.y.
Dinantian to Westphalian	355 Ma	300 Ma	55 m.y.
Upper Old Red Sandstone	~ 375 Ma	~ 355 Ma	20 m.y.
Post-Lower Old Red Sandstones, pre-quartz-dolerite intrusions	~ 390 Ma	295 Ma	< 95 m.y.
Post-Lower Old Red Sandstone, pre-Upper Old Red Sandstone	~ 390 Ma	~ 375 Ma	15 m.y.
Lower Old Red Sandstone	~ 410 Ma	~ 390 Ma	20 m.y.
Between date of Dalradian orogeny and date of uplift of Dalradian	520–490 Ma	460–440 Ma	~ 60 m.y.
Post-Lower Cambrian to date of Dalradian orogeny	~ 550 Ma	520–490 Ma	~ 50 m.y.

* Estimated without making allowance for compaction of sediments after deposition
† See footnote 8 and comments in final column of Table 10.3
‡ Assuming that 2000 m of the vertical difference is due to post-Westphalian folding (see Table 10.3).

Process	*Observation*	*Measurement*	*Reference*	*Rate of process*
Erosion	Thickness of U ORS and Carboniferous rocks eroded from western part of area	3000 m	Fig. 9.4	$\nless$ 0.01 mm/y
Crustal extension	Total width of east–west dykes and transgressive portions of Stirling Sill	~ 300 m	Sheet 39 W	~ 0.06 mm/y
Folding	Difference of elevation of base of Carboniferous from west of area to axis of Clackmannan Syncline	2000 m	Fig. 10.1	0.4 mm/y (vertical)
Crustal shortening	Measurement from vertical cross-sections	150 m	—	0.03 mm/y (horizontal)
Subsidence, sedimentation, and accumulation of volcanic rocks	Maximum thickness of Carboniferous sediments and volcanic rocks in the centre of the Clackmannan Syncline	2450+ m	Fig. 9.4	$\nless$ 0.04 mm/y*
Subsidence and sedimentation	Maximum thickness of U ORS sediments	700 m	Fig. 9.4	0.04 mm/y*
Movement of Highland Boundary Fault†	Juxtaposition of Dalradian schists (metamorphosed at depth ~ 10 km) against L ORS sediments (presumed to be formed near sea level)	10 000 m (Possible horizontal movement not determinable)	Table 10.2	> 0.1 mm/y (vertical)
Erosion	Thickness of L ORS eroded below pre-U ORS unconformity	$\nless$ 3500 m	Fig. 9.6	0.2 mm/y
Folding	Difference of elevation of L ORS lavas from axis of Strathmore Syncline to crest of Ochil Hills	4000 m‡	Fig. 10.1	0.3 mm/y (vertical)
Crustal shortening	Measurement from cross-sections	1200 m	—	0.08 mm/y (horizontal)
Accumulation of volcanic rocks and sedimentation	Average thickness of L ORS sediments plus thickness of L ORS volcanic rocks	5700 m	Fig. 9.4	0.3 mm/y*
Uplift and erosion	Depth of metamorphism of Dalradian schists in region of Highland Boundary Fault (schists presumed to have returned to near-surface)	10 000 m	Table 10.2	0.2 mm/y
Crustal thickening	Depth of metamorphism of Dalradian schists in region of Highland Boundary Fault (original sediments presumed to have formed close to sea level)	10 000 m	Table 10.2	0.2 mm/y

10.E ACTUALISTIC INTERPRETATION OF GEOLOGICAL MAPS

The principle in making an actualistic interpretation of a geological map is to use the results of all the preceding work (Sect. 9.A to 9.E, and 10.A to 10.D) to create a model of the geological evolution of the area. Present-day analogues of similar geological and geographical environments and processes are the key to interpretation (cf. Ch. 7).

The geological information is rarely complete enough to produce a model of an area in which every significant feature is accurately characterised and located in space and in time. A stylised cartographical or pictorial representation of the area is usually as much as can be achieved.

The interpretation of the geology can be represented:

(a) As a map (palaeogeographical map) showing the distribution of land and sea and of different environments of deposition (facies), deformation, erosion, etc. at selected stages in the development of the area.
(b) As block diagrams presenting the same features as in (a), and in addition showing the structure as vertical cross-sections on the sides of the block.

1. Select a time or time-period of interest.

2. Use the results of Sections 10.A and 10.B to review the geological history up to that time.

3. Identify the lithologies and facies of the rocks being deposited at that time and/or the structures being formed, together with any other contemporaneous processes (erosion, emplacement of igneous intrusions, etc.) (Sect. 10.C).

Fig. 10.3 The San Andreas Fault near Palmdale, California; a present-day situation broadly analogous to that of the Caledonide mountains, Highland Boundary Fault, and the Midland Valley of Scotland in Lower Old Red Sandstone time. Erosion of the upland areas to the left (southwest) produces detrital material for the sedimentary basin to the right; tectonic activity in the area maintains contemporaneous movement on the fault. Photograph: Aerofilms, Ltd.

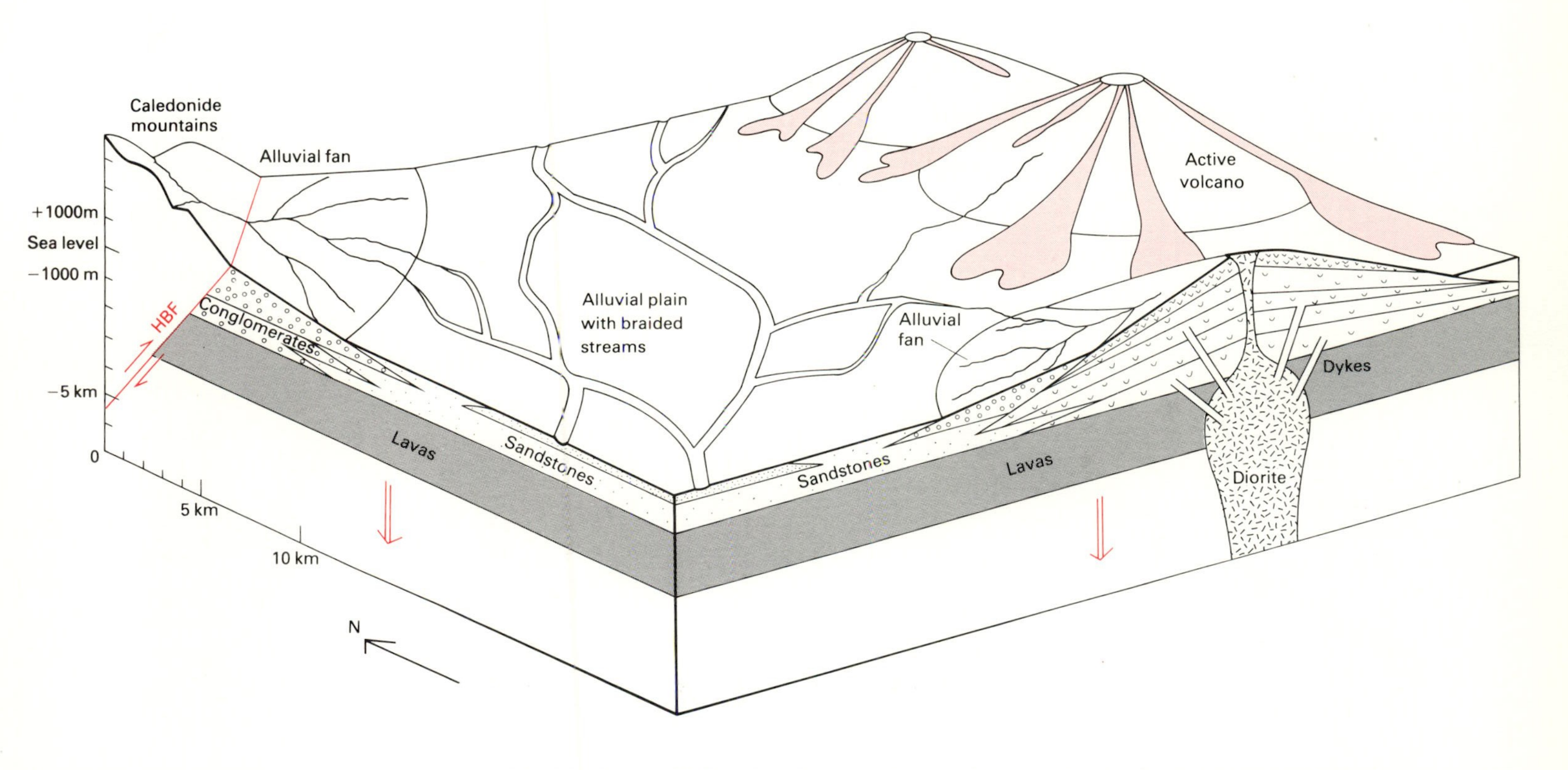

Fig. 10.2 Schematic representation of the area of B.G.S. Sheets 39W and 39E during Lower Old Red Sandstone time towards the end of volcanic activity. A thick pile of andesitic lavas was produced during the early part of the Lower Old Red Sandstone. The volcanoes have dioritic cores with radial dyke systems. As volcanoes become extinct, erosion produces alluvial fans of volcanic conglomerate interdigitated with lava flows from still active centres and with alluvial sediments. Meanwhile, erosion of the metamorphic rocks of the Caledonide mountains produces alluvial fans of conglomerate which form a pediment on the southeast side of the Highland Boundary Fault (HBF), and interdigitate with sandy sediments from a more distant source in a broad alluvial plain. Regional subsidence (including possible movement on the Highland Boundary Fault) maintains the area as a basin of deposition.

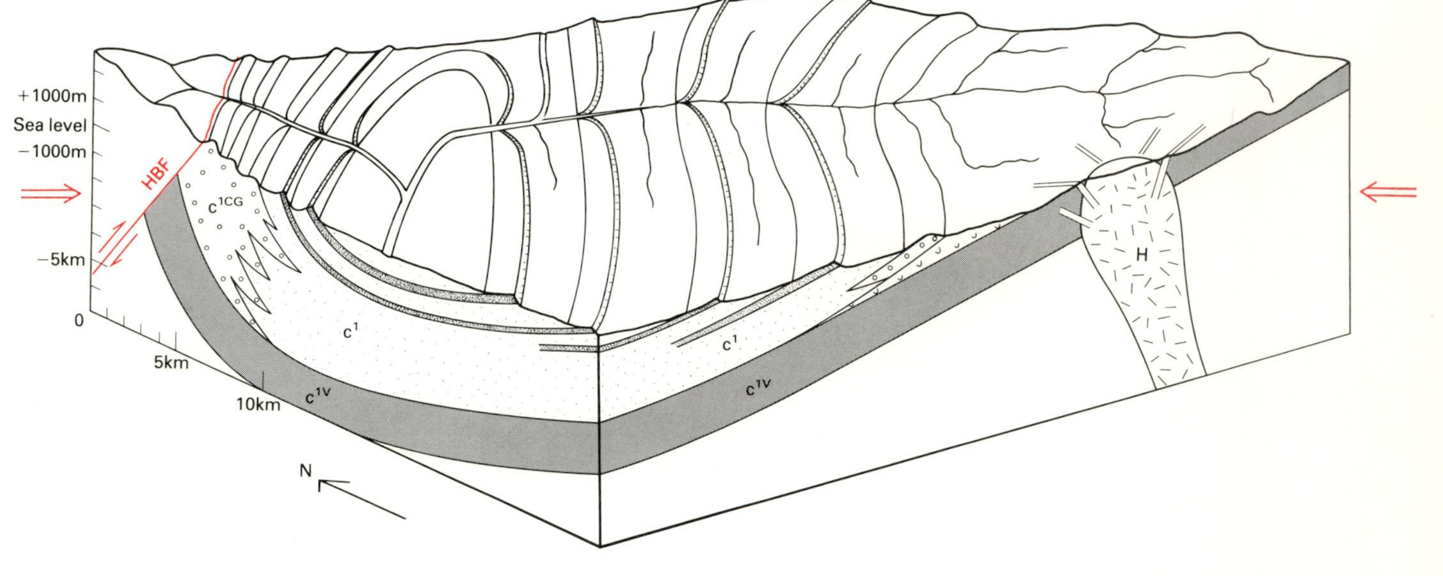

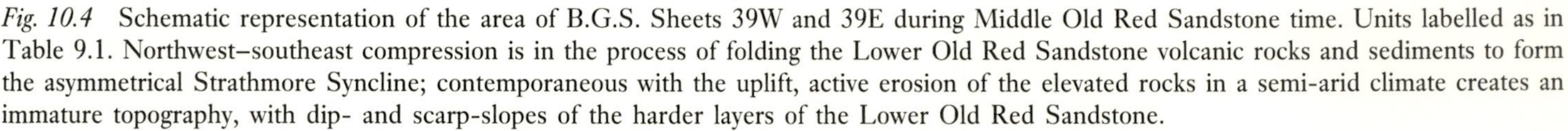

Fig. 10.4 Schematic representation of the area of B.G.S. Sheets 39W and 39E during Middle Old Red Sandstone time. Units labelled as in Table 9.1. Northwest–southeast compression is in the process of folding the Lower Old Red Sandstone volcanic rocks and sediments to form the asymmetrical Strathmore Syncline; contemporaneous with the uplift, active erosion of the elevated rocks in a semi-arid climate creates an immature topography, with dip- and scarp-slopes of the harder layers of the Lower Old Red Sandstone.

Fig. 10.5 A present-day recently uplifted fold belt of sandstones and shales, being vigorously eroded in a semi-arid climate (part of the Sulaiman Range, Pakistan); analogous to the area of the Midland Valley of Scotland in Middle Old Red Sandstone time. Photograph: Professor W. D. Gill.

4. Make use of the evidence from chronostructural units (Sect. 6.G) to establish the stage of structural evolution reached by that time.
5. Use general geological and geographical knowledge (aided by the information in Tables 1.1, 1.2, and 7.1) to form a picture of the area at the selected time.
6. Summarise all the information on a suitable map or block diagram.

Example Figs. 10.2 to 10.7.

10.F PLATE-TECTONIC INTERPRETATION

The identification of former plate distributions and environments draws upon a wider range of geological observations than are recorded on most published maps of geological-survey type. The inclusion of this section on plate-tectonic interpretation must not be taken to imply that identification of plate environments is necessarily either possible or desirable as a routine part of map interpretation. There are, though, features of the rocks or structures of an area that may indicate a particular location in relation to former plates or plate margins (Table 10.1), but it must be emphasised that plate-tectonic interpretation should only be attempted when it seems fully justified by the evidence and in the awareness of the consequences and implications for adjacent and for more distant areas.

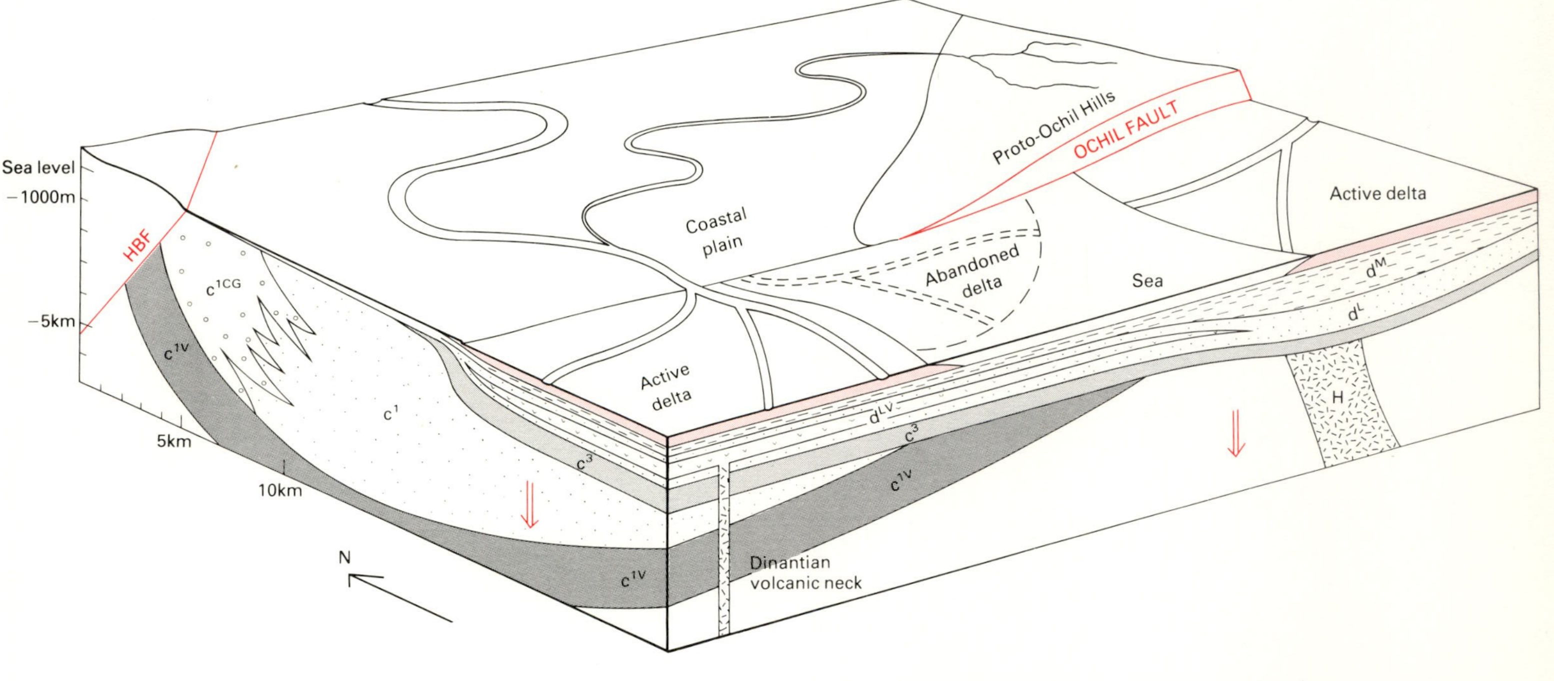

Fig. 10.6 Schematic representation of the area of B.G.S. Sheets 39W and 39E during the Carboniferous. Units labelled as in Table 9.1. Erosion of the Caledonide mountains and the Lower Old Red Sandstone volcanics of the proto-Ochil Hills has created a mature topography, with rivers flowing across the planed-off surface of the softer Lower Old Red Sandstone sediments. A basin of deposition has developed in the south of the area as a result of movement on the newly-formed Ochil Fault and its westward continuation into a monoclinal flexure. Deltas form along the coast and sweep around the base of the proto-Ochil Hills, with deposition of detrital sediments. A cover of vegetation will become future coal seams; in the adjoining sea areas marine sediments including limestones are deposited. Subsidence maintains the area as a sedimentary basin (the Kincardine Basin), and periodic switching of the delta produces a cyclic sequence of sediments. A period of volcanic activity occurred within the Calciferous Sandstone Measures.

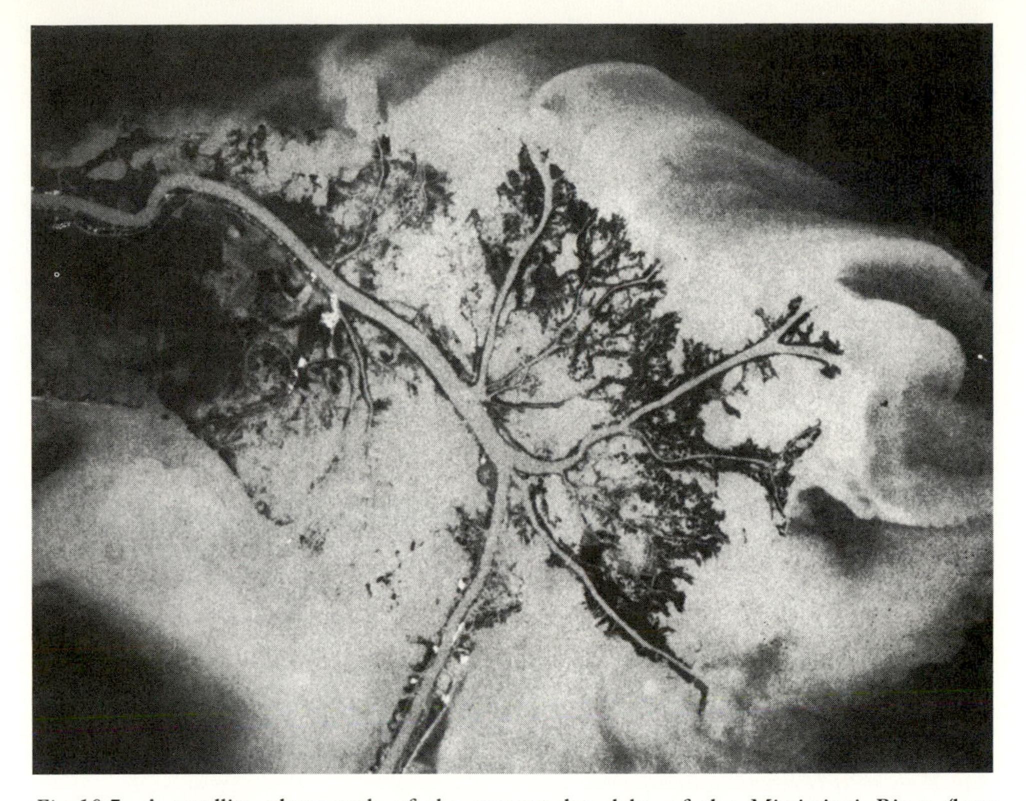

Fig. 10.7 A satellite photograph of the present-day delta of the Mississippi River (long dimension of the photograph about 120 miles); an area analogous to that of the Midland Valley of Scotland in Carboniferous time. The varied fluviatile, marine, and terrestrial environments within the delta area are indicated by the distributaries of the river and the interdistributary bays between. The plumes of sediment-laden water extend far out into the marine environment. Periodic avulsion of the river leads to the formation of a new delta system elsewhere along the coast-line, after which continued subsidence will lead to submergence of the present system and the ultimate preservation of the varied sediments as rocks of the distinctive deltaic facies. Photograph: NASA Astronaut, Hasselblad.

Relative horizontal movements of plates at divergent, transform, and convergent margins are an order of magnitude faster than any other major geological process (Table 1.2). Even within the time-span of a geological system plates originally thousands of kilometres apart can be brought into juxtaposition. If different parts of an area are recognised as belonging to originally separate plates they are likely to contain a record of fundamentally different geological histories, tectonics, and climates up to the date of juxtaposition, and thereafter to share a common history. This may indicate the need to reassess, and possibly to rewrite, preceding levels of interpretation.

Make use of Table 10.1 or any other information to recognise the critical evidence that is distinctive of particular regions of plates and plate margins.

Example See Table 10.5, which summarises the geological environments listed in Table 10.3 and suggests the plate-tectonic interpretation of the different parts of the area. A tentative reinterpretation of the geology of the area is shown in Fig. 10.8. Further interpretation is only possible using a much wider range of evidence than is available on the published maps (cf. Bluck *et al.* 1984).

Table 10.5 Summary of geological environments in the area of B.G.S. Sheets 39W and 39E, and plate-tectonic interpretation

	Precambrian	*Cambrian*	*Ordovician* *Arenig*	
Chronometric dates (Ma)	$\leqslant$ 570	$\geqslant$ 510	~ 480	435
Radiometric dates (Ma)		520——490	460——440	
Area northwest of Highland Boundary Fault				
Sedimentation				
Folding				
Metamorphism				
Intrusive igneous				
Uplift				
Erosion				
Highland Boundary Fault				
Highland Border Complex				
Sedimentation				
Ultrabasic igneous				
Deformation				
Erosion				
Area southeast of Highland Boundary Fault				
Sedimentation				
Extrusive igneous				
Intrusive igneous				
Folding		Early history not determinable		
Faulting				
Uplift				
Erosion				
Plate-tectonic interpretation				
Area NW of Highland Boundary Fault		?CONTINENTAL MARGIN	CONVERGENT PLATE MARGIN (OROGENY)	
Highland Border Complex			?OCEANIC PLATE	
Area SE of Highland Boundary Fault			Plate environment not known	

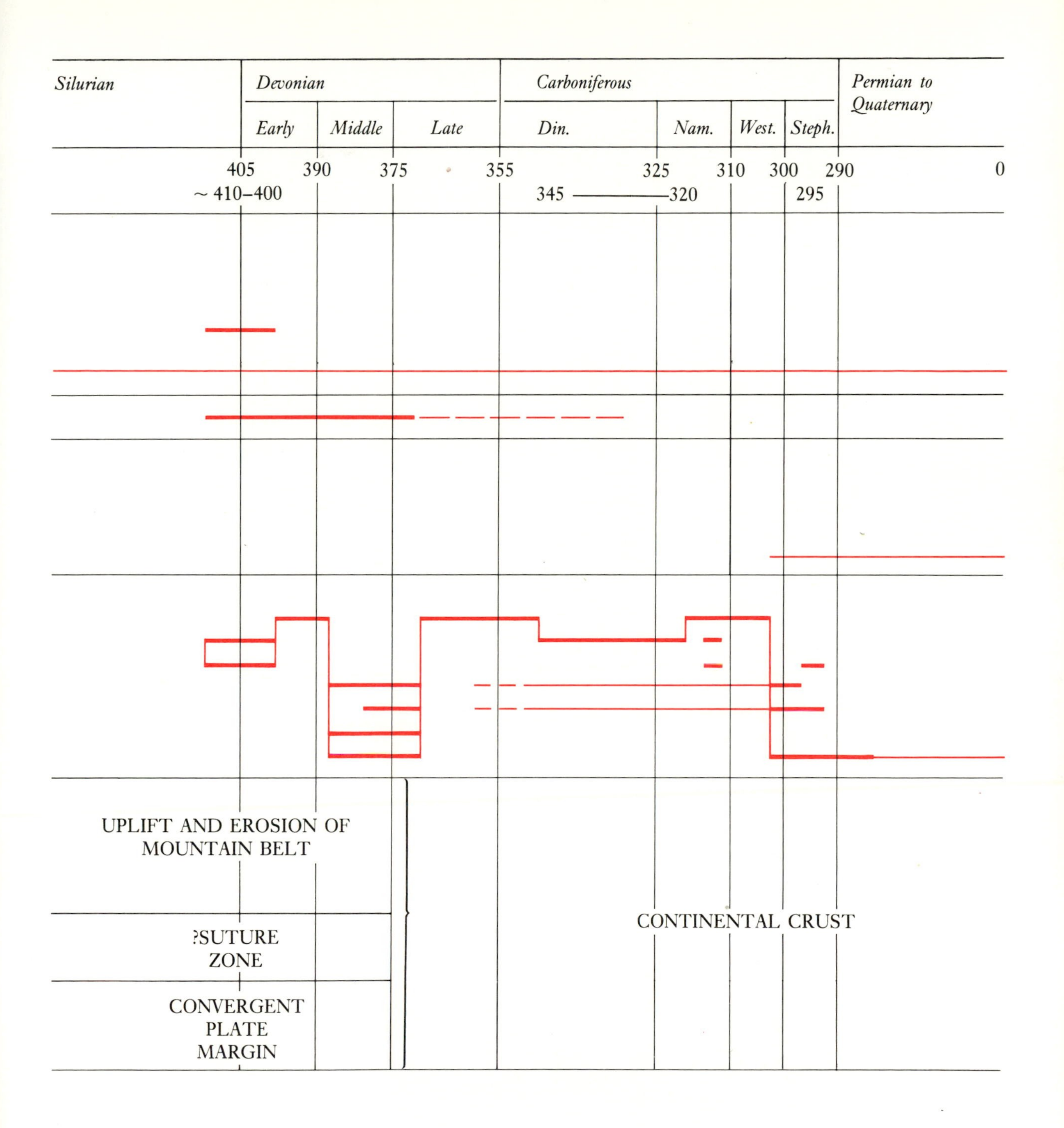

Silurian
Devonian
Early
Middle
Late
Carboniferous
Din.
Nam.
West.
Steph.
Permian to Quaternary
405
390
375
355
325
310
300
290
0
~ 410–400
345 —320
295
UPLIFT AND EROSION OF MOUNTAIN BELT
?SUTURE ZONE
CONVERGENT PLATE MARGIN
CONTINENTAL CRUST

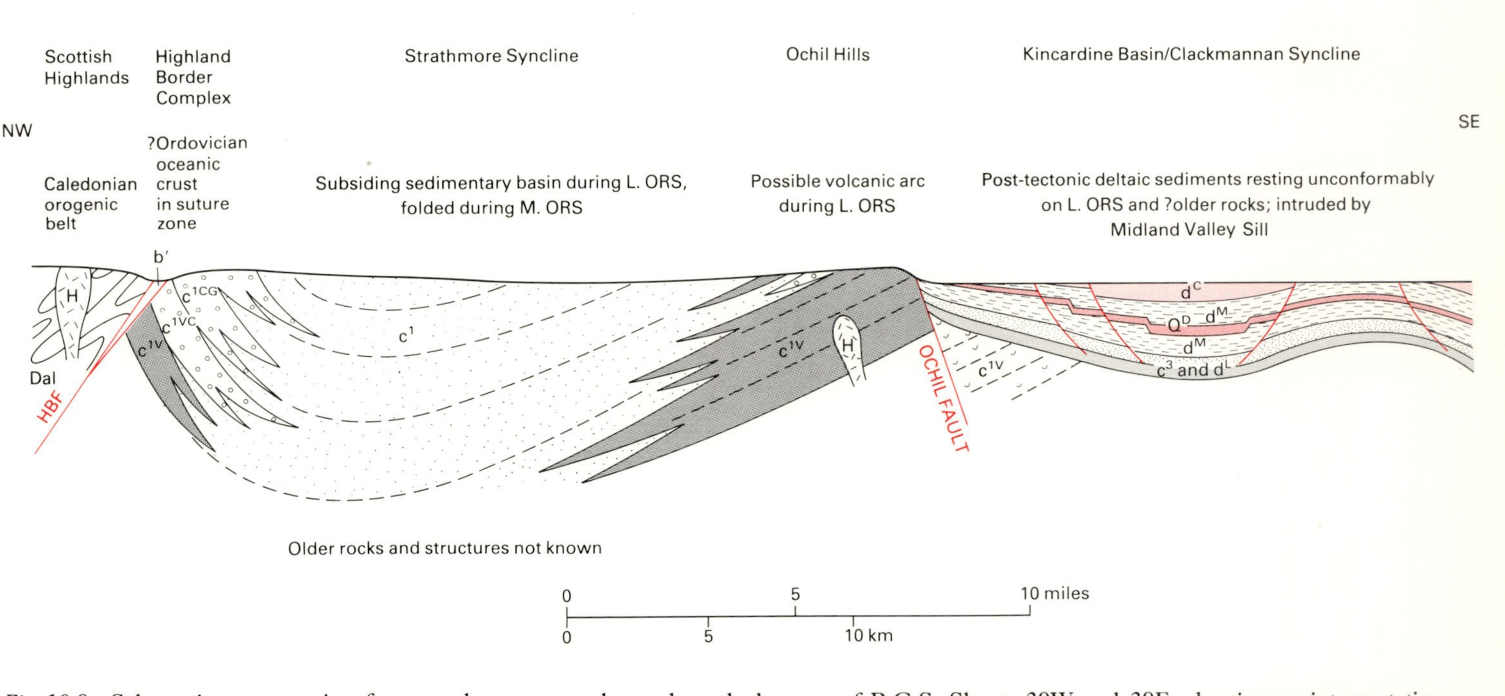

Fig. 10.8 Schematic cross-section from northwest to southeast through the area of B.G.S. Sheets 39W and 39E, showing an interpretation of the geology based on plate-tectonic concepts. The principal differences from the geometrically constructed sections in Figs 9.6 and 9.7 and from Fig. 10.1 are (i) facies and thicknesses in the sediments and volcanics of the Lower Old Red Sandstone are assumed to vary to match their presumed formation in the post-orogenic phase of a convergent plate margin. (ii) the Ochil Fault is assumed to have been a growth fault, forming the northern limit of subsidence of the Kincardine Basin and the boundary of the area of deposition of the Upper Old Red Sandstone and Carboniferous.

APPENDIX 1 RADIOMETRIC DATES

On many recent maps the radiometric ages of igneous and metamorphic rocks determined by isotopic analysis are included in the accompanying information. The age that is determined corresponds to the date in the thermal history of the rock when the temperature decreased to a level (the closure temperature) below which loss of the radiogenic isotope by diffusion ceased; different minerals and isotopes have different closure temperatures. The determined age of a rapidly-cooled shallow-level intrusion or of a lava is effectively the date of crystallisation. Slower-cooled deep-seated intrusions give cooling dates younger than the date of emplacement, and in some cases approximating to the date when the intrusion became exposed to erosion.

Metamorphic rocks can be dated in the same way as igneous rocks. The date of formation of sedimentary rocks can sometimes be determined from K-Ar dating of glauconites.

Table A.1 gives some details of the principal methods and materials that are used for dating.

Later metamorphic re-heating can flush radiogenic daughter isotopes out of certain rocks and minerals and re-set the original dates to a later date nearer to that of the metamorphism. For these reasons the most informative radiometric dating of intrusions uses different methods on the same rocks and minerals: for example, U-Pb zircon or Rb-Sr whole-rock dates on a granite give the date of intrusion and K-Ar dates on micas give a later, cooling or re-heating, date. During the interval between these two apparently discrete events other geological processes may have taken place.

The use of radiometric dates for the determination of stratigraphical ages is described in Appendix 2.

Table A.1 Isotopic methods used for dating rocks

Name of method	*U-Pb* (*Uranium-Lead*)	*Rb-Sr* (*Rubidium-Strontium*)	*K-Ar* (*Potassium-Argon*)	*Sm-Nd* (*Samarium-Neodymium*)
Parent → daughter isotope	(i) $^{238}U \rightarrow {}^{206}Pb$ (ii) $^{235}U \rightarrow {}^{207}Pb$	$^{87}Rb \rightarrow {}^{87}Sr$	$^{40}K \rightarrow {}^{40}Ar$	$^{147}Sm \rightarrow {}^{143}Nd$
Half-life	(i) 4.468×10^{9} y (ii) 7.04×10^{8} y	4.88×10^{10} y	1.28×10^{10} y*	1.06×10^{11} y
Suitable age range for method	10 m.y. and greater	10 m.y. and greater	50 000 years and greater; generally <200 m.y.	100 m.y. and greater; generally > 1000 m.y.
Geological materials commonly used (with closure temperature, where appropriate)	Zircon, galena, monazite, apatite, sphene, feldspar; coarse-grained acid to intermediate igneous rocks	Biotite (~ 200 °C), muscovite (~ 300 °C), potassium feldspar (~ 500 °C); coarse and fine-grained acid to intermediate igneous rocks; metamorphic rocks	Biotite (~ 200 °C), muscovite (~ 300 °C), nepheline, hornblende, and sanidine (~ 500 °C); fine-grained igneous rocks; contact-metamorphosed argillaceous rocks; authigenic glauconite in sediments	Plagioclase, pyroxene, garnet; coarse-grained acid to ultra-basic igneous rocks and metamorphic rocks
Comments	Zircon U-Pb is the most precise and accurate method for dating igneous rocks. Detrital zircon can be used to date the source rocks of sediments	Used to date time of emplacement (rather than time of cooling) of large intrusions. Sometimes gives ages 6–10% less than U-Pb zircon ages	The only method used for dating young basic extrusive and minor intrusive rocks. Ages are often lower than U-Pb or Rb-Sr ages and are in some cases cooling ages. Dates are affected by metamorphic re-heating. Date of sedimentation can be determined using authigenic glauconite	Used for dating Archaean rocks. Ages are less affected by later deformation and metamorphism than U-Pb, Rb-Sr and K-Ar ages

* Because ^{40}K also decays to ^{40}Ca the calculation of the radiometric date is more complex than for other methods

APPENDIX 2 THE GEOLOGICAL TIME-SCALE

Radiometric dates are determined from measurement of the amounts of radioactive and radiogenic elements in suitable minerals and rocks (Appendix 1). The dates of crystallisation of most igneous and metamorphic rocks can be satisfactorily determined in this way, but the radiometric determination of the date of formation of sedimentary rocks presents greater difficulties. However, by identifying igneous or metamorphic rocks whose ages relative to sedimentary rocks can be closely defined, it is possible to interpolate radiometric ages onto the stratigraphical column, and so to estimate the **chronometric age** of the stratigraphical epochs and their subdivisions. The most recent geological time-scale (Snelling, 1985) is given below. New data will lead to future revision of the dates of points on the time-scale, though the use to which the time-scale is put in this book – the determination of *intervals* of geological time – can be expected to remain approximately correct.

Radiometric and chronometric dates are given in Ma (millions of years before the present); the durations of intervals of geological time are given in m.y. (millions of years).

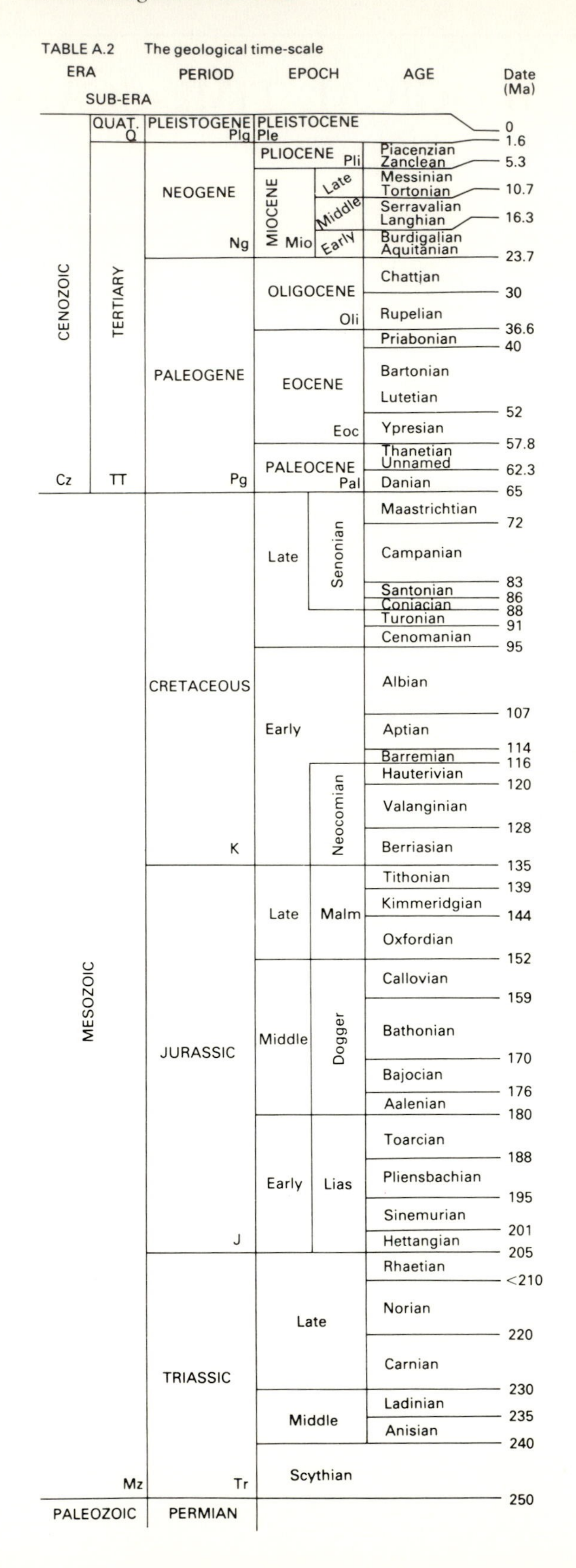

TABLE A.2 The geological time-scale

ERA	SUB-ERA	PERIOD	EPOCH		AGE	Date (Ma)
						0
CENOZOIC Cz	QUAT. Q	PLEISTOGENE Plg	PLEISTOCENE Ple			1.6
	TERTIARY TT	NEOGENE Ng	PLIOCENE Pli		Piacenzian	
					Zanclean	5.3
			MIOCENE Mio	Late	Messinian	
					Tortonian	10.7
				Middle	Serravalian	
					Langhian	16.3
				Early	Burdigalian	
					Aquitanian	23.7
		PALEOGENE Pg	OLIGOCENE Oli		Chattian	30
					Rupelian	36.6
			EOCENE Eoc		Priabonian	40
					Bartonian	
					Lutetian	52
					Ypresian	57.8
			PALEOCENE Pal		Thanetian	
					Unnamed	62.3
					Danian	65
MESOZOIC Mz		CRETACEOUS K	Late	Senonian	Maastrichtian	72
					Campanian	83
					Santonian	86
					Coniacian	88
					Turonian	91
					Cenomanian	95
			Early		Albian	107
					Aptian	114
					Barremian	116
				Neocomian	Hauterivian	120
					Valanginian	128
					Berriasian	135
		JURASSIC J	Late	Malm	Tithonian	139
					Kimmeridgian	144
					Oxfordian	152
			Middle	Dogger	Callovian	159
					Bathonian	170
					Bajocian	176
					Aalenian	180
			Early	Lias	Toarcian	188
					Pliensbachian	195
					Sinemurian	201
					Hettangian	205
		TRIASSIC Tr	Late		Rhaetian	<210
					Norian	220
					Carnian	230
			Middle		Ladinian	235
					Anisian	240
			Scythian			250
PALEOZOIC		PERMIAN				

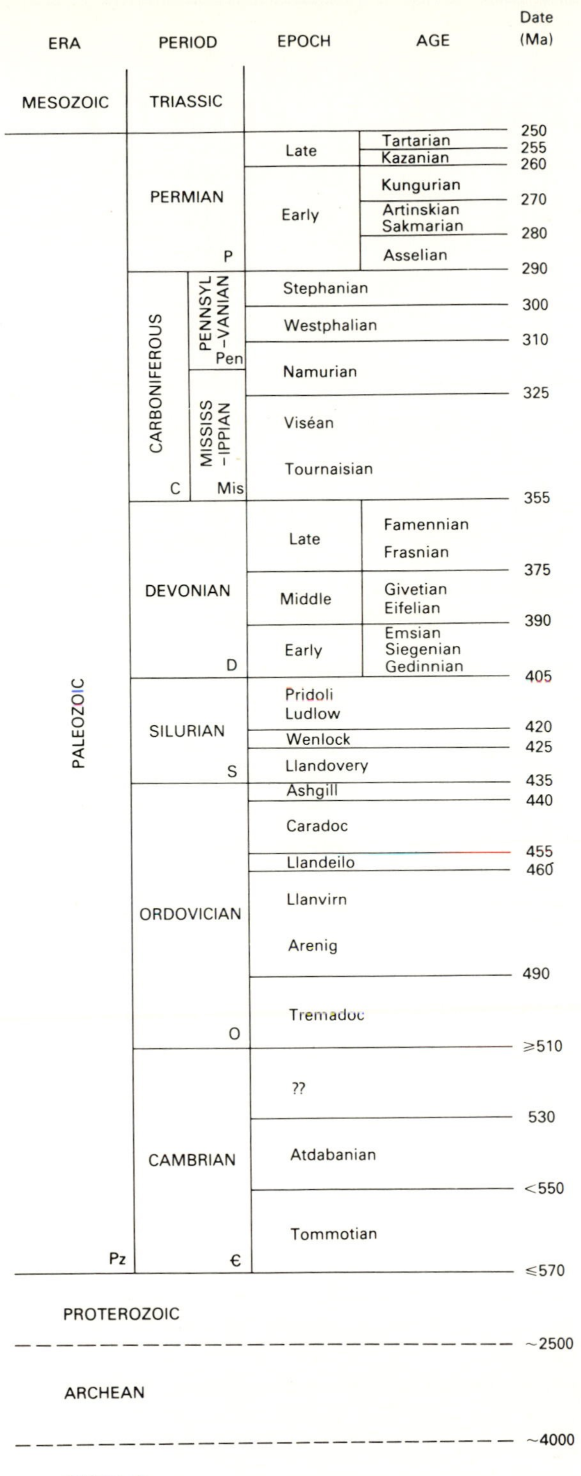
ERA
PERIOD
EPOCH
AGE
Date (Ma)
MESOZOIC
TRIASSIC
PERMIAN
P
Late
Tartarian
Kazanian
Early
Kungurian
Artinskian
Sakmarian
Asselian
250
255
260
270
280
290
CARBONIFEROUS
C
PENNSYL -VANIAN
Pen
MISSISS -IPPIAN
Mis
Stephanian
300
Westphalian
310
Namurian
325
Viséan
Tournaisian
355
DEVONIAN
D
Late
Famennian
Frasnian
375
Middle
Givetian
Eifelian
390
Early
Emsian
Siegenian
Gedinnian
405
PALEOZOIC
SILURIAN
S
Pridoli
Ludlow
420
Wenlock
425
Llandovery
435
Ashgill
440
ORDOVICIAN
O
Caradoc
455
Llandeilo
460
Llanvirn
Arenig
490
Tremadoc
≥510
CAMBRIAN
€
??
530
Atdabanian
<550
Tommotian
Pz
≤570
PROTEROZOIC
~2500
ARCHEAN
~4000
PRISCOAN

APPENDIX 3

CONSTRUCTION OF A TOPOGRAPHICAL PROFILE

1. Draw the line of section on the map (or, to avoid marking the map, on a tracing paper or film overlay).
2. Place the edge of a sheet of graph paper on the line. Mark the positions of the ends of the line of section. Hold the graph paper in this position.
3. Mark on the edge of the graph paper the points where topographical contours cut the line of section; label their heights.
4. Mark the positions of any additional topographical features (ridge crests, scarps, river valleys).
5. At either or both ends of the line of section mark the datum level (usually sea-level) and a scale of vertical heights, using the same scale as the map. Leave enough space – say at least 5 cm (2 in.) above the highest part of the section – for the insertion of geological data and constructions.
6. Mark on the section at the correct height above the datum level a point for each contour position from stage 3.
7. Mark at the appropriate heights the positions of the additional topographical features noted in stage 4.
8. Connect with a smooth line the points determined in stages 6 and 7.
9. Label the locations of the ends of the section and any named topographical features.
10. Erase the preliminary marks and labels (stages 3 and 4) so as to make space for subsequent geological constructions.

Notes:

(a) After a little practice it will be found possible to begin by constructing the vertical scale (stage 5), and to draw the topographical profile in one operation combining stages 2–4 and 6–8.

(b) If the topographical contours are not legible on the geological map, the topographical map of the same area should be used.

(c) In areas of gentle topography with steeply-dipping geological structures, and on small-scale maps (smaller than about 1 : 200 000) of non-mountainous regions the topographical profile can commonly be shown as flat.

(d) It may sometimes be necessary to adjust the topographical profile to obtain the best

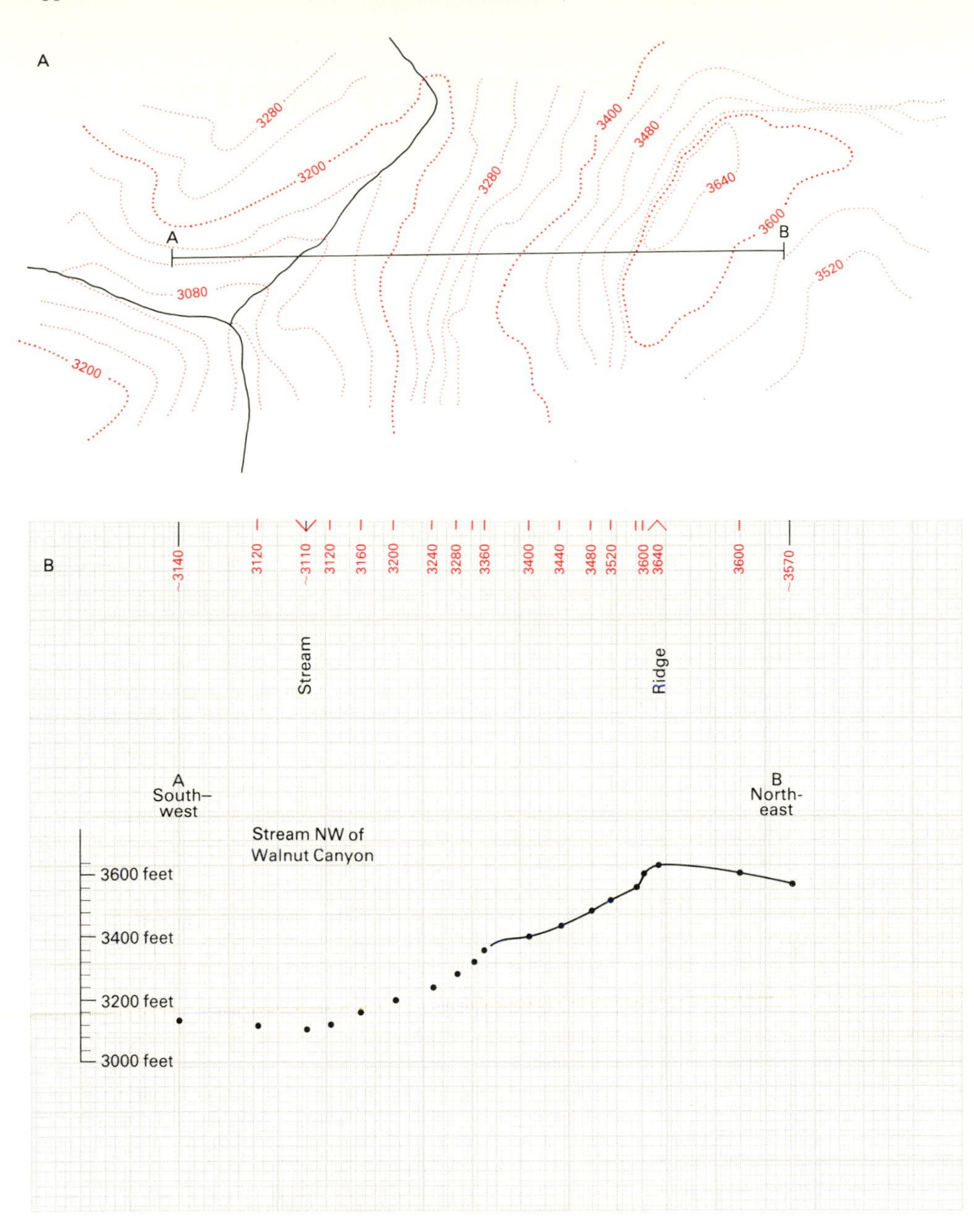

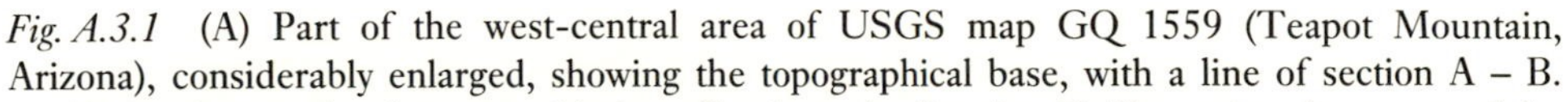

Fig. A.3.1 (A) Part of the west-central area of USGS map GQ 1559 (Teapot Mountain, Arizona), considerably enlarged, showing the topographical base, with a line of section A – B.
(B) Partly completed topographical profile along the line A – B, illustrating the stages of the procedure described above. The completed profile with the geology added is shown in Fig. 4.12.

fit of the topographical profile and the shapes of geological boundaries (cf. Section 5.B.7)

(e) Normally the vertical and horizontal scales should be the same. If there is good reason for using an exaggerated vertical scale, the appropriate scale factor should be applied at stage 5 (cf. Section 5.B.12)

APPENDIX 4

AN OUTLINE OF SPHERICAL TRIGONOMETRY

We are accustomed to think of distances between points on the surface of the Earth as being measured in kilometres or miles. However, distances over which the curvature of the Earth (mean radius 6371 km) becomes significant can alternatively be represented by the *angle* subtended by the two points at the centre of the Earth (as in the use of angular geographical co-ordinates measured relative to the Greenwich meridian and the equator). This forms the basis of the calculation of distances and directions on the surface of the Earth by the methods of spherical trigonometry.

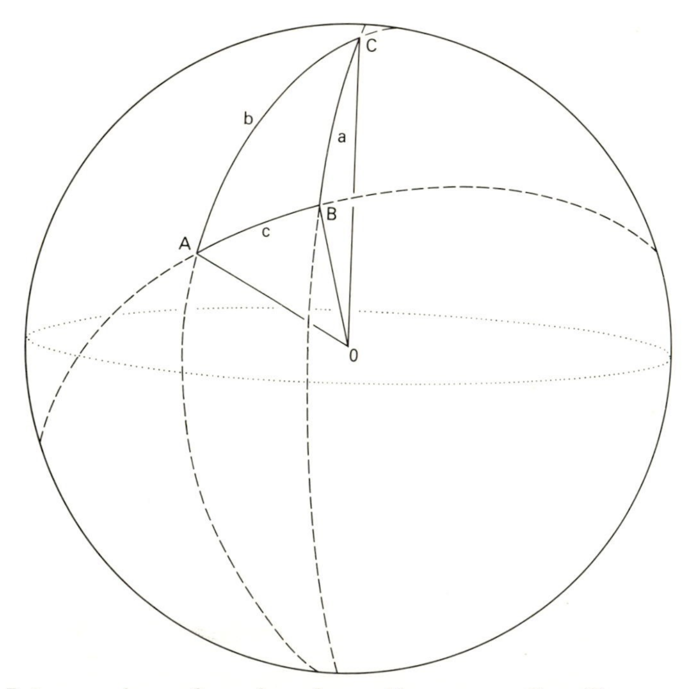

Fig. A.4.1 Points on the surface of a sphere with centre at O to illustrate the principles of spherical trigonometry. The lines AB, BC, and CA are segments of great circles, and the angles between the corners of the triangle are $c = \angle$ AOB, $a = \angle$ BOC, $b = \angle$ COA.

Note that if C is the north pole and the dotted line is the equator, and if points A and B are at geographical co-ordinates LATA, LONGA and LATB, LONGB, then $b = 90° -$LATA, $a = 90° -$ LATB, $\angle$ C = LONGB −LONGA; c represents the distance between points A and B, and $\angle$ A = compass direction of B from A, $\angle$ B = compass direction of A from B. In general $\angle A \neq 180° - \angle B$. (See Sect. 8.A.1.)

A **great circle** can be defined as a line on the surface of a sphere formed by the intersection of a plane passing through the centre of the sphere with its surface. On the surface of the Earth all lines of longitude are great circles and so also is the equator. The shortest distance between two points is measured along a great circle, and air and sea navigation routes are based on great circles.

A **small circle** can be defined as a line on the surface of a sphere such that all points on it are equidistant from a point which is the pole of the small circle. A great circle can be regarded as a special case of a small circle for which the distance of any point from its pole is 1/4 of the circumference of the sphere. Lines of latitude are small circles whose poles are the geographical north and south poles. Movement of plates on the surface of the Earth involves rotation about an imaginary axis through the centre of the Earth which emerges at a point on the surface which is the pole of rotation (Fig. 8.2).

A **spherical triangle** is formed by segments of three great circles (Fig. A.4.1). It consists of six elements – three sides (equivalent to the distances between the points A, B, and C, but measured in degrees), and three angles between the sides (equivalent to the differences in compass bearing between the directions AB and AC, BA and BC, and CA and CB). Sides and angles may have any value between 0° and 180°, and the sum of the angles is *not* constrained (as in plane triangles) to 180°. Any three elements define the shape of the triangle and can be used to calculate the values of any of the other three elements; the essential formulae to calculate any element from any three given data are:

$$\frac{\sin a}{\sin A} = \frac{\sin b}{\sin B} = \frac{\sin c}{\sin C} \qquad \ldots [1]$$

$$\cos \frac{A}{2} = \sqrt{\frac{(\sin s \cdot \sin (s - a))}{(\sin b \cdot \sin c)}} \quad \text{where } s = \frac{a + b + c}{2} \qquad \ldots [2]$$

$$\cos \frac{a}{2} = \sqrt{\frac{(\cos (S - B) \cdot \cos (S - C))}{(\sin B \cdot \sin C)}} \quad \text{where } S = \frac{A + B + C}{2} \qquad \ldots [3]$$

$$\cos c = \cos a \cdot \cos b + \sin a \cdot \sin b \cdot \cos C \qquad \ldots [4]$$

$$\cos C = \frac{\cos B \cdot \sin (A - x)}{\sin x} \quad \text{where } \tan x = \frac{1}{\tan B \cdot \cos c} \qquad \ldots [5]$$

$$\tan \frac{c}{2} = \frac{\cos \frac{1}{2} (A + B) \cdot \tan \frac{1}{2} (a + b)}{\cos \frac{1}{2} (A - B)} \qquad \ldots [6]$$

The geographical co-ordinate of latitude of a point is expressed as the angle from the equator (= 90° – the angle from the point to the north pole). So when equation [4] above is used to determine the distance and direction between two points defined by their geographical co-ordinates it is re-cast in the forms shown in Section 8.A.1, equations [1] and [3] (compare Figs. A.4.1 and 8.4).

The above formulae can only be used to calculate sides and angles of spherical triangles formed of arcs of great circles. Calculation of geographical distances and directions along small circles is described in Section 8.A.2.

APPENDIX 5

POLES OF ROTATION OF PLATES AND ANGULAR VELOCITIES

Table A.5 Poles of rotation of plates and angular velocities

Plate	*Pole of rotation*		Angular velocity
	Latitude	*Longitude*	*deg/m.y.*
Africa	+18.76°	−21.76°	0.139
Antarctica	+21.85°	+75.55°	0.054
Arabia	+27.29°	−3.94°	0.388
Caribbean	−42.80°	+66.75°	0.129
Cocos	+21.89°	−115.71°	1.422
Eurasia	+0.70°	−23.19°	0.038
India	+19.23°	+35.64°	0.716
Nazca	+47.99°	−93.81°	0.585
North America	−58.31°	−40.67°	0.247
Pacific	−61.66°	+97.19°	0.967
South America	−82.28°	+75.67°	0.285

(The data are from J. B. Minster and T. H. Jordan, 'Present-day plate motions', *J. Geophys. Res.*, **83**, 5331–5354, 1978.)

APPENDIX 6

GEOLOGICAL MAPS USED AS EXAMPLES IN THIS BOOK

The maps used as examples in this book have been selected to show a wide range of rock-types, structures, and regional environments. They can be used to form the core of a collection of maps for teaching purposes.

UNITED STATES GEOLOGICAL SURVEY

The following maps are obtainable from: US Geological Survey, Map Distribution, Federal Center, Bldg. 41, Box 25286, Denver, CO 80225.

1 : 24 000 Geological Quadrangle map GQ 1542, Honaker Quadrangle, Virginia (1981).

The boundary between the Valley and Ridge Province of the Appalachian Mountains with thrust sheets of Cambrian to Pennsylvanian sediments, and the Allegheny Plateau to the northwest with nearly horizontal Pennsylvanian. With the maps of the Calgary (Geol. Surv. of Canada) and Assynt (B.G.S.) areas it gives an impression of the process of development of large-scale thrusting.

1 : 24 000 Geological Quadrangle map GQ 1561, Goshen Quadrangle, Massachusetts (1981).

An area of strongly deformed metasedimentary and metavolcanic units with a particularly prominent dome structure. The map shows fold axial traces of more than one phase of folding and gives details of metamorphic effects including the kyanite isograd.

1 : 24 000 Geological Quadrangle map GQ 1465 (Surficial Geology), Palmer Quadrangle, Massachusetts (1978).

An area of Pleistocene glacial deposits and Holocene sediments with details of glacial lakes and spillways, positions of ice retreat, drumlins, eskers and other glacial features. Special explanatory sections show restored gradients of meltwater drainage systems.

1 : 24 000 Geological Quadrangle Map GQ 1559, Teapot Mountain Quadrangle, Arizona (1983).

The area is in the Basin and Range Province of the western United States, where crustal extension, with complex fault patterns and intrusive and extrusive igneous activity, has affected rocks from Precambrian to Tertiary in age.

1 : 24 000 Geological Quadrangle map GQ 1391, Hayden Quadrangle, Arizona (1977).

The map shows another area in the Basin and Range province of the western

United States where minor and major fault patterns and relationships between several phases of Mesozoic and Tertiary intrusions are clearly displayed. Country rocks range in age from Precambrian to Holocene. In the centre of the area is a porphyry copper deposit.

CALIFORNIA DIVISION OF MINES AND GEOLOGY

Obtainable from: California Division of Mines and Geology, PO Box 2980, Sacramento, California 95812, USA.

1 : 750 000 Geologic Map of California (1977)

A detailed map of the geology of the whole of the state, and including parts of the Modoc Plateau, Sierra Nevada batholith, Great Valley, the San Andreas Fault system, and the Franciscan rocks of the Coast Ranges.

VIRGINIA DIVISION OF MINERAL RESOURCES

Obtainable from: Branch of Distribution, US Geological Survey, 1200 South Eads Street, Arlington, Virginia 22202, USA.

1 : 62 500 map of Williamsville Quadrangle, Virginia (1979).

An area of northwest Virginia in which Palaeozoic shelf carbonates and clastic sediments are simply folded along northeast–southwest trending axes.

GEOLOGICAL SURVEY OF CANADA

Obtainable from: Geological Survey of Canada, 601 Booth Street, Ottawa K1A 0E8, Canada.

1 : 250 000 Map 1457A, Calgary, Alberta and British Columbia (1978), with a separate sheet of structural sections.

Part of the Main Ranges, Front Ranges, and Foothills of the Southern Rocky Mountains, with thrust sheets of rocks from late Precambrian to Tertiary age pushed eastwards over Tertiary rocks of the Great Plains. The area is economically important for its reserves of oil and gas.

GEOLOGICAL SURVEY OF JAPAN

The following maps are obtainable from: Maruzen Company Ltd, Import and Export Department, PO Box 5050, Tokyo International 100–31, Japan. (Appointed distributor authorised by the publisher, Geological Survey of Japan.)

1 : 500 000 Sheet 15, Kagoshima (1980).

A small-scale geological map covering the southern half of Kyushu; further detail is shown on the 1 : 200 000 map of Nobeoka (see below). The map symbols and the legend on each of these maps are in both Japanese and English.

1 : 200 000 Sheet N.I. 52–6, Nobeoka (1981)

A structural map of the east-central area of Kyushu, including the Median Tectonic Line and thrusts to the east (Pacific) side – a possible accretionary prism related to the subduction of the Pacific plate below the Japan island arc system. An inset map shows the zones of regional metamorphism and of thermal metamorphism around granite intrusions.

1 : 50 000 Sheet 15–23 (NJ–52–5–11), Taketa District (1977)

An area in central Kyushu in which calc-alkaline lavas and pyroclastic rocks from two volcanic centres (including Aso volcano) overlie Upper Palaeozoic metamorphic formations belonging to a major geotectonic belt in southwest Japan.

1 : 50 000 Sheet 6–89 (NJ–54–15–14), Matsushima District (1982)

A coastal area in northeast Honshu in which active faulting and folding are recorded and the diverse lithologies and morphological features of a fluviatile flood-plain and coastal plain are clearly shown.

GEOLOGICAL SURVEY OF SOUTH AUSTRALIA

Obtainable from: The Geological Survey of South Australia, Department of Mines, Adelaide.

1 : 250 000 South Australia Geological Atlas Series SH 54–9, Copley (1973).

The area is in the Flinders Ranges; the rocks are folded Proterozoic and Cambrian sediments of the Adelaide geosyncline intruded during the Palaeozoic by diapiric breccias; in the northeast there is the Mount Painter block of Lower Proterozoic metasediments with Precambrian and Ordovician granites. These rocks are unconformably overlain by thin Mesozoic and Tertiary continental deposits. On the west and east sides of the map area are Quaternary waterlaid and aeolian sediments fringing Lake Torrens and Lake Frome.

GOVERNMENT OF THE SULTANATE OF OMAN

Obtainable from: The Department of Earth Sciences, The Open University, Walton Hall, Milton Keynes, MK7 6AA, England.

1 : 100 000 Map 2 of the Oman Geological Ophiolite Project, Wadi Jizi (1980).

Part of the central area of the Oman ophiolite, a segment of oceanic crust and mantle thrust onto the northeast Arabian continental margin during the late Cretaceous.

GEOLOGICAL SURVEY OF WEST GERMANY

Obtainable from: Internationales Landkartenhaus GeoCenter, Postfach 80 08 30, D-7000 Stuttgart 80, West Germany.

1 : 1 000 000 Geologische Karte der Bundesrepublik Deutschland (1988).
A small-scale but detailed map showing the geology of a major part of Europe extending from the northern Alps to the North European Plain.

GEOLOGICAL SURVEY OF SWITZERLAND

Obtainable from: Kümmerly & Frey AG, Geographischer Verlag, CH-3001, Bern, Switzerland.

1 : 25 000 Geologischer Atlas der Schweiz, Blatt 1067, Arlesheim (1984)
An area at the south end of the Rhein graben, on the north side of the Jura Mountains, with molasse deposits of the Alpine orogeny and fluvial deposits of the Alpine glaciation.

SERVIZIO GEOLOGICO D'ITALIA

Obtainable from: Servizio Geologico d'Italia, Largo S. Susanna 13, Roma, Italy.

1 : 50 000 Carta Geologica d'Italia, Foglio 028, La Marmolada (1977).
A mountainous glaciated region in the western Dolomites in which folded and thrust Triassic carbonate rocks, some of reefal origin, are intercalated with volcanic formations. The map contains details of Pleistocene glacial deposits and Holocene sediments.

PARC NATUREL RÉGIONAL DES VOLCANS D'AUVERGNE (FRANCE)

Obtainable from: Parc Naturel Régional des Volcans d' Auvergne, Château Montlosier-Randanne, 63710 Rochefort-Montagne, France.

1 : 25 000 Carte de Volcanologie de la Chaîne des Puys, France (1983).
The classic Holocene volcanic chain, including the Puy de Dome, emplaced in crystalline basement rocks forming the eastern edge of the Massif Central near Clermont-Ferrand.

BRITISH GEOLOGICAL SURVEY

The following maps are obtainable from: British Geological Survey, Keyworth, Nottingham NG12 5GG.

1 : 63 360 Special Sheet of the Assynt area, Scotland (1965).
A classical area of British Geology. Lewisian (Archaean), Torridonian (Proterozoic), and Cambrian rocks, with intervening unconformities, and intruded by igneous rocks of various ages, form the foreland to the Caledonian orogenic belt. The Moine Thrust Zone makes a broad culmination in the Assynt area. A map of a complex area with many examples of important geological relationships.

1 : 50 000 Sheets 39W and 39E, Stirling and Alloa, Scotland (1974).

An area of the Midland Valley of Scotland, used in Chapters 9 and 10 for a detailed example of geological map interpretion.

1 : 50 000 Special Sheet of Arran, Scotland (1972).

Sills, dyke swarms, and volcanic vents of the British Tertiary igneous province intrude metasediments and sediments ranging in age from late Precambrian to Mesozoic. Particularly prominent is the North Arran Granite and the structural effects of its intrusion on the country rocks.

1 : 50 000 Sheet 111, Buxton, England (1978).

Marine and deltaic sediments of the Lower and Upper Carboniferous, with facies variation in both groups of rocks; post-Carboniferous folding and faulting and later Mississippi-Valley type mineralisation.

1 : 50 000 Sheet 233, Monmouth, England (1974).

Ten kilometres north of the area of the Bristol sheet, Upper Palaeozoic sediments subjected to several episodes of folding with intervening unconformities. In the east is the synclinal Forest of Dean coalfield whose structure is clearly displayed by outcrop pattern.

1 : 63 360 Special Sheet of Bristol District, England (1962).

Includes the area of the 1 : 25 000 Sheet ST45 of Cheddar (see below), and the area of periclines and basins of Palaeozoic rocks to the north, including the Somerset and Bristol coalfields, overlain unconformably by Triassic and Jurassic rocks.

1 : 25 000 Sheet ST45, Cheddar, England (1983).

A simple asymmetrical Hercynian anticline and thrust structure in Carboniferous and Devonian sediments of the Mendip Hills, unconformably overlain by Mesozoic sediments.

1 : 63 360 Sheet 281, Frome, England (1965).

Twenty kilometres east of the Cheddar map, in an area where the Hercynian folds in Palaeozoic rocks had a topographical and structural influence on the deposition of unconformably overlying Mesozoic sediments.

1 : 250 000 Sheet 48N 10W and part of 48N 11W, Little Sole Bank and part of Austell Spur, Sea Bed Sediments (1985).

One of a series of maps showing the composition of sea-bed sediments around the British Isles. The area is on the edge of the continental shelf, and shows also the bathymetry of the continental slope with submarine canyons.

GEOLOGICAL WORLD ATLAS

Obtainable from: Service Géologique National, B.P.6009, 45018 Orléans Cedex, France.

1 : 36 000 000 Sheet 20, Pacific Ocean (1976).

The whole of the Pacific Ocean, showing the ages of basement rocks, magnetic anomalies, thickness of sediments, fracture zones, earthquake epicentres, and volcanoes. Methods of interpretation of this map are given in Chapter 8.

LIST OF SYMBOLS ON MAPS

	Topographical contours *or* lines of latitude and longitude
	Lithological boundary
	Lithological boundary (conjectural)
30	Direction and amount of dip (longer line shows strike direction)
60	Direction and amount of overturned dip
	Structure contour
	Fault
	Direction of downthrow of fault
	Direction of relative movement of strike-slip fault
	Anticline
	Syncline
	Direction of plunge of fold
	Unconformity
	Direction and velocity of plate movement
	Relative direction and velocity of plate movement

REFERENCES

Anderton, R., Bridges, P. H., Leeder, M. R. & Sellwood, B. W., 1979, *A Dynamic Stratigraphy of the British Isles*. George Allen & Unwin, London.

Badgley, P. C., 1959, *Structural Methods for the Exploration Geologist*. Harper, New York.

Barnes, J. W., 1981, *Basic Geological Mapping*. Open University Press, Milton Keynes.

Blatt, H. 1982, *Sedimentary Petrology*. Freeman, San Francisco.

Blatt, H., Middleton, G. V. & Murray, R. C., 1980, *Origin of Sedimentary Rocks* (2nd edn). Prentice-Hall, Englewood Cliffs, New Jersey.

Bluck, B. J. *et al.* (eds), 1984, 'Deep geology of the Midland Valley of Scotland and adjacent regions', *Trans. R. Soc. Edinburgh*, Vol. 75, Pt 2.

Cox, A. & Hart, R. B., 1986, *Plate Tectonics: How It Works*. Blackwell, Oxford.

Craig, G. Y. (ed.), 1983, *Geology of Scotland* (2nd edn). Scottish Academic Press Ltd, Edinburgh.

Cronan, D. S., 1980, *Underwater Minerals*. Academic Press, London.

Dinham, C. H. & Haldane, D., 1932, 'The Economic Geology of the Stirling and Clackmannan Coalfield'. *Memoir of the Geological Survey, Scotland*. HMSO, Edinburgh.

Elder, J., 1976, *The Bowels of the Earth*. OUP, Oxford.

Evans, A. M., 1987, *An Introduction to Ore Geology* (2nd edn). Blackwell, Oxford.

Faure, G., 1986, *Principles of Isotope Geology* (2nd edn). Wiley, New York.

Francis, E. H., 1982, 'Magma and sediment – I: emplacement mechanism of late Carboniferous tholeiite sills in northern Britain', *J. Geol. Soc. London*, **139**, 1–20.

Francis, E. H., Forsyth, I. H., Read, W. A., & Armstrong, M., 1970, 'The geology of the Stirling District', *Memoir of the Geological Survey of Great Britain*. HMSO, Edinburgh.

Gillen, C., 1982, *Metamorphic Geology*. George Allen & Unwin, London.

Greenwood, H. J., 1976, 'Metamorphism at moderate temperatures and pressures', Ch. 2 of Section A in D. K. Bailey & R. Macdonald (eds) *'The Evolution of the Crystalline Rocks'*. Academic Press, London.

Hallam, A., 1981, *Facies Interpretation and the Stratigraphic Record*. Freeman, Oxford.

Hobbs, B. E., Means, W. D. & Williams, P. F., 1976, *An Outline of Structural Geology*. Wiley, New York.

Hobson, G. D. & Tiratsoo, E. N., 1985, *Introduction to Petroleum Geology* (2nd edn). Gulf Publishing Co., Houston, Texas.

Holmes, A., 1978, *Principles of Physical Geology* (3rd edn by D. L. Holmes). Nelson, London.

Hutchison, C. S., 1983, *Economic Deposits and their Tectonic Setting*. Macmillan, London.

Hyndman, D. W., 1985, *Petrology of Igneous and Metamorphic Rocks* (2nd edn). McGraw-Hill, New York.

Keller, E. A., 1982, *Environmental Geology* (3rd edn). Charles E. Merrill Publishing Company, Columbus.

Kennett, J. P., 1982, *Marine Geology*. Prentice-Hall, Englewood Cliffs, New Jersey.

Lahee, F. H., 1961, *Field Geology* (6th edn). McGraw-Hill, New York.

Leeder, M. R., 1982, *Sedimentology: Process and Product*. George Allen & Unwin, London.

Link, P. K., 1982, *Basic Petroleum Geology*. Oil and Gas Consultants Inc., Tulsa, Oklahoma.

Middlemost, E. A. K., 1985, *Magmas and Magmatic Rocks*. Longman, London.

Miyashiro, A., 1973, *Metamorphism and Metamorphic Belts*. George Allen & Unwin, London.

North, F. K., 1985, *Petroleum Geology*. Allen & Unwin, Boston.

Ollier, G. D., 1981, *Tectonics and Landforms*. Longman, London.

Owen, H. G., 1983, *Atlas of Continental Displacement: 200 Million Years to the Present*. CUP., Cambridge.

Park, R. G., 1983, *Foundations of Structural Geology*. Blackie, Glasgow.

Press, F. & Siever, R., 1982. *Earth* (3rd edn). Freeman, San Francisco.

Ragan, D. M., 1985, *Structural Geology: An Introduction to Geometrical Techniques* (3rd edn). Wiley, New York.

Reading, H. G. (ed.), 1986, *Sedimentary Environments and Facies* (2nd edn). Blackwell, Oxford.

Roberts, J. L., 1982, *Geological Maps and Structures*. Pergamon, Oxford.

Selley, R. C., 1985a, *Elements of Petroleum Geology*. Freeman, New York.

Selley, R. C., 1985b, *Ancient Sedimentary Environments and their Sub-surface Diagnosis* (3rd edn). Chapman & Hall, London.

Smith, A. G., Hurley, A. M. & Briden, J. C., 1980, *Phanerozoic Paleocontinental World Maps*. CUP, Cambridge.

Snelling, N. J. (ed.), 1985, 'The Chronology of the Geological Record'; *Mem. 10, Geol. Soc. of London*. Blackwell, Oxford.

Stanton, R. L., 1972, *Ore Petrology*. McGraw-Hill, New York.

Turner, F. J., 1981, *Metamorphic Petrology*. McGraw-Hill, New York.

Vernon, R. H., 1976, *Metamorphic Processes*. George Allen & Unwin, London.

Ward, C. R. (ed.), 1984, *Coal Geology and Coal Technology*. Blackwell, Melbourne.

Williams, H. & McBirney, A. R., 1979, *Volcanology*. Freeman, Cooper, & Co., San Francisco.

Ziegler, P. A., 1982, *'Geological Atlas of Western and Central Europe'*. Shell Internationale Petroleum Maatschappij B.V., Elsevier, Amsterdam.

INDEX